METHODS IN
MICROBIOLOGY

A complete list of other titles in this series is available from the Publisher on request.

METHODS IN
MICROBIOLOGY

Volume 20
Electron Microscopy in Microbiology

Edited by

F. MAYER

Institüt für Mikrobiologie der Georg-August Universität-Göttingen,
Grisebachstr. 8, D-3400 Gottingen, FRG

1988

ACADEMIC PRESS
Harcourt Brace Jovanovich, Publishers
London   San Diego   New York   Berkeley   Boston
Sydney   Tokyo   Toronto

ACADEMIC PRESS LIMITED
24–28 Oval Road
London NW1

United States Edition published by
ACADEMIC PRESS INC.
Orlando, Florida 32887

British Library Cataloguing in Publication Data
Methods in microbiology. — Vol. 20:
Electron microscopy in microbiology.
1. Microbiology. Laboratory techniques
I. Mayer, Frank
576'.028

ISBN 0-12-521520-6

Printed in Great Britain by Galliard (Printers) Ltd, Great Yarmouth

CONTRIBUTORS

G. Acker Universität Bayreuth, Abteilung Elektronenmikroskopie, NW I D-8580 Bayreuth, Federal Republic of Germany

R. Bauer c/o Carl Zeiss Oberkochen, Application for Transmission Electron Microscopy, D-7082 Oberkochen, Federal Republic of Germany

H. Engelhardt Max-Planck-Institut für Biochemie, D-8033 Martinsried, Federal Republic of Germany

H. Gerberding Georg-August-Universität Göttingen, Institut für Mikrobiologie, Abteilung Mikromorphologie, D-3400 Göttingen, Federal Republic of Germany

H. J. Geuze University of Utrecht, Department of Cell Biology, Medical School, Utrecht, The Netherlands

J. R. Golecki Universität Freiburg, Institut für Biologie II, Mikrobiologie, D-7800 Freiburg, Federal Republic of Germany

A. Holzenburg Universität Basel, Maurice E. Müller-Institut für hochauflösende Elektronenmikroskopie am Biozentrum, CH-Basel, Switzerland

W. Johannssen c/o E. Merck, Sparte Reagenzien, D-6100 Darmstadt, Federal Republic of Germany

G.-W. Kohring Georg-August-Universität Göttingen, Institut für Mikrobiologie, Abteilung Mikromorphologie, D-3400 Göttingen, Federal Republic of Germany

J. Lalucat Universidad de las Islas Baleares, Departamento de Biologia, Facultad de Ciencias, 07071 Palma de Mallorca, Spain

R. Lurz Max-Planck-Institut für Molekulare Genetik, D-1000 Berlin 33, Federal Republic of Germany

F. Mayer Georg-August-Universität Göttingen, Institut für Mikrobiologie, Abteilung Mikromorphologie, D-3400 Göttingen, Federal Republic of Germany

P. Messner Universität für Bodenkultur, Zentrum für Ultrastrukturforschung und Ludwig-Boltzmann-Institut für Ultrastrukturforschung, A-1180 Wien, Austria

M. Müller Eidgenössische Technische Hochschule Zürich, Laboratorium für Elektronenmikroskopie I, Institut für Zellbiologie, Universitätstrasse 2, CH-8092 Zürich, Switzerland

Th. Mund Georg-August-Universität Göttingen, Institut für Mikrobiologie, Abteilung Mikromorphologie, D-3400 Göttingen, Federal Republic of Germany

D. Pum Universität für Bodenkultur, Zentrum für Ultrastrukturforschung und Ludwig-Boltzmann-Institut für Ultrastrukturforschung, A-1180 Wien, Austria

M. Rohde Georg-August-Universität Göttingen, Institut für Mikrobiologie, Abteilung Mikromorphologie, D-3400 Göttingen, Federal Republic of Germany

U. B. Sleytr Universität für Bodenkultur, Zentrum für Ultrastrukturforschung und Ludwig-Boltzmann-Institut für Ultrastrukturforschung, A-1180 Wien, Austria

J. W. Slot University of Utrecht, Department of Cell Biology, Medical School, Utrecht, The Netherlands

E. Spiess Deutsches Krebsforschungszentrum, Institut für Zellforschung, D-6900 Heidelberg, Federal Republic of Germany

H.-D. Tauschel   Hinterdorfstrasse 29, D-7637 Ettenheim 3, Federal Republic of Germany

A. H. Weerkamp   University of Groningen, Department of Oral Biology, Dental School, Groningen, The Netherlands

K. Zierold   Max-Planck-Institut für Systemphysiologie, Rheinlanddam 201, D-4600 Dortmund 1, Federal Republic of Germany

PREFACE

Volume 20 of "Methods in Microbiology" is concerned with electron microscopic techniques applied for the elucidation of microbial structures and structure–function relationships at cellular, subcellular, and macromolecular levels. Many of the recent findings on the ultrastructural features of microorganisms have been obtained with newly developed methods; cryotechniques are of major importance in this respect. Nevertheless, classical approaches have not lost their validity. Both conventional and newer methods have therefore been incorporated into this volume. The topics dealt with are meaningful not only in bacterial cytology, but also in physiology, enzymology, biochemistry, and molecular biology, as well as aspects of medical and biotechnological application.

At the cellular level the topics covered include the analysis of bacterial surface structures (polysaccharides, S-layer), membranes, and special inclusions (R-bodies), and approaches for the qualitative and quantitative investigation of low and high molecular weight cellular components. At the subcellular level, the analysis of isolated membranes and of proteoliposomes is described. At the macromolecular level, nucleic acids and selected enzymes are treated. A more general chapter on cryopreparation of microorganisms for transmission and scanning electron microscopy provides an introduction to the field; a chapter on computerized image evaluation and reconstruction was put at the end of the volume because the quantitative extraction of structural data and their handling are prerequisites for the improved understanding of what is seen on electron micrographs. Some of the individual chapters may contain descriptions of procedures also presented in other chapters. In part, this was unavoidable; it was also desirable in that the reader may discover that specific procedures work in a number of variations. This appears to be a feature of the majority of electron microscopic approaches, which often demand a specific skill and patience, and which have developed in different directions after their basic features had become common knowledge.

It is hoped that this volume will encourage workers in a variety of fields of microbiology, and that it contributes to a wider application of modern preparation, imaging, and image evaluation techniques.

The editor wishes to express his gratitude to the authors for their co-operation, to G. Gottschalk for initiating this volume and for his advice, to the staff of Academic Press in London, and to the Chairman of the Advisory Board, John R. Norris.

Frank Mayer

CONTENTS

1

Cryopreparation of Microorganisms for Electron Microscopy

M. MÜLLER

Laboratorium für Elektronenmikroskopie, Institut für Zellbiologie,
Eidgenössiche Technische Hochschule Zürich, Zürich, Switzerland

I. Introduction

The aim of an electron microscopical study is to provide a structural basis for the correlation of structure and function. Structural information is therefore more valuable when the structures are preserved more completely; preparation techniques should aim to preserve the smallest significant details.

Electron microscopy is the only tool with the inherent power to observe structural details at macromolecular dimensions within the context of complex biological systems. It is therefore important to continue the development of this unique feature of electron microscopy to complement the biochemical and molecular biological techniques available. In practice, however, electron microscopy is used extensively in its "instamatic function", i.e. to illustrate subjective expectations derived from *a priori* information

1

Copyright © 1988 by Academic Press Limited
All rights of reproduction in any form reserved

obtained by other techniques. With the exception of immunocytochemical methods, biological electron microscopy is rarely used as a source of primary information. This use of electron microscopy can be understood when conventional preparative procedures based on chemical fixation and dehydration are used; it is no longer justified, or at least to a much lesser extent, with the advent of appropriate cryofixation techniques.

The main problem encountered during preparation of biological specimens for electron microscopy arises from the necessity to transform the aqueous biological sample into a solid state in which it can resist the physical impact of the electron microscope (high vacuum, electron beam irradiation). The visualization of intracellular structures by transmission electron microscopy requires very thin sections. Biological samples embedded in plastic compounds, epoxy or methacrylate resins, yield the requisite thin sections. Dehydration of the samples in graded series of an organic solvent, usually acetone or ethanol, is required since these resins are not totally miscible with water. Some effects of the organic solvent, e.g. extraction of lipids (Weibull *et al.*, 1983), gross conformational changes, collapse phenomena, aggregation of proteins and other macromolecules due to complete dehydration (Kellenberger, 1987), are reduced by previous chemical fixation. Chemical fixatives react relatively slowly and cannot preserve all cellular components. Most of the diffusible ions are lost or redistributed during sample preparation. Fixation influences the diffusion properties of the membranes and therefore results in alterations of shape, volume, and content of the cell and its components. Dehydration of fixed samples results in further dimensional changes (shrinkage) (Lee, 1984), which are by no means isotropic. It becomes evident from this that the initial potential of electron microscopy cannot be approached by specimen preparation procedures based on chemical fixation and dehydration.

Ideally one would like specimen preparation procedures that simultaneously guarantee absolute preservation of the dimensions and of the spatial distribution of diffusible elements. Antigens, receptors, lectin-binding sites, etc., should become demonstrable through immunocytochemical techniques. A universal specimen preparation procedure will perhaps remain a dream. One must nevertheless attempt to realize it so that the integrating potential of electron microscopy can further develop into a complementary tool in modern biological research.

Dehydration is the most limiting step with respect to the preservation of the structural integrity. It affects the complex interactions of macromolecules, membranes, cellular water, bound and diffusible ions, in a way that cannot be controlled by means of chemical fixation. Cryofixation represents the alternative; it halts the physiological processes very rapidly. Chemically unpretreated biological material can be immobilized in its natural environment; only minor dimensional changes are introduced (when frozen the volume of water increases only by 9%: Ushiyama *et al.*, 1979).

II. Cryofixation

The structural integrity of the biological material is guaranteed only if cryofixation brings about solidification of water or solutes in a vitreous or microcrystalline state in the absence of any chemical treatment (aldehyde prefixation, for example, reduces possible effects of cryoprotectants: Rash and Hudson, 1979).

Freezing cellular water in the vitreous, amorphous state would be ideal, since the basic nature of the liquid would be preserved. Vitrification of cellular water, however, requires very high cooling rates, present only in very thin layers at the specimen surface. These layers are usually too thin (<1 µm) to represent the bulk of the sample or to be processed further for electron microscopy. Their thickness is, however, related to the composition of the cellular fluid (i.e. the concentration of components that act as natural cryoprotectants). Dubochet *et al.* (1987) claim vitrification of thicknesses up to 5–10 µm of non-cryoprotected tissue of liver or kidney. True vitrification of biological solutions has been demonstrated by low temperature electron diffraction (Dubochet *et al.*, 1982) in a transmission electron microscope equipped with a cold stage using the "bare grid" technique (Adrian *et al.*, 1984). By this technique thin (~ 100 nm) aqueous layers of suspensions are formed within the meshes of an electron microscope grid and vitrified by immersion into liquid ethane or propane at $\sim -180°C$. Thus true vitrification is currently limited to biological samples (suspensions of viruses, phages, liposomes, macromolecules, etc.) that can be prepared in this way.

Heat can be extracted only through the surface of the sample. Heat transfer from deeper within the specimen is limited by the low thermal conductivity of the water and the developing solid layer. Very high cooling rates can be achieved which may lead to the immobilization of a thin layer in the vitreous state at the surface of the sample. Insufficient cooling rates allow ice crystals to form deeper in the sample. More heat is produced by ice crystal formation with increasing depth than is transferred through the ice to the cooled surface. This progressively reduces the cooling rate and results in increasing ice crystal dimensions. Solutes are excluded from the crystal lattice and concentrated between neighbouring ice crystals to form a eutectic as ice crystals form. In electron micrographs of biological material cryofixed with insufficient cooling rates, the eutectic appears as a network of segregation compartments. In practice, the absence of a segregation pattern is a good indication of adequate cryoimmobilization with perhaps only very small ice crystals, if any, and is referred to as *microcrystalline*. A correct analysis of the state of the cryoimmobilized cellular water is, however, only possible by electron diffraction of frozen hydrated cryosections (Dubochet *et al.*, 1983).

Under optimized freezing conditions (sample mass: Bachmann and Schmitt, 1971; Gulik-Krzywicki and Costello, 1978; optimized application of the cold: Müller *et al.*, 1980b; Costello and Corless, 1978; Dempsey and Bullivant, 1976a,b) and optimized specimen geometry the first detectable segregation patterns appear at a depth of 5–10 μm. This depth depends on the composition of the cellular fluid, as mentioned above. It cannot be significantly increased by any increase in cooling rates at the specimen surface.

Phase separation occurs whenever ice crystals form. Other physiological effects within the cells, such as increase in local solute concentration or change in pH, must therefore follow. These may affect smaller structural details even if no visible segregation compartments are present. It is important to be aware of such effects in order to avoid mis- or overinterpretation of the micrographs.

The above discussion of cryofixation is perfunctory at best. A few aspects relevant to practical application should be stressed:

(1)   Cryoimmobilization (by very high cooling rates), in the microcrystalline state, of untreated aqueous samples is only possible in very thin (~ 10 μm) superficial layers.

(2)   The term "microcrystalline" refers to the usable layer, in which no compartmentalization due to excessive ice crystal formation can be detected. Minor structural alterations due to effects of phase separation may only be excluded if true vitrification can be demonstrated.

(3)   True vitrification of untreated biological samples is obtained only in extremely thin superficial layers, and at present is therefore, with the exception of the "bare grid technique" (Dubochet *et al.*, 1982), of limited practical value.

(4)   The very high cooling rates (e.g. 10^4 K s^{-1}) required for adequate cryofixation bring about a rapid arrest of physiological processes and therefore allow the study of dynamic events with an estimated time resolution of 0.1 ms (Knoll *et al.*, 1987).

(5)   During vitrification, water expands linearly by about 2% (Dubochet *et al.*, 1982). During the change of water from a liquid to a solid phase an increase in *volume* of about 9% has been reported (Ushiyama *et al.*, 1979). These very small changes indicate that cryoimmobilization in the microcrystalline state may well preserve the dimensional relations close to the living state, thus forming the basis of a technique to

measure significant structural details as a function of the physiological state, rather than of the applied preparative procedure (as may be the case in the conventional procedures based on chemical fixation and dehydration).

During the last few years our knowledge of the complex behaviour of cellular water during freezing has improved greatly. A comprehensive introduction to cryofixation is given by Robards and Sleytr (1985) and by Bachmann and Mayer (1987). These reviews deal with the fundamental aspects of freezing and are highly recommended to everybody already using, or planning to enter the field of cryofixation-based electron microscopy. In depth knowledge about the physics of water and ice is essential for correct interpretation of results, i.e. to what extent our results reflect the true living state.

Rapid cooling techniques guarantee optimal structural preservation in a thin ($\sim 10\,\mu$m) layer; the very high cooling rates immobilize the cellular water in the vitreous, or in the microcrystalline state, and concomitantly rapidly arrest physiological processes, i.e. dynamic events at membranes (Heuser et al., 1979; Knoll et al., 1987). Rapid cooling techniques are therefore only applicable to samples that can be prepared in a thin layer (e.g. suspensions of macromolecules, isolated cell organelles, viruses, microorganisms, fungal hyphae, etc.) or to natural or cut surfaces of biological tissue samples. Thicker, more complex systems (e.g. plant or animal tissue, fungus–host interactions, root nodules) can be studied by cryofixation-based electron microscopy only if the physical properties of the cellular water are influenced in such a way that cryoimmobilization in the vitreous or microcrystalline state is achieved with much slower cooling rates. The impregnation of larger samples with cryoprotectants, usually in combination with aldehyde prefixation, is frequently used (Skaer, 1982). Numerous artefacts, however, have been shown to be introduced by this procedure (Plattner and Bachmann, 1982), for example the loss of diffusible ions and the redistribution of intramembraneous particles.

The aim of cryofixation is solely to immobilize the specimen physically. A method of cryofixation based on the application of high hydrostatic pressure was developed by Moor and coworkers (Moor and Höchli, 1970; Riehle and Höchli, 1973; Moor et al., 1980; Müller and Moor, 1984b). The effect of the high pressure can be elucidated by applying the principle of Le Chatelier: freezing increases the volume of water. This expansion, and consequently the crystallization, is hindered by high pressure. This effect is demonstrated by a lowering of the freezing point and by reduced rates of nucleation and ice crystal growth (Riehle, 1968). Consequently, less heat is produced by crystallization and has to be extracted per unit time by cooling. This means that vitrification can be achieved with reduced cooling rates. As deduced from the phase diagram of water (Kanno *et al.*, 1975) the most profitable pressure zone is at 2045 bar, where the melting point of water is lowered to the minimum of 251 K. Samples with dimensions of up to 500 μm can be adequately cryoimmobilized by high pressure freezing. The optimal sample dimensions again depend on the composition of the cellular fluid, i.e. on the presence of components that bring about cryoprotective activity. The suitable specimen thickness for plant material is limited to about 300 μm, whereas animal tissue blocks up to 600 μm thick are frequently successfully cryofixed (Hunziker *et al.*, 1984; Moor *et al.*, 1980). High pressure freezing thus permits structural analysis of more complex systems, i.e. fungus–host interactions or, in the centre of a tissue sample, of cells that have not suffered from traumatic excision. These advantages are somewhat reduced by the relatively slow cooling rates (approx. 500 K s^{-1}) achieved in the centre of the sample. These rates may be too slow to catch dynamic events at membranes or to prevent structural alterations due to lipid phase transition and segregation phenomena. On the other hand, the transition temperature of membrane lipids is raised by about 20 K kbar^{-1} (Macdonald, 1984). This means that by applying a pressure of more than 2 kbar (which is attained in $\sim$ 15 ms) the membrane lipids may be immobilized very quickly, purely by the action of high pressure. Experimental data supporting this assumption as well as on other short-lived high pressure effects on biological material are not yet available. Possible reactions of biological specimens to high pressure have been discussed by Müller and Moor (1984b) and Moor (1987).

III.   Follow-up procedures

Successfully cryoimmobilized samples have to be further processed for electron microscope analysis by various follow-up procedures, each of which yields different information and poses different technical problems. Subsequent processing has to be performed at sufficiently low temperatures such that devitrification and secondary ice crystal growth are avoided. The devitrification range for vitreous, amorphous water was found at $\sim$ 140 K by means of low temperature electron diffraction (Dubochet *et al.*, 1983); vitreous water recrystallizes into cubic ice at higher temperatures (Dubochet *et al.*, 1983). In this state no effects of phase separation due to ice crystal formation are yet visible when employing the most frequently used follow-up procedures (freeze-fracturing and freeze-substitution). At higher temperatures, i.e. above 190 K, cubic ice may be transformed into hexagonal ice, in which modification ice crystals may rapidly grow and alter the specimen.

A. Physical procedures

Low temperature electron microscopy of vitrified thin aqueous layers and of cryosections (Stewart and Vigers, 1986), as well as freeze-fracturing, are considered to be direct, purely physical procedures. They provide reliable structural information, most closely related to the living state (Dubochet et al., 1987). Cryosectioning of untreated cryofixed biological material at present is still a very demanding technique (Dubochet and McDowall, 1984) and the sections are generally too thick to provide structural identification at high resolution. On the other hand, cryosections observed in the microscope at low temperatures, either frozen hydrated or freeze-dried (Zierold, 1987), represent the best, if not the only, way towards qualitative and quantitative information of the spatial distribution of diffusible ions by X-ray microanalysis. None of these direct physical procedures is suitable for immuncytochemical work unless mild chemical fixation and cryoprotection precedes cryofixation [compare the cryosectioning technique of Tokuyasu (1984) for the labelling of intracellular antigens and the label-fracture techniques of Pinto da Silva and Kan (1984)].

Freeze-fracturing represents the simplest and best established physical technique for obtaining a "safe" representation of structural details (down to ~5 nm). Robards and Sleytr (1985), among others, have reviewed this technique in detail, and its application to microorganisms has been considered by Chapman and Staehlin (1986). Freeze-fracturing allows the description of specific structural aspects which depend on the fracturing behaviour of the sample and its components. It is especially suited to characterizing membranes, since, for energetic reasons, the fracture plane proceeds through the hydrophobic interior of the membranes, thus providing information about size and distribution of intramembraneous particles (IMP). The pattern formed by the IMPs is characteristic for each specific membrane fracture face. Alterations of these specific patterns may reflect dynamic processes at membranes (Knoll et al., 1987) or, if the sample was cooled too slowly, the occurrence of phase transitions and segregation phenomena of the lipid phase. The nature of the IMPs is still under discussion. They may indicate the positions of transmembrane or intramembraneous proteins but, as pointed out by Verkleij (1984), may also be of lipidic nature.

B. Hybrid techniques

Freeze-fracturing is a straight-forward, easily handled technique which provides reliable structural information. Its major disadvantages are that it is rarely applicable to any purpose other than structural description and that the fracture plane proceeds at random. These problems can be partially overcome by hybrid techniques which combine the advantages of cryofixation with those of the conventional plastic embedding and thin sectioning procedures. *Freeze-substitution* and *freeze-drying* are frequently used for these purposes (for review see Steinbrecht and Müller, 1987). Both procedures are essentially dehydration processes. Freeze-substitution dissolves the ice in a cryoimmobilized specimen by an organic solvent, and freeze-drying eliminates the frozen water by sublimation in a vacuum chamber. Freeze-drying and freeze-substitution must be performed well above the devitrification range of amorphous water (~ 140 K). Due to the low vapour pressure, freeze-drying at temperatures below about 170 K would lead to impractically long drying times. The temperature limits during freeze-substitution are set by the melting point of the solvent used and the amount of water the solvent can take up at low temperatures (Humbel and Müller, 1986). Generally, temperatures of 180–190 K are considered to be "safe" for cryofixed biological samples because of the rather high natural cryoprotective activity of many cellular components (see Steinbrecht, 1980). After completion of the dehydration process the samples are warmed to room temperature, infiltrated with the embedding resin, and heat polymerized. With respect to the preservation of the structural integrity, these hybrid techniques are much more obscure than the purely physical follow-up procedures such as cryosectioning or freeze-fracturing outlined above, since effects of the organic solvents (e.g. lipid extraction: Weibull *et al.*, 1984) and the embedding chemistry (Causton, 1986; Weibull and Christiansson, 1986) are not excluded. The structural description provided by the physical procedures, in which the water remains in the specimen, thus represents the standard by which all the other procedures have to be measured. Freeze-substitution and freeze-drying may allow an accurate control of the dehydration process, but, due to our incomplete knowledge of cellular water, they are still insufficiently understood.

Our present knowledge about the role of water in the cell and the effects of its removal are summarized by the following statements (for details see Bachmann and Mayer, 1987; Kellenberger *et al.*, 1986; Kellenberger, 1987; Steinbrecht and Müller, 1987):

(1) Water in the cell exhibits different physicochemical properties and is classified into two major groups, namely, bulk, or free water, in contrast to anomalous water referred to as "bound water", "non-freezable water", etc. This anomalous water is thought to be closely associated with surfaces of macromolecules, membranes and ions, and is sometimes also termed hydration shell, surface-modified water, or vicinal water.

(2) This surface-modified water is extremely important for metabolism (Clegg, 1979; Negendank, 1986) as well as for the maintenance of the

structural integrity of proteins and other cell constituents (Tanford, 1980).

(3) One may conclude that the bulk water is more easily removed and affects the preservation of the structural integrity less than the water of the hydration shells during the dehydration process in biological electron microscopy.

The above assumptions are supported by the non-linear shrinkage behaviour of cells and tissues during conventional dehydration by organic solvents at room temperature. The cells start to shrink when $\sim 70\%$ of the cellular water is replaced by the organic solvent. Fully dehydrated, they shrink up to 30–70% of their initial volume (Lee, 1984). This is a first indication that part of the cellular water can be removed which does not introduce gross dimensional changes. However, there is some water closely associated with the cellular structures. Removing this residual water may lead to conformational changes of cellular components (collapse) and aggregation (Kellenberger, 1987).

The temperatures above which different types of macromolecules collapse when exposed to dehydrating agents such as organic solvents and vacuum were determined by MacKenzie (1972). These temperatures range from 215 K to 263 K and seem to depend only on the temperature and the polarity of the dehydrating agent. Wildhaber et al. (1982) and Gross (1987) studied the freeze-drying of test specimens containing deuterium oxide D_2O instead of H_2O and followed its escape with a mass spectrometer. They observed a first peak of D_2O evaporating in the temperature range 180–190 K which approached zero after 2 h at 190 K. A second peak of D_2O was observed only after heating the specimen further and had a maximum between 220 K and 230 K. This suggests that the water in the different groups is held in the tissue by different forces, and it may be concluded that some of the specimen water is bound, and therefore needs a higher energy for evaporation than the free solvent. The temperature at which the second peak of D_2O was observed by Wildhaber et al. (1982) is in the temperature range of the collapse temperature of MacKenzie (1972), and it may be speculated that it corresponds to the release of the water of the hydration shell.

C. Freeze-drying and freeze-substitution

Ideally, freeze-drying and freeze-substitution could be used to control the residual water content, i.e. how much water has to be left so that the cells maintain their structural and functional integrity, and how much water has to be removed to allow successful plastic embedding. Experiments, however, have shown that the hydration shells can prevent an efficient copolymerization between biological material and resin (Humbel and Müller, 1986).

Strongly hydrated organelles may therefore not become embedded at all, or resin and biological material may separate very easily along membranes. Freeze-drying, in contrast to freeze-substitution, has not yet found very wide application in the dehydration of cryofixed samples for subsequent plastic embedding, except in the preparation of samples for ion measurements by X-ray microanalysis (e.g. Ingram and Ingram, 1984). It is, however, very successfully used in combination with metal shadowing in transmission as well as in scanning electron microscopy (e.g. Giddings and Wray, 1986; Gross, 1987; Walther et al., 1984b). A detailed overview of the theoretical and experimental data on freeze-substitution and freeze-drying, together with the most generally used procedures, has been recently given by Steinbrecht and Müller (1987) and Robards and Sleytr (1985).

The organic solvents used to freeze-substitute cryofixed biological samples frequently contain fixatives (e.g. OsO_4, uranylions, aldehydes). These are assumed to stabilize the biological structures at the ice–solvent interface, or during the gradual or stepwise rise in temperature, but little is known about their reactivity at these low, subzero temperatures. There is, however, experimental evidence that uranylions react with and prevent the extraction of lipids by the solvent even at the lowest temperatures (180 K). A reaction of OsO_4 with the double bonds of unsaturated fatty acids has been reported to occur at 203 K (White et al., 1976), while glutaraldehyde starts to cross-link proteins at 223 K (Humbel and Müller, 1986). Fixatives may be necessary to reduce solvent effects, e.g. the loss of lipids and other low molecular weight constituents. In addition, they might help to reduce the effects of conformational changes and aggregations of macromolecules and supramolecular structures which inevitably occur as the hydration shells are removed at higher temperatures (cf. collapse temperatures of MacKenzie, 1972). The much more homogeneous finer grained appearance of the cytoplasma as compared to conventionally dehydrated samples supports this assumption. The presence of fixatives in the substitution medium is essential if freeze-substitution is followed by conventional embedding at room temperature and heat polymerization at 335 K. Samples prepared in this way permit excellent structural description with preservation of the dimensions comparable to freeze-fracturing (Menco, 1986), and have yielded valuable new ultrastructural information. *Escherichia coli* cells, freeze-substituted in acetone OsO_4, show new aspects of the bacterial nucleoid (Hobot et al., 1985; Kellenberger et al., 1986) which is now recognized as a "ribosome free" space rather than as a vacuole-like empty space containing black particles which represent DNA aggregates (Fig. 8). Furthermore, the plasma membranes of the *E. coli* cells no longer appear irregular and undulated (Hobot et al., 1984). Essentially the same appearance of the nucleoid was reported for *Bacillus subtilis* by Amako and Takade (1985) after freeze-substitution. These authors in addition found the cell surface covered with a fibrous layer probably representing glycoproteins, and intracytoplasmic membranous structures (mesosomes). Mesosomes of Bacillus sp. have previously been characterized as an artefact of the conventional fixation procedure by rapid freezing followed by freeze-fracturing, freeze-substitution (Ebersold et al., 1981) and cryosectioning (Dubochet et al., 1983). Freeze-substitution has become an important tool in mycological cytology (Howard and Aist, 1979; Hoch and Howard, 1980) and has led to a partial revision of fungal ultrastructure. Freeze-substitution of fungi has been comprehensively reviewed recently by Hoch (1986). Besides the ultrastructure of the fungal hyphae, its interaction with the host cell is of interest. Study of these interactions using freeze-substitution is severely limited by the thin superficial zone in which adequate cryofixation can be obtained. Freezing under high pressure (Moor, 1987) will therefore open new perspectives in phytopathology.

Excellent structural preservation is obtained by freeze-substitution in acetone or methanol containing fixatives like OsO_4, uranylions or aldehydes, either alone or in combination (Inoué et al., 1982; Ito and Ischikawa, 1982; Müller et al., 1980b; Zalokar, 1966), followed by conventional embedding and heat polymerization. Samples prepared in this way, however, are rarely useful for cytochemical studies. Specimens which permit both an optimal structural description and the labelling of intracellular antigens are obtained by combining freeze-substitution with low temperature embedding in Lowicryl (Humbel and Müller, 1983, 1986; Hunziker et al., 1984). Greatly improved labelling efficiency was demonstrated by immunolabelling of the protein groE in E. coli after freeze-substitution followed by low temperature embedding in Lowicryl K4M, as compared to conventional low temperature embedding by the PLT (progressive lowering of temperature) technique (Carlemalm et al., 1986). Freeze-substitution followed by low temperature embedding is currently being investigated in many laboratories (e.g. Hunziker and Herrmann, 1987) and will undoubtedly find increasing applications, especially with the prospect of using high pressure freezing to cryofix thicker and more complex samples. The label efficiency of freeze-substituted and low temperature embedded samples is mainly affected by OsO_4 as a stabilizing additive: it should therefore be avoided. Uranylions and aldehydes at low concentration show no undesired effects with respect to label efficiency; they again help to minimize effects of the organic solvent and the Lowicryls (Weibull et al., 1983, 1984). Furthermore, they may improve the stainability of the biological structures. Under carefully controlled conditions, however, freeze-substitution in a pure solvent can be combined with low-temperature embedding at very low temperatures. A detailed description of the technique, its problems and the instrumentation needed, are found in Humbel and Müller (1986).

IV.   Cryotechniques in scanning electron microscopy

The factors affecting the preservation of the structural integrity are the same in both scanning and transmission electron microscopy. Shrinkage due to complete dehydration and drying may be even more pronounced in scanning electron microscopy since the removed water is usually not replaced by an embedding resin. Techniques based on cryofixation again help to overcome the major problems.

A.   Low temperature scanning electron microscopy (LTSEM)

Low temperature scanning electron microscopy represents the direct, physical approach. Modern LTSEM equipment consists of a high vacuum preparation chamber directly attached to the scanning microscope. In the preparation chamber the specimen can be kept at a controlled temperature. It may be retained intact, fractured or dissected; and kept either fully frozen hydrated, partially freeze-dried ("etched"), or fully freeze-dried. The samples may be coated for subsequent observation in the SEM. The gate valve between preparation chamber and microscope is then opened and the sample is transferred onto a temperature controlled stage in the scanning electron microscope where it is examined at very low temperatures (e.g. 100 K), Uncoated samples may be repeatedly dissected or "etched" if necessary. They may also be partially freeze-dried under visual control in the microscope. A comprehensive overview on LTSEM has been recently given by Beckett and Read (1986). LTSEM experiments by Read and Beckett (1983), Read et al. (1983), Beckett et al. (1984) have illustrated the various dimensional and structural changes that occur in partially or completely freeze-dried specimens. They conclude that the traditional classification of "free" or "bulk" water, as opposed to "bound water" or "water of hydration shells", might provide a greatly simplified view of cellular water (Clegg, 1982). Models that describe more comprehensively the complex interactions of macromolecules, membrane surfaces, ions and water are supported (e.g. Clegg, 1979; Negendank, 1986). LTSEM is currently restricted to lower magnifications (e.g. up to 10 000 ×), mainly due to technical limitations of the equipment (stability of cold stages, moderate resolving power of conventional SEM-instruments). A reliable representation of structural facts is, however, always more valuable than the high resolution detection of insignificant structural details. LTSEM has found wide application, especially in the field of mycology (Beckett and Read, 1986). It may be the method of choice to analyse electrolytes within cells and tissues by X-ray microanalysis of frozen-hydrated bulk samples (Hall and Gupta, 1984).

LTSEM will undoubtedly develop towards improved resolution. Reasonably priced, easily handled high resolution SEM instruments, equipped with reliable field emission guns as well as more sophisticated cryopreparation attachments, are now commercially available. Furthermore, high pressure freezing (Moor, 1987) offers adequate cryofixation of samples of significant size (e.g. a round disc of 2 mm diameter of a fungus-infected apple leaf which is 300 μm thick is frozen in the microcrystalline state throughout the entire sample; Fig. 6).

B. Freeze-drying and freeze-substitution for SEM observation at room temperature

Freeze-drying of cryofixed samples can be performed under well controlled conditions only if the temperature is constant throughout the sample. This is only achieved in very thin samples that stay in perfect thermal contact with the stage of the freeze-drier (e.g. membranes, frozen hydrated cryosections, macromolecular solutions). The experiments of Wildhaber et al. (1982) and Gross (1987) support the observation of Frederik et al. (1984) and of Dubochet et al. (1983) that cryosections can be kept partially freeze-dried at 193 K without apparent shrinkage. At this temperature—independent of the drying time—a certain amount of water is removed from the specimen. More water is only released when the specimen is heated to the temperature range between 223 and 243 K, and is referred to as the water of hydration shells. Shrinking now occurs. This temperature range is well in accordance with the collapse temperatures (263–223 K) determined by MacKenzie (1972) for freeze-dried model solutions. Removal of the hydration shells most certainly leads to conformational changes of proteins and macromolecules. Whether or to what extent it is the main factor responsible for dimensional alterations (e.g. shrinkage) is not yet clear. In a cell, water may play a much more complex role than in the above-mentioned model experiments, and its controlled removal may be very difficult. Only fully freeze-dried samples can be examined in a SEM at room temperature. The specimens therefore have always suffered from shrinkage and collapse. Nevertheless, fully freeze-dried samples can provide useful information since, in contrast to the conventional critical-point drying procedure, any interaction with organic solvents is avoided.

Such a solvent effect is illustrated by the different appearance of the yeast cell surface after conventional and cryofixation-based preparation for electron microscopy. The cells reveal a smooth surface (Fig. 1d) after chemical fixation, dehydration in graded ethanol series followed by critical-point drying. Hair-like structures, termed fimbriae, are detected in SEM after cryofixation followed by complete freeze-drying (Fig. 1a) and in TEM after partial freeze-drying ("deep etching") (Fig. 1b) or freeze-substitution (Fig. 1c) (Walther et al., 1984b; Tokunaga et al., 1986; Baba and Osumi, 1987). The fimbriae represent a dynamic structure, the size and distribution of which

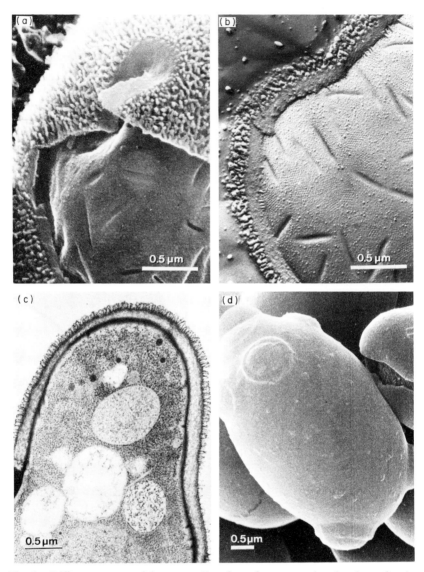

Fig. 1. Different aspects of the yeast cell surface after conventional and cryofixation-based preparation for transmission and scanning electron microscopy. (a) Freeze-drying for SEM, (b) "deep-etching" (partial-freeze-drying), and (c) freeze-substitution (Müller *et al.*, 1980a) for TEM preserve the fimbriae, while fimbriae at the cell surface are lost after conventional preparation for SEM by chemical fixation, dehydration in ethanol and critical-point drying (d). Note the collapsed overall structure of the fully freeze-dried cell shown in (a).

depend on physiological parameters such as growth conditions (medium, growth phase). They seem to be involved in cell–cell (agglutination, mating, cell division) and cell–substrate interactions (cf. Fig. 4). The finding that the fimbriae are lost during the dehydration step in graded ethanol series suggested a useful way of detaching them from the cell surface for subsequent biochemical analysis. When cells are incubated in an aqueous solution containing 30% ethanol or more, the fimbriae are no longer apparent in SEM micrographs (Fig. 1d). When the material present in the alcohol solution is lyophilized, electrophoresed on SDS-gels and stained for proteins and carbohydrates, slowly migrating glycoproteins and low molecular weight proteins can be detected. Acid phosphatase, a well characterized glycoprotein, was identified by immunoblotting using a monoclonal antibody against the protein moiety. It could be demonstrated *in situ* by immuno-scanning electron microscopy that at least some of the fimbriae contain acid phosphatase (Fig. 2) (Walther *et al.*, 1984a). Weinstock and Ballou (1986) suggested that about 60 kD of α-agglutinin, a glycoprotein larger than 200 000 kD, protrudes out of the yeast cell wall. It is conceivable that part of the fimbriae might correspond to agglutinin-like structures. At present little is known about the role of the fimbriae (Maurer and Walther, 1987) since they are detected routinely only after cryopreparative procedures.

Immunocytochemistry at the ultrastructural level relies on complexing the specific, primary antibody with a marker system that can be identified in the electron microscope. Colloidal gold particles of various diameters attached to the primary antibody either directly, via protein A from *Staphyloccus aureus*, or via a second antibody, are widely used. Small gold markers (5–15 nm) are easily detected in thin sections by transmission electron microscopy and, due to recent instrumental improvements (Walther *et al.*, 1984; Walther and Müller, 1986), by scanning electron microscopy as well. SEM immunocyto-chemistry permits the analysis of surface antigens (for a recent review see Hodges *et al.*, 1987). It is especially useful in localizing antigens which are present only in few copies, or which form clusters at the cell surface. Such antigens are easily missed by thin sectioning. Figure 2 shows cells of *Schizosaccharomyces pombe*, labelled with an antibody against acid phosphatase. The primary antibody is visualized by a second antibody coupled to 10 nm colloidal gold. In Fig. 2a a secondary electron signal is used to describe the surface topography. The gold particles are identified by the material-dependent signal of the backscattered electrons (Fig. 2b). The combination of the two images permits an accurate localization of acid phosphatase on some of the fimbriae. The specimen was immunolabelled in suspension, rapidly frozen by plunging into liquid propane (88 K) and freeze-dried at 188 K for 1 h. It was then coated with 8 nm carbon and, still under vacuum, warmed to room temperature. Examination of the specimen was

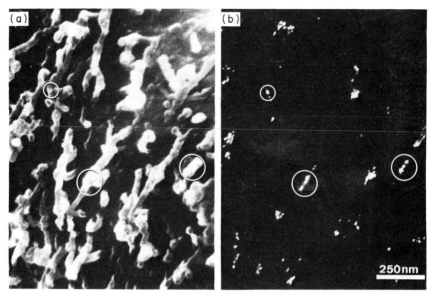

Fig. 2. Immunolabelling of acid phosphatase on the surface of *S. pombe*. (a) Surface structures visualized by the secondary electron signal. (b) Distribution of the gold particles in the same area visualized by the material-dependent backscattered electron signal. The circles indicate corresponding structures. Most of the gold label is localized on the fimbriae.

performed in a conventional high resolution scanning electron microscope, equipped with a field emission gun and a highly sensitive detector for backscattered electrons (Walther and Müller, 1986).

More recently, ultra-high resolution scanning electron microscopes of the "in lens" type (Koike *et al.*, 1971) have become commercially available. Equipped with a field emission gun, they enable the examination of samples with an electron probe of diameter less than 1 nm. Adequate structural preservation is required to make full use of this resolving potential. Experiments using the T_4-polyhead mutant as a model specimen (Laemmli *et al.*, 1976) have shown that only rapid freezing followed by freeze-drying reveals the ring-like structure of the capsomers, which have a diameter of 8 nm (Laemmli *et al.*, 1976) (Fig. 3). Conventional critical-point drying after chemical dehydration of the aldehyde-fixed samples by ethanol or after freeze-substitution failed to preserve structural details at this level. Figure 3 shows a secondary electron image of a rapidly frozen, freeze-dried T_4-polyhead preparation. Freeze-dried at 188 K for 30 min, it was coated with a thin (1–2 nm) continuous film of chromium by "double axis rotary shadowing" (Shibata *et al.*, 1984), warmed to room temperature and examined in an ultra-high resolution scanning electron microscope (Hitachi S-900) at a primary

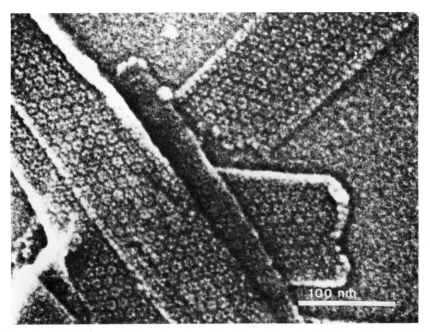

Fig. 3. TEM-like resolution of surface details is achieved by high resolution SEM only after freeze-drying. The ring-like capsomeres of a T_4-polyhead (diameter 8 nm) are clearly shown (Laemmli *et al.*, 1976).

magnification of 200 000 × . It is essential that the very thin, continuous metal film is of high quality for the generation of high topographic resolution; it should be fine grained, continuous and sufficiently thin to avoid levelling of small topographic details, and it should be made of a metal in which the generation of backscattered electrons is minimized. "Double axis rotary shadowing" allows the production of thin continuous metal films at low temperatures by electron gun evaporation. During shadowing, the electron gun is moved from 0–90° while the cooled specimen is rotated. The necessary prerequisites for high resolution scanning electron microscopy (instrumentation, signal generation and detection) of biological objects have been recently reviewed by Peters (1985). The intention of Fig. 3 is to illustrate the potential that scanning electron microscopy, optimized with respect to both structural preservation and instrumentation, might achieve in future.

Freeze-drying should be performed out of a clean solvent, e.g. distilled water, in order to avoid the deposition of solutes onto the specimen surface during drying. Aldehyde prefixation therefore is frequently required to render the specimen resistant against the treatment with distilled water. Freeze-substitution followed by critical point drying partially overcomes this problem in the medium resolution range. Barlow and Sleigh (1979) systematically tested various freeze-substitution media for stabilizing the metachronal wave of the cilia of *Paramecium* in the SEM. An acceptable preservation of the fimbriae of the yeast cell surface by freeze-substitution and critical point drying was demonstrated by Baba and Osumi (1987).

V. Methods of cryofixation

Several techniques for the rapid cryoimmobilization of thin aqueous layers of suspensions and tissues are currently employed, either by using home-made or commercially available equipment including spray-freezing (Bachmann and Schmitt, 1971), propane-jet freezing (Müller *et al.*, 1980b; Müller and Moor, 1984a), metal mirror freezing (impact-freezing, slam-freezing: Van Harreveld and Crowell, 1964; Fernandez-Moran, 1960; Heuser *et al.*, 1979) and plunge-freezing (Costello *et al.*, 1984; Handley *et al.*, 1981).

The high degree of acceptance achieved by these techniques is best documented by the way in which they are treated in recent books and reviews, e.g. Robards and Sleytr (1985), Chapman and Staehlin (1986), Sitte *et al.* (1987), minor differences in judging the relative merits depending mainly on the scope of application and whether or to what extent the authors have been engaged in the development of a particular procedure. A comprehensive overview of the currently available cryofixation techniques together with a helpful, however subjective, discussion on the choice of the most suitable method has been recently given by Sitte *et al.* (1987).

The extraction of the maximum information from a cryofixed sample, as well as an accurate interpretation of results, often depend on correlated information from several follow-up procedures. The choice of the most appropriate freezing method therefore depends on the size of the sample under investigation and the follow-up procedure by which one wants to analyse it. Sandwich techniques are recommended to cryofix cell suspensions, since they are equally suited to further processing by freeze-fracturing, freeze-substitution and freeze-drying for scanning electron microscopy. Heat can be extracted from the specimen sandwich either by controlled rapid plunging into a suitable coolant (Costello *et al.*, 1984; Sitte, 1987) or by shooting the coolant simultaneously from opposite directions onto the specimen surface (propane-jet: Müller *et al.*, 1980b; Müller and Moor, 1984a). Successful cryofixation with high reproducibility depends on the same prerequisites for both techniques: the total mass of the sandwich must be as low as possible (Handley *et al.*, 1981) and the assembly of the sandwich has to be carefully controlled, since sufficient freezing is achieved only in a very thin layer. A conventional EM grid (gold or copper) is dipped into an adequately diluted pellet, such that the meshes of the grid are filled. This grid is then placed between two low mass copper platelets where it serves as a spacer, and hence limits the thickness of the aqueous layer to be frozen (Müller and Moor, 1984a). Adequately manufactured copper platelets are now commercially available (Balzers Union, Balzers FL). The frozen sandwiches are fractured under liquid nitrogen for further processing by freeze-substitution or in a freeze-fracture apparatus for freeze-drying or replication. Sandwiches always split between either side of the spacer grid and a copper platelet. The fracture plane is therefore forced through the best frozen part of the sample. For replication, the copper platelet with the adhering grid is convenient, since the replica usually adheres to the grid. This greatly facilitates handling of the replica during the cleaning process. The platelet without the grid is used for SEM observations, after freeze-drying and coating.

The complementary information exhibited by the follow-up procedures mentioned above is shown in Fig. 4. The micrographs originate from a study of Käppeli et al. (1984) on the relationship between the structure of the cell surface of the yeast *Candida tropicalis* and hydrocarbon transport. At the surface of *C. tropicalis* cells grown on hexadecane as a carbon source, scanning electron microscopy of freeze-dried samples reveals a distinct fringe of long fimbriae (Fig. 4a). This result is confirmed in thin sections of the same cells after freeze-substitution (Fig. 4b). Strongly contrasted areas in the fimbrial fringe of the hexadecane-grown cells relate the fimbriae to the binding of hydrocarbons. The fimbriae are much shorter on the surface of cells grown on glucose as a carbon source (Figs 4d and e) and can be almost completely removed by proteolytic digestion, as revealed by SEM (Fig. 4c). Freeze-fracturing of hydrocarbon-grown cells exhibits large crystalline structures in the cell interior (Fig. 5). Only the contours of these crystals are detected in thin sections after freeze-substitution in methanol. This indicates dissolution by the organic solvent, and therefore suggests a lipidic nature of the crystals.

High pressure freezing is at present the only known practical way of cryofixing larger samples. Its development started about 20 years ago (Moor, 1987) but adequate instrumentation became commercially available (Balzers Union, Balzers, FL) only recently. The technique is not yet widely used due to the high price of the equipment and the small number of stimulating results. This is not expected to change very rapidly, despite the fact that the commercial high pressure freezer works reliably well with respect to the physical performance. It provides high cooling rates and reaches 2100 bar within 15 ms with a perfect coordination of the rise in pressure with the drop in temperature. Major problems seem to arise from the way pressure and cold are transferred to the sample. From our experience the yield in well cryofixed specimens is greatly enhanced if the sample matches exactly the cavities of the metal planchets between which it is sandwiched for high pressure freezing

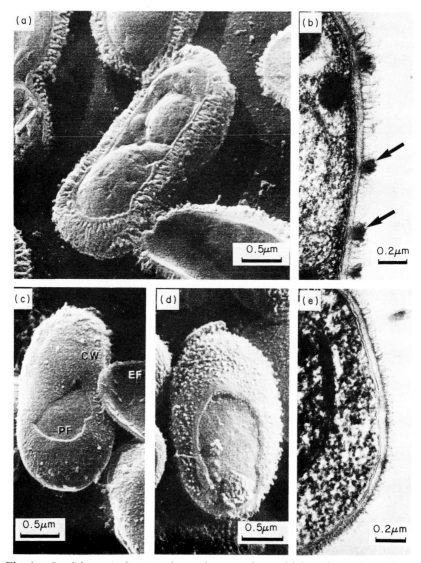

Fig. 4. *Candida tropicalis* grown in continuous culture with hexadecane (a and b) and glucose (d and e) as the carbon sources. Cells sandwiched between low mass copper platelets and rapidly frozen by the propane-jet technique (Müller *et al.*, 1980b, 1984) were freeze-dried for SEM (a, c and d) or freeze-substituted for TEM thin section analysis (b and e). Note the much longer fimbriae of the hexadecane-grown cells (a and b). The arrows in (b) mark contrasted regions which occur as a consequence of incubation of the cells with hexadecane prior to rapid freezing. Protease treatment removed the fimbriae (c). In (c) the plasma membrane is fractured, exposing the plasmic fracture face (PF) and an exoplasmic fracture face (EF) of a neighbouring cell.

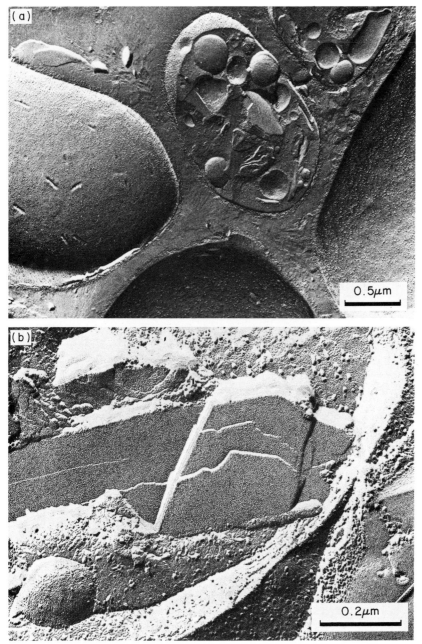

Fig. 5. *Candida tropicalis* grown with hexadecane as a carbon source. Freeze-fracturing of cells cryofixed identically to Fig. 4 reveal large crystalline structures in the interior (b). Their fracturing behaviour, as well as the fact that they are dissolved by the organic solvent during freeze-substitution, suggests a lipidic nature of the crystals.

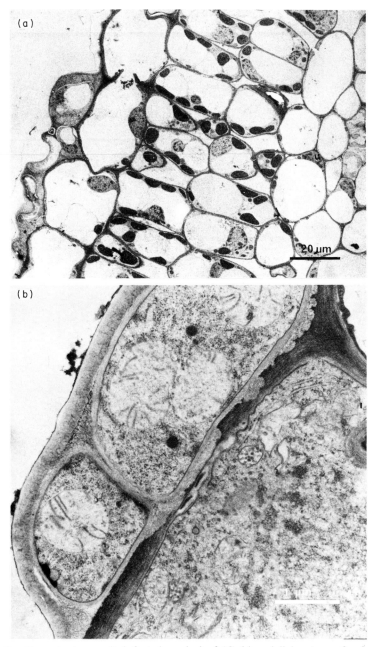

Fig. 6. *Venturia inaequalis*-infected apple leaf (Golden delicious) cryofixed under high pressure, freeze-substituted in acetone/OsO$_4$ and embedded in Araldite/Epon. Note the well preserved cellular organization (a), the absence of an ice crystal induced segregation network and the distorted cell wall, as well as the undulated plasma membrane of the host cell at the site of interaction with the fungus (b).

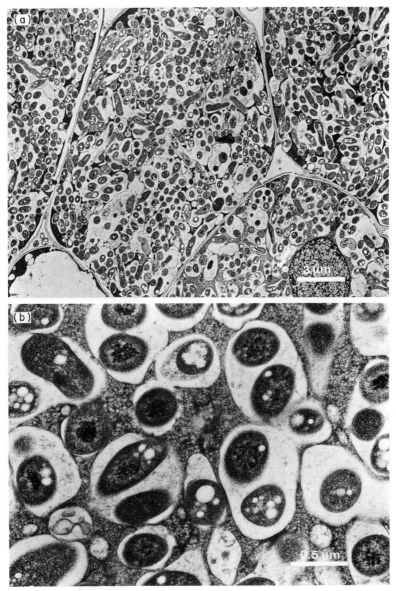

Fig. 7. Nodules of soybean roots infected with *Bradyrhizobium japonicum* strain 110 were sliced to a thickness of ~200 μm and mildly prefixed in 0.2% glutaraldehyde and 0.3% formaldehyde. Discs of diameter 2 mm were punched out of the slices and cryofixed under high pressure. Freeze-substitution in acetone containing uranylacetate (2%) and glutaraldehyde (0.2%) was followed by low temperature embedding into Lowicryl HM 20 at 243 K. Note the well preserved overall structure (a) and the preservation of the peribacteroid membrane. Aldehyde prefixation, however, induces aggregation of the nuclear material of the bacteroids and, due to osmotic effects, the shape of the peribacteroid membrane may have been altered (b).

(Müller and Moor, 1984b). Adequate cryofixation of thick samples (300 μm) is achieved with a high yield (60–80%) of plant material, e.g. *Venturia*-infected apple leaves (Fig. 6) and root nodules (Fig. 7) as well as of thick suspensions of microorganisms that fulfil the above geometrical requirement (Fig. 8). Discs of 2 mm diameter are punched out of the leaves or out of thick microtome slices of root nodules. The tissue discs are then sandwiched between two aluminium planchets which contain a cavity that matches the specimen dimensions as precisely as possible. As it is compressible, the air in the intercellular space of the leaves must be replaced by water. Locally different compressibility not only reduces the cryoprotective effect of high pressure, but also results in the destruction of the cellular organization. Intercellular air is replaced with water by mild evacuation of the punched specimen discs in tap water.

The yield in adequately cryofixed samples by high pressure freezing can be further increased by mild aldehyde fixation or by a short incubation in 3–5% methanol as a cryoprotectant. The root nodule in Fig. 7, for example, was fixed with 0.2% glutaraldehyde and 0.3% formaldehyde prior to freezing.

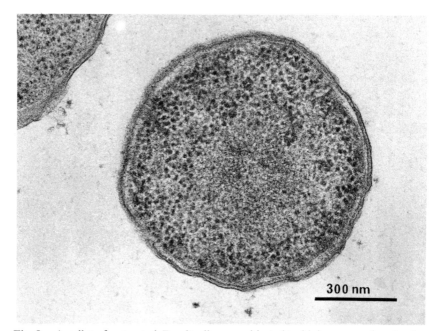

300 nm

Fig. 8. A pellet of untreated *E. coli* cells was subjected to high pressure freezing in a 200 μm thick layer, freeze-substituted in acetone containing 1% OsO_4 and embedded in Araldite/Epon. The figure shows a section through a cell situated approximately in the centre of the frozen pellet. Note the quality of cryofixation and the absence of aggreggated nuclear material in the "ribosome free" area.

High pressure freezing yields adequate cryofixation but osmotic effects, for example on the peribacteroid membrane, cannot be excluded. Furthermore, the nuclear material of the *Bradyrhizobium japonicum* cells appears aggregated and condensed. Condensation of the nuclear material cannot be detected in untreated, high pressure frozen and freeze-substituted *E. coli* cells (Fig. 8). The bacterial nucleoid appears as a "ribosome free" space (Hobot *et al.*, 1985).

VI. Conclusions

Cryofixation is alone in being able to immobilize biological structures close to the living state. The application of cryofixation-based preparative procedures is mandatory for the correlation of structure and function at the ultrastructural level. Purely physical techniques provide safe structural description and preserve the spatial distribution of diffusible elements. Hybrid techniques, for example freeze-substitution, facilitate thin sectioning and, in combination with low temperature embedding, immunocytochemical studies of intracellular antigens. The quality of structural preservation, however, has to be gauged by purely physical procedures such as cryosectioning and freeze-fracturing. Adequate procedures to cryofix suspensions and thin superficial layers of tissues are available. High pressure freezing offers the prospect of being able to cryofix samples up to a thickness of approximately 0.5 mm.

Acknowledgements

I wish to thank my colleagues Dr P. Walther, Dr D. Studer and M. Michel for kindly providing some of the micrographs. Special thanks are given to M. Horvath and M. L. Yaffee for their help with the manuscript.

References

Adrian, M., Dubochet, J., Lepault, J. and McDowall, A. W. (1984). *Nature* **308**, 32–36.
Amako, K. and Takade, A. (1985). *J. Electron Microsc. (Japan)* **34**, 13–17.
Baba, M. and Osumi, M. (1987). *J. Electron Microsc. Techn.* **5**, 249–261.
Bachmann, L. and Mayer, E. (1987). In *Cryotechniques in Biological Electron Microscopy* (R. A. Steinbrecht and K. Zieroid, eds), pp. 3–34. Springer-Verlag, Berlin.
Bachmann, L. and Schmitt, W. W. (1971). *Proc. Natl. Acad. Sci. (USA)* **68**, 2149–2152.
Barlow, D. I. and Sleigh, M. A. (1979). *J. Microsc. (Oxford)* **115**, 81–95.
Beckett, A. and Read, N. D. (1986). In *Ultrastructure Techniques for Microorganisms* (H. C. Aldrich and W. J. Todd, eds), pp. 45–86. Plenum Press, New York and London.
Beckett, A., Read, N. D. and Porter, R. (1984). *J. Microsc. (Oxford)* **136**, 87–95.
Carlemalm, E., Villiger, W., Acetarin, J.-D. and Kellenberger, E. (1986). In *The Science of Biological Specimen Preparation 1985* (M. Müller, R. P. Becker, A. Boyde and J. J. Wolosewick, eds), pp. 147–154. SEM, AMF O'Hare, IL.

Causton, B. (1986). In *The Science of Biological Specimen Preparation 1985* (M. Müller, R. P. Becker, A. Boyde and J. J. Wolosewick, eds), pp. 209–214. SEM, AMF O'Hare, IL.

Chapmann, R. L. and Staehlin, L. A. (1986). In *Ultrastructure Techniques for Microorganisms* (H. C. Aldrich and W. J. Todd, eds) pp. 213–240. Plenum Press, New York and London.

Clegg, J. S. (1982). In *Biophysics of Water* (F. Franks and S. F. Mathias, eds), pp. 365–383. Wiley, New York.

Clegg, J. S. (1979). In *Cell-associated Water* (W. Drost-Hansen and J. S. Clegg, eds), pp. 363–413. Academic Press, London and New York.

Costello, M. J. and Corless, J. M. (1978), *J. Microsc.* **112**, 17–37.

Costello, M. J., Fetter, R. and Corless, J. M. (1984). In *The Science of Biological Specimen Preparation 1984* (J.-P. Revel, T. Barnard and G. H. Haggis, eds), pp. 105–115. SEM, AMF O'Hare, IL.

Dempsey, G. P. and Bullivant, S. (1976a). *J. Microsc. (Oxford)* **106**, 251–260.

Dempsey, G. P. and Bullivant, S. (1976b). *J. Microsc. (Oxford)* **106**, 261–271.

Dubochet, J. and McDowall, A. W. (1984). In *The Science of Biological Specimen Preparation 1984* (J. P. Revel, T. Barnard and G. H. Haggis, eds), pp. 147–152. SEM. AMF O'Hare, IL.

Dubochet, J., Lepault, J., Freeman, R., Berriman, J. A. and Homo, J. C. (1982). *J. Microsc.* **128**, 219–237.

Dubochet, J., McDowall, A. W., Menge, B., Schmid, E. N. and Lickfeld, K. G. (1983), *J. Bacteriol.* **155**, 381–390.

Dubochet, J., Adrian, M., Chang, J-J., Lepault, J. and McDowall, A. W. (1987). In *Cryotechniques in Biological Electron Microscopy* (R. A. Steinbrecht and K. Zierold, eds), pp. 114–131. Springer-Verlag, Berlin.

Ebersold, H. R., Cordier J.-L. and Lüthy, P. (1981). *Arch. Microbiol.* **130**, 19–23.

Fernandez-Moran, H. (1960). *Ann. N.Y. Acad. Sci.* **85**, 689–698.

Frederik, P. M., Busing, W. M. and Hax, W. M. A. (1984). In *Electron Microscopy 1984*, Vol II (A. Csanády, P. Röhlich and D. Szabó, eds), pp. 1411–1412. Proceedings of the 8th European Congress on Electron Microscopy, Budapest, Hungary.

Giddings, T. H. and Wray, G. P. (1986). In *Ultrastructure Techniques for Microorganisms* (H. C. Aldrich and W. J. Todd, eds), pp. 241–265. Plenum Press, New York and London.

Gross, H. (1987). In *Cryotechniques in Biological Electron Microscopy* (R. A. Steinbrecht and K. Zierold, eds), pp. 205–215. Springer-Verlag, Berlin.

Gulik-Krzywicki, T. and Costello, M. J. (1978). *J. Microsc.* **112**, 103–113.

Hall, T. A. and Gupta, B. L. (1984). *J. Microsc. (Oxford)* **136**, 193–208.

Handley, D. A., Alexander, J. T. and Chien, S. (1981). *J. Microsc. (Oxford)* **121**, 273–282.

Heuser, J. E., Reese, T. S., Dennis, M. J., Jan, Y., Jan, L. and Evans, L. (1979). *J. Cell. Biol.* **81**, 275–300.

Hobot, J. A., Carlemalm, E., Villinger, W. and Kellenberger, E. (1984). *J. Bacteriol.* **160**, 143–152.

Hobot, J. A., Villiger, W., Escaig, J., Maeder, M., Ryter, A. and Kellenberger, E. (1985). *J. Bacteriol.* **162**, 960–971.

Hoch, H. C. (1986). In *Ultrastructure Techniques for Microorganisms* (H. C. Aldrich and W. J. Todd, eds), pp. 183–212, Plenum Press, New York.

Hoch, H. C. and Howard, R. J. (1980). *Protoplasma*, **103**, 281–297.

Hodges, G. M., Southgate, J. and Toulson, E. C. (1987). *Scanning Microscopy* **1**, 301.

Howard, R. J. and Aist, J. R. (1979). *J. Ultrastruct. Res.* **66**, 224–234.

Humbel, B. M., Marti, T. and Müller, M. (1983). *Beitr. Elektronmikrosk. Direktabb. Oberfl.* **16**, 585–594.
Humbel, B. M. and Müller, M. (1986). In *The Science of Biological Specimen Preparation 1985* (M. Müller, R. P. Becker, A. Boyde and J. J. Wolosewick, eds), pp. 175–183. SEM, AMF O'Hare, IL.
Hunziker, E. B. and Herrmann, W. (1987). *J. Histochem. Cytochem.* **6**, 647–655.
Hunziker, E. B., Herrmann, W., Schenk, R. K., Müller, M. and Moor, H. (1984). *J. Cell. Biol.* **98**, 267–276.
Ingram, F. D. and Ingram, M. J. (1984). In *The Science of Biological Specimen Preparation 1984* (J.-P. Revel, T. Barnard and G. H. Haggis, eds), pp. 167–174. SEM, AMF O'Hare, IL.
Inoué, K., Kurosumi, K. and Deng, Z. P. (1982). *J. Electron Microsc.*, **31**, 93–97.
Ito, T. and Ischikawa, A. (1982). *J. Electron Microsc.*, **31**, 235–248.
Kanno, H., Speedy, R. J. and Angell, C. A. (1975). *Science* **189**, 880–881.
Käppeli, O., Walther, P., Müller, M. and Fiechter, A. (1984). *Arch. Microbiol.* **138**, 279–282.
Kellenberger, E. (1987). In *Cryotechniques in Biological Electron Microscopy* (R. A. Steinbrecht and K. Zierold, eds), pp. 35–63. Springer-Verlag, Berlin.
Kellenberger, E., Carlemalm, E. and Villiger, W. (1986). In *The Science of Biological Specimen Preparation 1985* (M. Müller, R. P. Becker, A. Boyde and J. J. Wolosewick, eds), pp. 1–20. SEM, AMF O'Hare, IL.
Knoll, G., Verkleij, A. J. and Plattner, H. (1987). In *Cryotechniques in Biological Electron Microscopy* (R. A. Steinbrecht and K. Zierold, eds), pp. 258–271. Springer-Verlag, Berlin.
Koike, H., Ueno, K. and Suzuki, M. (1971). In *Proceedings of the 29th Annual Meeting of the Electron Microscopical Society of America*, Baton Rouge/L.A., pp. 28–29.
Laemmli, U. K., Amos, L. A. and Klug, A. (1976). *Cell.* **7**, 191–203.
Lee, R. M. K. W. (1984). In *The Science of Biological Specimen Preparation 1984* (J-P. Revel, T. Barnard and G. H. Haggis, eds), pp. 61–70. SEM, AMF O'Hare, IL.
Macdonald, A. G. (1984). *Phil. Trans. R. Soc. Lond., Ser. B.* **304**, 47–68.
MacKenzie, A. P. (1972). *Scanning Electron Microsc.* **2**, 273–280.
Maurer, A. and Walther, P. (1987). In *Membranous Structures*, pp. 39–63. Academic Press, London.
Menco, B. P. M. (1966). *J. Electron. Microsc. Tech.* **4**, 177–240.
Moor, H. (1987). In *Cryotechniques in Biological Electron Microscopy* (R. A. Steinbrecht and K. Zierold, eds), pp. 175–191. Springer-Verlag, Berlin.
Moor, H. and Höchli, M. (1970). In *Proceedings of the 7th International Congress on Electron Microscopy, Grenoble,* Vol. 1 (P. Favard, ed.), pp. 449–450. Société Française de Microscopie Electronique, Paris.
Moor, H., Bellin, G., Sandri, C. and Akert, K. (1980). *Cell. Tissue Res.* **209**, 201–216.
Müller, M. and Moor, H. (1984a). In *Proceedings of the 42nd Annual Meeting of the Electron Microscopical Society of America* (G. W. Bailey, ed), pp. 6–9. San Francisco Press, San Francisco.
Müller, M. and Moor, H. (1984b). In *The Science of Biological Specimen Preparation 1984* (J.-P. Revel, T. Barnard and G. H. Haggis, eds), pp. 131–138. SEM, AMF O'Hare, IL.
Müller, M., Marti, T. and Kriz, S. (1980a). In *Electron Microscopy 1980,* Proceedings of the 7th European Congress on Electron Microscopy, Leiden (P. Brederoo and W. de Priester, eds), Vol. 2, pp. 720–772.
Müller, M., Meister, N. and Moor, H. (1980b). *Mikroskopie (Wien)* **36**, 129–140.
Negendank, W. (1986). In *The Science of Biological Specimen Preparation 1985* (M.

Müller, R. P. Becker, A. Boyde and J. J. Wolosewick, eds), SEM, AMF O'Hare, IL.
Peters, K.-R. (1985). *Scanning Electron Microsc.* **4**, 1519–1544.
Pinto da Silva, P. and Kan, F. W. K. (1984). *J. Cell Biol.* **99**, 1156–1161.
Plattner, H. and Bachmann, L. (1982). *Int. Rev. Cytol.* **79**, 237.
Rash, J. E. and Hudson, C. S. (eds) (1979). *Freeze-fracture: Methods, Artefacts, and Interpretation* Raven Press, New York.
Read, N. D. and Beckett, A. (1983). *J. Microsc. (Oxford)* **132**, 179–184.
Read, N. D., Dorter, R. and Beckett, A. (1983). *Can. J. Bot.* **61**, 2059–2078.
Riehle, U. (1968). Ueber die Vitrifizierung verdünnter wässriger Lösungen. Dissertation no. 4271, Eidgenössiche Technische Hochschule, Zürich.
Riehle, U. and Höchli, M. (1973). In *Freeze-etching Technique and Applications* (E. L. Benedetti and P. Favard, eds), pp. 31–61. Sociéte Française de Microscopie Electronique, Paris.
Robards, A. W. and Sleytr, U. B. (1985). *Low Temperature Methods in Biological Electron Microscopy.* Elsevier, Amsterdam.
Shibata, Y., Arima, T. and Yamamoto, T. (1984). *J. Microsc. (Oxford)* **136**, 121–123.
Sitte, H., Edelmann, L. and Neumann, K. (1987). In *Cryotechniques in Biological Electron Microscopy* (R. A. Steinbrecht and K. Zierold, eds), pp. 87–113. Springer-Verlag, Berlin.
Skaer, H. le B. (1982). *J. Microsc.* **125**, 137.
Steinbrecht, R. A. (1980). *Tissue Cell* **12**, 73–100.
Steinbrecht, R. A. and Müller, M. (1987). In *Cryotechniques in Biological Electron Microscopy* (R. A. Steinbrecht and K. Zierold, eds), pp. 149–172. Springer-Verlag, Berlin.
Stewart, M. and Vigers, G. (1986). *Nature* **319**, 631–636.
Tanford, C. (1980). *The Hydrophobic Effect: Formation of Micelles and Biological Membranes*, 2nd edn. John Wiley & Sons, New York.
Tokunaga, M., Kusamichi, M. and Koike, H. (1986). *J. Electron Microsc.* **3**, 237–246.
Tokuyasu, K. T. (1984). In *Immunolabelling for Electron Microscopy* (J. M. Polak and I. M. Varndell, eds), pp. 71–82. Elsevier, Amsterdam.
Ushiyama, M., Cravalho, E. G. and Levin, R. L. (1979). *J. Membr. Biol.* **6**, 112.
Van Harreveld, A. and Crowell, J. (1964). *Anat. Rec.* **149**, 381–385.
Verkleij, A. J. (1984). *Biochim. Biophys. Acta.* **779**, 43–63.
Walther, P., Kriz, S., Müller, M., Ariano, B. H., Bordbeck, U., Ott, P. and Schweingruber, M. E. (1984a). *Scanning Electron Microsc.* **3**, 1257–1266.
Walther, P., Müller, M. and Schweingruber, M. E. (1984b). *Arch. Microbiol.* **137**, 128–134.
Walther, P. and Müller, M. (1986). In *The Science of Biological Specimen Preparation 1985* (M. Müller, R. P. Becker, A. Boyde and J. J. Wolosewick, eds), pp. 195–201. SEM, AMF O'Hare, IL.
Weibull, C. and Christiansson, A. (1986). *J. Microsc.* **142**, 79–86.
Weibull, C., Christiansson, A. and Carlemalm, E. (1983). *J. Microsc.* **129**, 201–207.
Weibull, C., Villiger, W. and Carlemalm, E. (1984). *J. Microsc. (Oxford)* **134**, 213–216.
Weinstock, K. and Ballou, C. E. (1986). *J. Biol. Chem.* **34**, 16174–16179.
White, D. L., Andrews, S. B., Faller, J. W. and Barrnett, R. J. (1976). *Biochim. Biophys. Acta* **436**, 577–592.
Wildhaber, I., Gross, H. and Moor, H. (1982). *J. Ultrastruct. Res.* **80**, 367–373.
Zalokar, M. (1966). *J. Ultrastruct. Res.* **15**, 469–479.
Zierold, K. (1987). *Cryotechniques in Biological Electron Microscopy* (R. A. Steinbrecht and K. Zierold, eds), pp. 132–148. Springer-Verlag, Berlin.

2

Analysis of Crystalline Bacterial Surface Layers by Freeze-etching, Metal Shadowing, Negative Staining and Ultra-thin Sectioning

UWE B. SLEYTR, PAUL MESSNER and DIETMAR PUM

Zentrum für Ultrastrukturforschung und Ludwig Boltzmann-Institut für Ultrastrukturforschung, Universität für Bodenkultur, A-1180 Wien, Austria

METHODS IN MICROBIOLOGY
VOLUME 20   ISBN 0 12 521520 6

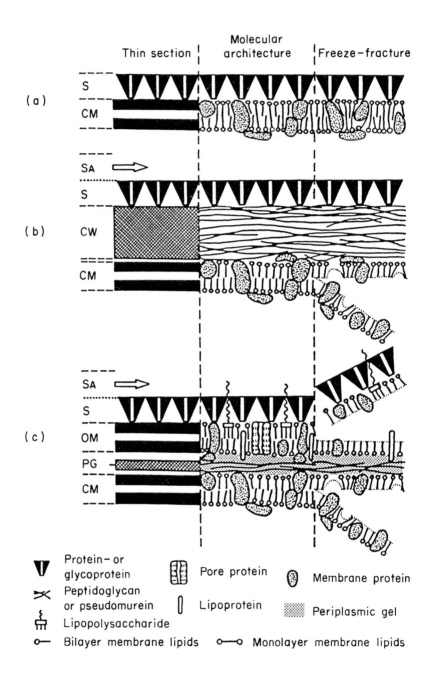

	Thin section	Molecular architecture	Freeze-fracture

(a) S, CM

(b) SA, S, CW, CM

(c) SA, S, OM, PG, CM

▼ Protein- or glycoprotein

⚏ Pore protein

◉ Membrane protein

✄ Peptidoglycan or pseudomurein

▯ Lipoprotein

▒ Periplasmic gel

🖁 Lipopolysaccharide

o— Bilayer membrane lipids   o—o Monolayer membrane lipids

I. Introduction

Most prokaryotic cells possess layered assemblies of polymers and heteropolymers external to the cytoplasmic membrane. On the basis of structural and biochemical studies cell envelope profiles can be classified into three main categories (Fig. 1):

(1) cell envelopes composed only of a crystalline array of macromolecules which is in association with the cytoplasmic membrane;

(2) Gram-positive cell envelopes frequently having a relatively thick rigid sacculus (e.g. peptidoglycan or pseudomurein) external to the cytoplasmic membrane; and

(3) Gram-negative envelopes with a thin peptidoglycan sacculus and an outer membrane.

Although not a universal feature, many prokaryotic organisms possess a crystalline array of proteinaceous subunits as the outermost component of the cell envelope, commonly referred to as the S (surface) layer (for reviews see Sleytr, 1978; Sleytr and Messner, 1983, 1988; König and Stetter, 1986; Sleytr *et al.*, 1986a; Smit, 1987; Baumeister and Engelhardt, 1987).

S-layers are found on members of nearly every taxonomic group in Gram-positive and Gram-negative eubacteria and archaebacteria (Sleytr and Messner, 1983, 1988; Sleytr *et al.*, 1986a) and some green algae (Roberts *et al.*, 1982; Shaw and Hills, 1982). Crystalline layers similar to S-layers have also been detected in bacterial sheaths (Shaw *et al.*, 1985; Stewart *et al.*, 1985; König and Stetter, 1986) and spore coats (Holt and Leadbetter, 1969). The location and ultrastructure of S-layers of a great variety of organisms has been investigated by electron microscopy of thin sections of freeze-etched, freeze-dried, shadowed, negatively stained or frozen hydrated preparations. S-layers of archaebacteria have predominantly hexagonal (p6) symmetry (Sleytr *et al.*, 1986a), whereas in eubacteria square (p4) and oblique (p2) lattices are also commonly observed (Sleytr and Messner, 1983). Although p1 and p3

Fig. 1. Schematic drawing of the three main categories of bacterial cell envelopes containing S-layers. Left: thin section profiles as revealed by transmission electron microscopy. Centre: molecular structure showing major components. Right: freeze-fracture behaviour of cell envelopes. (a) Cell envelope structure of archaebacteria which lack a rigid cell wall component; (b) Gram-positive cell envelope; (c) Gram-negative cell envelope. CM, cytoplasmic membrane; CW, rigid cell wall layer in Gram-positive organisms (peptidoglycan in eubacteria, pseudomurein in archaebacteria); OM, outer membrane; PG, peptidoglycan; S, crystalline surface layer; SA, indicates location for possible additional S-layers. Redrawn and modified diagram from Sleytr and Glauert (1982) and Sleytr and Messner (1983).

symmetry are both feasible, such lattices have not been convincingly demonstrated. Some bacteria have been shown to produce more than one S-layer on their surface (Remsen *et al.*, 1970; Beveridge and Murray, 1976; Stewart and Murray, 1982; Tsuboi *et al.*, 1982; for reviews see Sleytr and Messner, 1983, 1988).

Chemical and structural analyses have indicated that most S-layers are composed of a single, homogeneous protein or glycoprotein species. The subunits have molecular weights ranging from approximately 40 000 to 250 000, being arranged into morphological units (which consist of two, four or six monomers with centre-to-centre spacings ranging from 10 to 32 nm) (Sleytr and Messner, 1983; Sleytr *et al.*, 1986a). Comparative studies on the distribution and uniformity of S-layers have shown that individual strains of a species can exhibit a remarkable degree of heterogeneity regarding the structure of the crystalline arrays and the molecular weight of the constituent subunits (Messner *et al.*, 1984).

Various methods have been developed for the isolation and purification of S-layers (for reviews see Sleytr and Messner, 1983; Koval and Murray, 1984; Messner and Sleytr, 1988).

The constituent subunits of most S-layers interact with each other and with the underlying cell envelope components (Fig. 1) through non-covalent forces; common procedures for solubilization involve treatment with chaotropic agents, detergents, or metal chelating agents, or changes in pH and/or ionic strength. A few S-layers, particularly from extremely thermophilic archaebacteria, are highly resistant to common denaturants, suggesting that most of their protein subunits are covalently linked (König and Stetter, 1986; Messner *et al.*, 1986b).

Isolated subunits from various organisms have the ability to reassemble *in vitro* into regular arrays with the same lattice dimensions as those observed in intact cells upon removal of the disintegrating agent. Depending on the S-layer symmetry, the reassembly conditions and the morphology of the S-layer subunits, flat sheets, cylinders or vesicles are formed (Aebi *et al.*, 1973; Hastie and Brinton, 1979a; Sleytr, 1981; Sleytr and Messner, 1983; Koval and Murray, 1984; Jaenicke *et al.*, 1985; Sleytr *et al.*, 1986b; Messner *et al.*, 1987).

S-layers are involved in constant interaction between the cell and its environment; as porous proteinaceous membranes completely covering the cell surface, S-layers have the potential to function as: (1) protective coats, molecular sieves, and molecule and ion traps; (2) promoters for cell adhesion and surface recognition; and (3) a framework which determines and maintains cell shape or envelope rigidity (for reviews see Sleytr and Messner, 1983, 1988; Smit, 1987; Koval and Murray, 1986).

Due to their surface location, structural regularity, and the fact that S-layer subunits are produced in larger amounts than any other class of proteins in the cell, crystalline surface layers are eminently suitable subjects for studies of the processes involved in the synthesis, glycosylation and secretion of proteins, as well as structural aspects of recognition and assembly in biological macromolecules.

This chapter reviews the electron microscopical preparation and evaluation procedures most often used to study S-layers.

II. Freeze-etching

A. General principles

Both freeze-fracture replication and freeze-etching techniques involve the production of a vacuum-deposited replica from a fracture plane in a frozen, fully hydrated specimen (for review see Robards and Sleytr, 1985). Structural details are revealed by two different processes, namely fracturing and etching. The precise fracture path within a bacterial cell is generally unpredictable. The fracture passes through the regions of the cell envelope which offer the least resistance. The preferential cleavage planes for the three major types of bacterial cell envelopes are illustrated in Fig. 1. The fractured cell suspension can be replicated immediately after fracturing. Surfaces of single cells or isolated envelope components (e.g. S-layers) which are not favoured by the fracture plane can be exposed by controlled sublimation of ice from the fracture face (etching). For many bacteria, it has been shown that extended areas of S-layers on the cell surface can be visualized by deep etching.

B. Sampling, pretreatment and freezing

When using freeze-etching techniques for studying S-layers, the main purpose of the sampling, pretreatment and cryofixation procedures is to preserve the regular array as closely as possible to that existing *in vivo*. These requirements are common to all cryotechniques and have been reviewed in detail (Lickfeld, 1976; Robards and Sletyr, 1985).

Prokaryotic cells are almost ideal specimens for evaluation by freeze-etching techniques. It is possible to freeze the specimen directly in such a way that it retains good ultrastructure, solute distribution and viability. Plate cultures are used if at all possible. With liquid cultures, centrifugation should be as short as possible to prevent unpredictable modifications of the bacterial fine structure (Lickfeld, 1976). Cells are best frozen as small samples (1.0 μl) in suspension, or as pellets, without cryoprotection, using standard freezing methods (Plattner and Bachmann, 1982; Robards and Sleytr, 1985; Sitte *et al.*, 1987). Due to the relatively low average water content, little damage from growth of ice crystals will occur within most prokaryotic cells, but extracellular ice crystals can cause distortions to the cell. Such extracellular ice crystals can also dislocate highly hydrated slime and capsular material, which is then trapped in the eutectic network surrounding the ice crystals. Thus, with some organisms the detection of S-layers depends on the development of suitable methods for the selective disintegration and solubilization or mechanical removal (e.g. using shear forces) of masking material before freezing. If a cryoprotectant is required, then the cells are infiltrated prior to freezing in 20–30% buffered glycerol solution with a pH and a final ionic strength equivalent to those of the growth environment (Lickfeld, 1976). However, it should be remembered that with cryoprotected specimens extended areas of S-layers cannot be revealed upon deep etching due to the unetchable pseudo-eutectic network of the glycerol–water matrix generated during freezing.

There are many different designs of holders for mounting specimens. As a general rule, the smaller the specimen volume and the lower the thermal capacity of the specimen support, the more likely it is that cells can be frozen without damaging ice crystal formation occurring. Freezing methods suitable for prokaryotic organisms are described in another chapter of this volume. There are also many publications reviewing rapid freezing methods (for references see Plattner and Bachmann, 1982; Robards and Sleytr, 1985; Sitte *et al.*, 1987).

C.   Fracturing and etching

Several freeze-etching apparatus or devices are commercially available. In practice, fracturing of the specimen is achieved either by applying tensile stress, which produces a single cleavage plane, or by cutting with a cold scalpel or microtome knife (for reviews see Lickfeld, 1976; Sleytr and Robards, 1977; Robards and Sleytr, 1985). After fracturing, cell surfaces not exposed by the fracturing process can be revealed by controlled sublimation (etching) of ice under vacuum conditions. To avoid specimen contamination, it is necessary that the temperature of the specimen and the vacuum are monitored and controlled precisely (Umrath, 1977; Robards and Sleytr, 1985). In practice, etching is generally performed at specimen temperatures between -95 and $-100°C$ (178 and 173 K). For studying bacterial cell surface structures, the etching process should not lower the surface of the ice exposed by the fracture plane by more than 1 µm; for geometrical reasons exposed cells cannot be completely replicated (for an extensive compilation of freeze-drying times at different temperatures see Umrath, 1983; Robards and Sleytr, 1985).

D.   Shadowing and coating

The replication process has to copy the relief produced by specimen fracturing and etching as faithfully as possible, and consequently involves a two-step process. First, an electron-opaque metal is evaporated onto the specimen at an oblique angle (usually 45°). The patches of shadowing material are subsequently bound together and reinforced by depositing a low electron scattering supporting layer of carbon.

For high resolution (fine grain) shadowing, the most common elements are platinum and iridium, which are evaporated together with carbon and a tantalum–tungsten alloy (for reviews see Robards and Sleytr, 1985; Gross, 1987). For routine studies of S-layers on intact cells, a heavy metal film 1.5–2 nm thick is used. However, for the highest resolution shadowing, a metal film (Ta–W) thickness of 0.7–1.0 nm will give sufficient contrast.

With unidirectional shadowing it is not possible to provide contrast in all directions. To overcome this problem the specimen may be rotated during shadowing. Since images of rotary shadowed structures can vary dramatically, depending on the shadowing angle and the thickness of the metal film, special care is required to avoid misinterpretation (Neugebauer and Zingsheim, 1979). Such considerations are of particular importance if attempts are made to interpret the subunit structure of S-layers.

E.   Thawing and replica cleaning

The heavy metal/carbon replicas of the required thickness are extremely fragile and brittle. Consequently, the thawing and replica cleaning procedures are among the most delicate steps in the procedure. Replicas of most prokaryotic cells can be cleaned successfully with a 30–40% solution of chromic acid within 1 to 2 days at 20°C. Other cleaning procedures have been suggested (Lickfeld, 1976; Robards and Sleytr, 1985). Replicas of suspensions of bacteria contain considerable redundancy of information and consequently can be picked up on uncoated low transmission grids with wide bars.

F.   Applications and interpretation

Unless excessive, unstructured amorphous material is present on the cell surface, freeze-etching will give the closest picture of the *in vivo* structure and orientation of the S-layers (Sleytr and Glauert, 1975, 1982; Sleytr, 1978). Most S-layers completely cover the cell surface, leaving no gaps (Figs 2a–c). Some organisms have been shown to possess more than one S-layer (for reviews see Sleytr and Messner, 1983, 1988). In such instances the fracturing process may separate adjacent layers (Watson and Remsen, 1970; Beveridge and Murray, 1974, 1975).

On rod-shaped cells S-layer patterns are generally uniform over large areas of the cylindrical part of the cells. Oblique and square lattices (Figs 2a,b) are usually arranged with one axis parallel to the long axis of the cell, but with

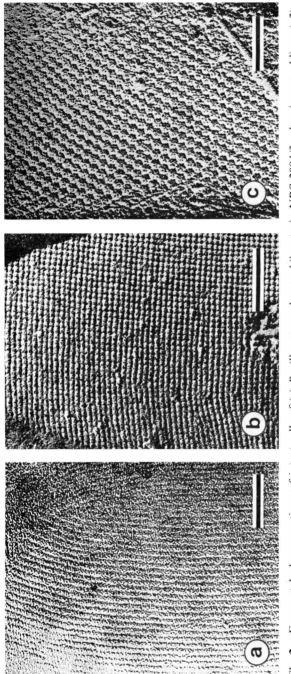

Fig. 2. Freeze-etched preparations of intact cells of (a) *Bacillus stearothermophilus* strain NRS 2004/3a showing an oblique (p2) surface lattice (from Messner *et al.*, 1986b, with permission). (b) *Desulfotomaculum nigrificans*, B200-71 with a square (p4) lattice (from Sleytr *et al.*, 1986b, with permission). (c) *Archaeoglobus fulgidus* strain z with a hexagonal (p6) array of subunits (G. Zellner and P. Messner, unpublished). Bars, 100 nm.

some organisms a strain-specific skew angle may be observed (Sleytr and Glauert, 1975; Sleytr, 1978; Messner *et al.*, 1986b). More variations are observed in the alignment of hexagonal lattices, but there is a tendency for one axis to be approximately parallel to the long axis of the rod. In contrast to the cylindrical part, there is an accumulation of lattice faults at the poles of the cell (Fig. 2a) and at sites of constriction (Sleytr and Glauert, 1976; Messner *et al.*, 1986b). In these areas, the orientation of the lattice vectors frequently changes, giving the appearance of a mosaic composed of small crystallites. These observations are in accordance with theoretical considerations which indicate that it is only a cylinder, and not a spherical surface, that can be covered with a crystalline array without any lattice faults occurring (Caspar and Klug, 1962; Deatherage *et al.*, 1983; Messner *et al.*, 1986a).

Evaluation of the local frequency of lattice faults at cell poles and septation sites has given a strong indication for a dynamic process of assembly of S-layers involving a continuous entropy-driven rearrangement of the constituent subunits on a growing cell surface (Sleytr and Glauert, 1975; Sleytr, 1981). Both dislocations and disclinations have been demonstrated on S-layers of intact cells (Sleytr and Glauert, 1975), supporting the suggestion that during lattice growth new subunits may be added at such sites (Harris and Scriven, 1970).

Freeze-etching has also been shown to be suitable for evaluating S-layers labelled with morphologically detectable marker molecules. Labelling with polycationic ferritin, a marker for negative surface charges, has been used for probing the net charge of native and chemically modified S-layers (Sleytr and Friers, 1978; Sleytr and Glauert, 1982; Sára and Sleytr, 1987a,b).

III. Freeze-drying and metal shadowing

A. General principles

Although there are methods available for studying S-layers in a fully hydrated frozen state, for the examination of non-frozen material the water must be removed. In electron microscopy the following drying methods can be used: (1) evaporation (air) drying from distilled water, volatile buffer or organic solvents which have replaced water in a previous preparation step; (2) freeze-drying of hydrated specimens or specimens substituted with non-aqueous volatile solvents; and (3) critical-point drying (after replacing the specimen water with an organic solvent) from CO_2 or a suitable halocarbon (for reviews see Nermut, 1977; Umrath, 1983; Robards and Sleytr, 1985; Steinbrecht and Müller, 1987). Comparative studies have shown that the preservation of biological material is strongly dependent on the dehydration procedure used. Studies comparing air drying with other methods for removing water have demonstrated that air drying can lead to serious alterations of the fine structure such as dimensional changes, disruption, aggregation and collapse (Kistler and Kellenberger, 1977; Kellenberger and Kistler, 1979; Dubochet *et al.*, 1982; Robards and Sleytr, 1985; Kellenberger, 1987). The main, although not the exclusive, reason for these artefacts is the enormous force exerted at the air–water interface during drying by surface tension. With some specimens the structural preservation can be improved by chemical fixation before air drying, but for many specimens the surface tension forces are so great that the resulting artefacts are unacceptable for high resolution studies. Freeze-drying procedures may help to overcome many of these artefact problems. On the other hand, artefacts specifically related to freeze-drying are known (for a review see Robards and Sleytr, 1985). Delicate structures such as S-layers liberated from the surrounding ice in the course of etching vibrate under vacuum as a result of their thermal energy, and may collapse irreversibly upon the support (Kistler and Kellenberger, 1977; Robards and Sleytr, 1985; Kellenberger, 1987).

Numerous high resolution studies on S-layer structures have been made on freeze-dried or air-dried S-layer preparations or cell wall fragments. In practice, specimens are first adsorbed onto a supporting film from a drop of suspension. The specimen is then rapidly frozen by immersion into a cryogen such as liquid nitrogen or nitrogen slush, and subsequently freeze-dried before being metal shadowed and carbon stabilized. Isolated S-layers or bacterial cell walls and fragments with S-layers attached to them tend to curve and buckle when forced flat onto the surface of a supporting layer. Thus, it is advantageous to use "recrystallized" S-layers, formed by a self-assembly process of isolated subunits. Such assemblies commonly possess a high degree of regularity and good long-range order, as required for detailed analysis by image processing procedures (see below) (Aebi *et al.*, 1973; Sleytr and Glauert, 1982; Baumeister and Engelhardt, 1987; Messner *et al.*, 1986b).

B. Specimen mounting and freezing

For studying S-layers by freeze-drying, or air drying and heavy metal shadowing, grids with carbon-coated support films are used for specimen mounting. For most mounting procedures it is advantageous to make the surface of the support hydrophilic. This is best done by glow discharge in air at low pressure, or by radiation with ultra-violet light. Many S-layers have been shown to possess a net negative surface charge, at least on one of their surfaces (Sleytr and Sára, 1986; Messner *et al.*, 1986b; Sára and Sleytr, 1987a). To achieve a better adsorption of negatively charged S-layers, the support can be positively charged with poly-L-lysine or alcian blue (Nermut and Williams, 1977, 1980; Robards and Sleytr, 1985).

Specimen mounting can be done by two different methods: (1) spray-freezing (Williams, 1953; Lickfeld, 1976); or (2) adsorption (Nermut, 1977, 1982). In the spraying technique, the specimen is directly sprayed with an atomizer onto coated grids firmly attached to a liquid nitrogen-cooled metal block (for a review see Robards and Sleytr, 1985). This is achieved by mounting the specimen support directly onto the temperature-controlled specimen table of a freeze-etching or freeze-drying unit. To prevent hoar frost formation, the specimen support is mounted on the specimen table and then cooled down under vacuum. After venting with dry nitrogen, the cold surface of the support is maintained free from hoar frost before spraying the specimen by continuing to purge with dry nitrogen gas. For spraying, S-layer preparations are best suspended in distilled water or volatile buffer (e.g. 0.1–0.3 M ammonium acetate or ammonium bicarbonate).

The adsorption technique is the more suitable technique for mounting S-layer preparations. In contrast to the spray-freezing procedure, S-layer fragments come into contact with the support and are adsorbed onto it before freezing. Adsorption of the specimen onto the support is best done by floating the grid on a drop of the S-layer preparation. The required adsorption time is strongly dependent on the concentration of the S-layer fragments in the suspension and the affinity of the S-layers to the support. Negative staining techniques (see below) may be used to check the success of this preparation step. Using these mounting procedures it is only possible to obtain good results if the liquid surrounding the specimen is free of non-sublimable (non-volatile) components. For removing such masking material, specimen supports loaded by the adsorption technique can be passed through several washes in distilled water or volatile buffer. Most of the excess liquid is then removed by contact with wet blotting paper and the specimen is immediately coated by immersion in a cryogen. Freezing must be achieved within 2–3 s to avoid air drying. In order to produce a thin film of water prior to freezing, it has been suggested that the grids are placed for a few seconds with the specimen side down on a stack of wet filter paper in a Petri dish. With hydrophilic supports, the filter paper only needs to be kept slightly wet, whereas for hydrophobic specimen carriers the paper must be soaked (Würtz et al., 1976; Kistler et al., 1977; Robards and Sleytr, 1985).

Because specimens mounted by the adsorption technique are embedded in a layer of water only a few μm thick, excellent cooling rates can be obtained by rapid immersion of the grid into liquid cryogens (e.g. nitrogen or nitrogen slush, propane) (for a review see Robards and Sleytr, 1985).

C. Freeze-drying and shadowing

Frozen specimens are either stored under liquid nitrogen or directly transferred to the precoated ($-150°$C, 123 K) specimen stage of a freeze-etching unit. The chamber is then evacuated and the specimen brought up to the freeze-drying temperature. To avoid ice crystal damage during the specimen transfer and the freeze-drying period, the specimen temperature should be kept below the recrystallization temperature. To avoid such artefacts it is recommended that the specimen temperature during freeze-drying is maintained between -85 and $-95°$C (188 and 178 K). Under optimal vacuum conditions (e.g. 1.33×10^{-3} Pa) a 50 μm thick layer of ice sublimes at $-85°$C (188 K) within *c.* 25 min; at $-95°$C (178 K) the same layer would take two and a half hours to sublime. Such calculations show that in general relatively short drying times are sufficient (for detailed data on sublimation rates see Umrath, 1983; Robards and Sleytr, 1985).

After freeze-drying the specimen is shadowed (unidirectional or rotary shadowing), following the procedure applied in freeze-etching. To reduce or avoid secondary structural changes caused by the uptake of atmospheric moisture, the shadowed specimens can be stabilized by subliming a thin (0.5 nm) layer of carbon from above. To reduce the thermal load, both shadowing and carbon coating should be done with the specimen still at low temperature. The specimen is then warmed to ambient temperature before venting the chamber, thus avoiding condensation of atmospheric moisture.

Artefact problems (shrinkage, wrinkling, folding, collapse) related to freeze-drying procedures have been described (for reviews see Nermut, 1982; Robards and Sleytr, 1985). Delicate structures, such as cylindrical S-layer self-assembly products (Kistler and Kellenberger, 1977; Sleytr, 1978), may collapse during adsorption, during freeze-drying, or during shadowing and replication. Cylindrical S-layer self-assembly products (liberated from the surrounding ice) frequently show fewer structural details than areas adsorbed onto the support. Such artefacts are more frequently observed with specimens mounted by spray freezing.

D. Applications and interpretation

Freeze-drying combined with heavy metal shadowing is the most straight-forward approach for obtaining information about the surface structure of S-layers. High resolution unidirectional or rotary shadowing can depict characteristic topographical differences between both S-layer surfaces (Kistler *et al.*, 1977; Smith and Kistler, 1977; Baumeister *et al.*, 1981; Messner *et al.*, 1986a, 1986b). Although in general the structural resolution revealed is not better than 2 nm (Baumeister and Engelhardt, 1987; Messner *et al.*, 1986b), such differentiations are of particular importance when the orientation (sidedness) of S-layer self-assembly products, or the reattachment process of isolated S-layer fragments or subunits to artificial surfaces (e.g. charged surfaces) or cell wall fragments are to be studied. A structural feature seen with most S-layers is a smooth outer and a more structured (sculptured) inner surface. The freeze-dried and shadowed preparation of S-layer sacculi of *Thermoproteus tenax* (Fig. 3a) illustrates such characteristic differences in the topographies of the inner and outer surfaces. While the outer surface of the layer appears relatively smooth, the inner surface is highly sculptured. A characteristic feature seen on the inner side of the sacculus is dome-shaped protrusions (Messner *et al.*, 1986a; Wildhaber and Baumeister, 1987).

Images of the outer S-layer surface obtained by freeze-drying and high resolution shadowing are generally comparable to those observed in freeze-etching preparations of intact cells. With some specimens differences in the topography of the outer surface have been observed which may be related to collapse phenomena (Messner *et al.*, 1986a). High resolution shadowed air- or freeze-dried S-layers reveal structural information complementary to that obtained by negative staining (Kistler *et al.*, 1977; Baumeister and Engelhardt, 1987; Messner *et al.*, 1986a, 1986b). Particularly for high resolution studies, one should be aware of the fact that the shadowed image does not necessarily correlate with the "real" surface relief of the specimen. Even at a favourable shadowing angle, the metal deposition on the crystalline array depends not only on the local slope angle with respect to the evaporation source, but also on differences in the physicochemical properties of the specimen. Despite the uncertainties in interpreting the heavy metal deposit on S-layers due to specific decoration phenomena, attempts have been made to compute a topographical map of metal-shadowed crystalline arrays (Smith and Kistler, 1977; Smith and Ivanov, 1980; Guckenberger, 1985).

Freeze-drying and heavy metal shadowing may also be used for evaluating labelling experiments. For demonstrating surface-located charged groups, S-layer sacculi and self-assembly products have been labelled with polycationic ferritin (PCF) (Messner *et al.*, 1986a, 1986b). Such experiments have shown that in many S-layers differences exist in the net charge of both surfaces. Using *T. tenax* (Fig. 3b) it has been demonstrated that one PCF molecule is bound per morphological unit at the six-fold axis of the hexagonal S-layer lattice. Labelling with PCF has also been used to demonstrate lattice faults (Fig. 3b). By evaluating the location and distances of the wedge disclinations on S-layer sacculi collapsed onto the support at different orientations, it is possible to calculate that six disclinations per cell pole are present (Messner *et al.*, 1986a). This is in agreement with the theoretical consideration that, as was found for the capsids of icosaeder viruses (Caspar and Klug, 1962; Sleytr *et al.*, 1982a), at least six pentameric units are required in a hexagonal lattice to provide a continuous covering over a hemispherical cell pole.

Freeze- or air-dried and heavy metal shadowed preparations with known orientation in the microscope have been used to characterize the handedness

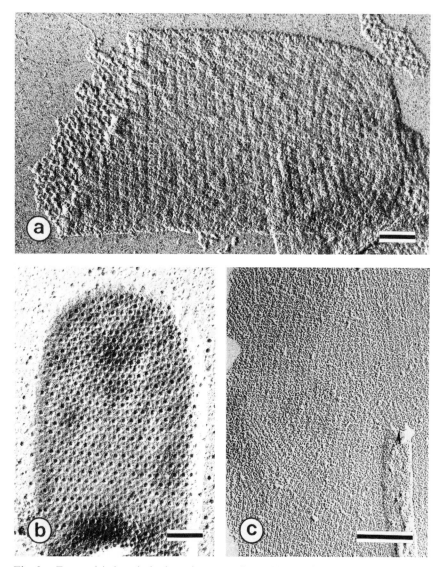

Fig. 3.   Freeze-dried and shadowed preparations. (a) Envelopes of *Thermoproteus tenax* showing the hexagonal S-layer lattice on the smooth outer and the more sculptured inner surface. (b) The same preparation as in (a), but labelled at pH 5.7 with polycationic ferritin. The marker binds to the extracellular surface in a regular fashion. (c) Randomly oriented S-layer crystallites of *Bacillus stearothermophilus* strain NRS 2004/3a attached to a carbon layer positively charged with alcian blue. Bars, 100 nm. (a) and (b) from Messner *et al.* (1986a), with permission; (c) from Messner *et al.* (1986b), with permission.

of the morphological units of S-layers (Messner *et al.*, 1986a). S-layer subunits or fragments frequently have the ability to reattach to charged surfaces. Although such reconstituted S-layers generally have the appearance of crazy-paving, the small crystallites exhibit an identical orientation and consequently, as seen in oblique (p2) lattices (Fig. 3c), an identical handedness of the base vectors (Sleytr *et al.*, 1982b; Messner *et al.*, 1986b).

IV. Thin sectioning

A. General principles

Thin sectioning is the electron microscopical preparation technique most suitable for identifying the major groups of bacterial cell envelope profiles (Fig. 1). Certain information, such as whether there is more than one S-layer or whether the S-layers are involved in the cell division process, can only be obtained unambiguously from thin sections.

Conventional thin sectioning includes the following general preparation steps: (1) fixation; (2) dehydration; (3) embedding and sectioning; and (4) post-staining.

Several excellent articles have been published on the methodology of ultra-thin sectioning (Hayat, 1971; Plattner, 1971; Glauert, 1974; Robinson *et al.*, 1986; Plattner and Zingsheim, 1987). A detailed description of thin sectioning, especially as applied to the investigations of the fine structure of prokaryotic cells, has been given by Lickfeld (1976) in this series.

B. Fixation

For thin sectioning of bacteria, Kellenberger *et al.* (1958) introduced a fixation procedure using 1% osmium tetroxide in veronal buffer at pH 6.0. A disadvantage of this fixative is the slow penetration rate (Glauert, 1974). Proteinaceous materials are best stabilized with bifunctional aldehydes like glutaraldehyde (Sabatini *et al.*, 1963), but several other aldehydes can also be applied (e.g. formaldehyde or acrolein).

One of the most commonly used procedures is the glutaraldehyde–osmium tetroxide double fixation method. The convenience and accuracy of this procedure has made it the standard fixation method in many laboratories. Care should be taken that the fixative matches the natural environmental conditions of the microorganism in terms of pH, osmolarity and ionic state. The fixatives are normally dissolved in buffers [cacodylate, 4-(2-hydroxy-ethyl)-piperazine-1-ethane-2-sulphonic acid (HEPES), veronal, phosphate] and adjusted to the appropriate pH and osmotic conditions (Beveridge and Murray, 1974; Glauert, 1974; Lickfeld, 1976; Sleytr and Glauert, 1976). The prefixation with 2.5–5% glutaraldehyde in 0.1 M cacodylate or HEPES buffer at pH 6.8–7.2 can be performed either in the liquid culture for 20 min at room or growth temperature with vigorous shaking, or "en bloc" on the pellet after harvesting the cells by centrifugation. If the prefixation was carried out in liquid cultures, the cells are then pelleted for the main fixation which is performed with 1% osmium tetroxide in the described buffer systems for 1 h at 20°C. Prior to staining with 0.5–2% uranyl acetate for 1 h at room temperature, the pellet can be embedded in 2% Noble agar and, after solidification, cut into small cubes ($< 1 \, mm^3$).

For contrast enhancement of S-layer structures ruthenium red (Luft, 1971) and/or tannic acid (Mizuhira and Futaesaku, 1971) can be added to the glutaraldehyde and/or osmium tetroxide (Beveridge and Murray, 1975; Hayat, 1981). Ruthenium red (0.15%) probably acts not only as a stain, but also as a fixative together with OsO_4. If applied, post-staining with uranyl acetate and lead citrate for contrast enhancement is not required. The reaction mechanism of tannic acid, which is generally applied at a concentration of *c*. 4%, is not yet fully understood. It is thought to react preferentially with the amino groups of proteins by cross-linking, so forming electron-dense precipitates (Hayat, 1981).

C.   Dehydration

Removal of water from the sample is accomplished by passing the fixed specimen through a graded series of solutions of increasing concentrations of the dehydrating agent in water (either ethanol or acetone) (Glauert, 1974). The monomers of the embedding resins should be soluble in the dehydration medium. If the resins are not miscible in the latter, an intermediate such as propylenoxide should be used. Dehydration can also be performed with 2,2-dimethoxypropane (Muller and Jacks, 1975), which is converted into methanol and acetone by interaction with the specimen water.

D.   Embedding and sectioning

Different resins are commercially available for embedding specimens, e.g. epoxy resins (Epon 812), Araldite or "Spurr's low viscosity medium" (Spurr, 1969), polyesters (Vestopal) or various methacrylates. Overviews of standard embedding media including remarks on specific properties, such as structural preservation and toxicity, are given by Plattner (1971) and Luft (1973).

During the last few years new resins have been developed for particular applications. "Lowicryl" resins (Carlemalm *et al.*, 1982) can be polymerized by UV-radiation even at low temperatures ($< -30°C$, 240 K). The most frequently used Lowicryl resins are the polar K4M and the apolar HM20, but no investigations on bacterial S-layer structures have yet been performed with these substances. Another newly developed resin not yet applied to studies on S-layers is the water-soluble melamin, which provides high contrast of biological samples despite low mass loss (Bachhuber and Frösch, 1983; Westphal and Frösch, 1984).

Ultra-thin sections are cut in ultramicrotomes with either glass or diamond knives. A detailed description of the principle and function of the microtomes has been given by Reid (1974) and Sitte and Neumann (1983).

E. Post-staining

Thin sections (<100 nm) generally do not possess enough contrast to show fine structural details. However, the contrast of biological samples can be enhanced by post-staining of the sections with solutions of heavy metal salts like uranyl or lead salts. The most common method is post-staining with lead citrate at pH 12.0 (Reynolds, 1963). Combined uranyl and lead staining is possible, but for good results the uranyl salt must be applied prior to the lead (Glauert, 1974) to prevent precipitation on the section.

F. Applications and interpretation

On sections of prokaryotic cell envelopes (Figs 4a–c) the cytoplasmic membrane and the outer membrane (only present in Gram-negative eubacteria) appear as a triple-layered "unit-membrane" structure, due to the accumulation of electron-dense reaction products of the fixative (Plattner, 1971; Sleytr and Glauert, 1982). Structural details up to $c.$ 2 nm can be resolved by this method. The peptidoglycan or the pseudomurein layer can be seen as an electron-dense layer of variable thickness which correlates reasonably well with the Gram-staining reaction of the investigated bacteria. On sections of Gram-positive and Gram-negative cell envelopes, S-layers are either seen as separate layers or in close contact with the rigid wall component (peptidoglycan or pseudomurein layer) or outer membrane, respectively (for reviews see Sleytr, 1978; Beveridge, 1981; Sleytr and Glauert, 1982). If the cells are cut tangentially very close to the surface, both the lattice type and the lattice constants of the S-layer can be determined. In cross-section the globular S-layer subunits frequently exhibit an electron-dense shell and a less dense core structure, but often the distinction of a coherent S-layer is not possible and only a fuzzy, unstructured S-layer can be detected.

Thin sectioning allows us to demonstrate how S-layers are involved in the cell division process. In some organisms a surplus of S-layer subunits can be seen on the septation site. This material can be incorporated in the newly formed cell poles which remain completely covered throughout the

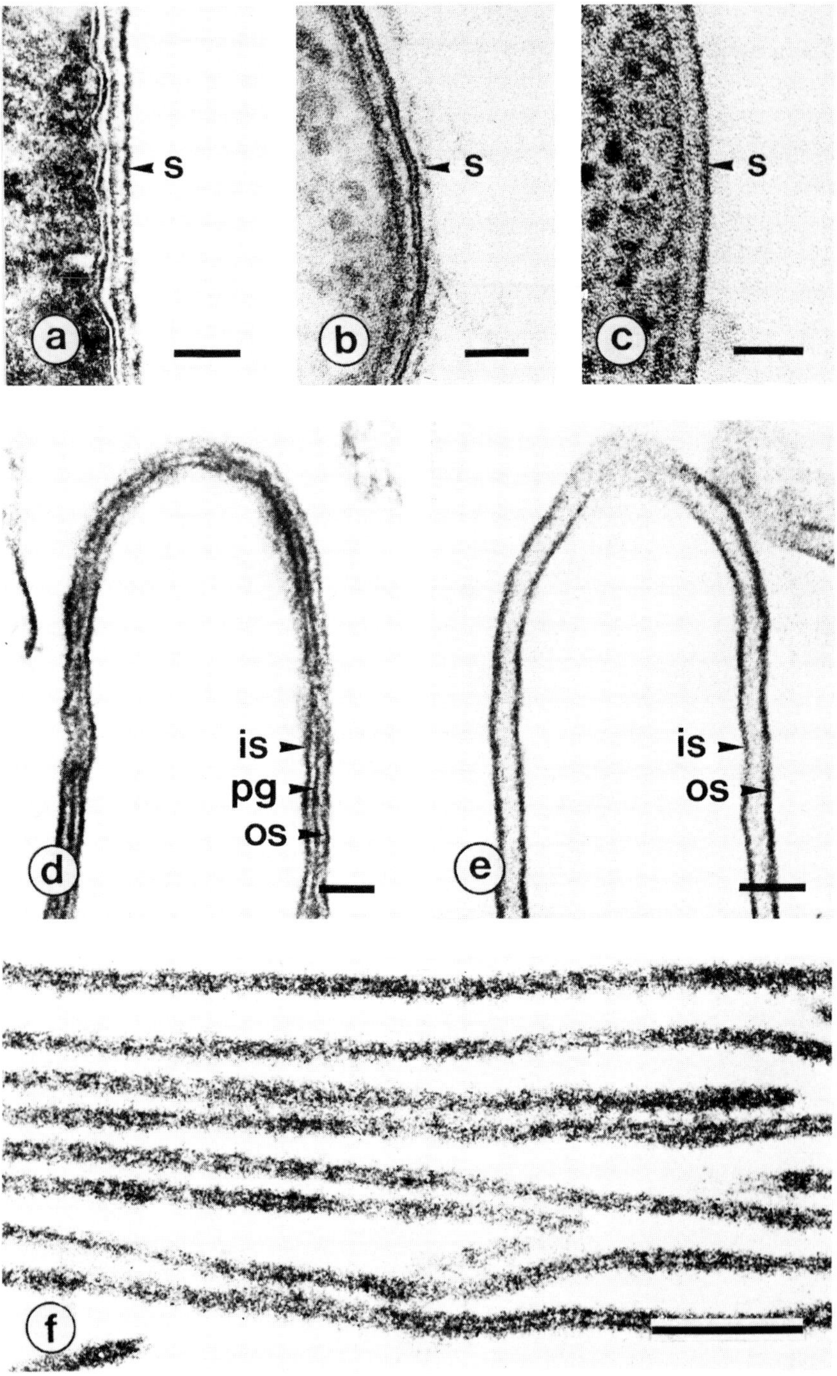

a
s

b
s

c
s

d
is
pg
os

e
is
os

f

separation process (Sleytr and Glauert, 1976). In the case of Bacillaceae lysed cells or isolated cell wall, preparations can exhibit a second S-layer on the inner side of the peptidoglycan sacculus (Sleytr, 1978; Sára and Sleytr, 1987b) (Fig. 4d). Upon digestion of the electron-dense murein layer with lysozyme, a double-layered envelope structure composed of an outer and an inner S-layer can be seen (Fig. 4e).

Sometimes isolated S-layers appear only weakly contrasted in thin sections after standard fixation procedures. The application of ruthenium red and/or tannic acid, both as stain and fixative (see above), can enhance the contrast of S-layers considerably (Fig. 4f).

As an alternative to the standard procedures for thin sectioning other methods for studying bacterial envelope structures may be used. Low temperature embedding techniques with Lowicryl resins (Carlemalm et al., 1982) have recently provided new concepts for the ultrastructure of the cell envelope of Gram-negative bacteria. The space between cytoplasmic and outer membrane seems to be filled by a "periplasmic gel" (Hobot et al., 1984). Cryoelectron microscopy of frozen hydrated sections is now an established method (Dubochet et al., 1983; McDowall et al., 1983). Both techniques have the potential to contribute significantly to the study of S-layers at a structural level close to the native state.

V. Negative staining

A. General principles

Due to its convenience and extreme rapidity, the negative staining technique is prominent amongst EM preparation methods for studying bacterial cell envelope structures. Using this technique S-layers can be demonstrated on intact bacteria and preparations of cell envelopes (for reviews see Sleytr, 1978; Beveridge, 1981; Sleytr and Messner, 1983; Smit, 1987). Negative staining is a particularly powerful technique for high resolution studies on isolated S-layer

Fig. 4. Ultra-thin sections of intact cells of (a) *Bacillus stearothermophilus*, PV72; (b) *Aeromonas salmonicida*; and (c) *Methanococcus voltae*. s, S-layer. (d)–(f) Thin sections of Triton X-100 treated cell wall sacculi of *Bacillus stearothermophilus*, PV72. (d) Native sacculus after removal of the cytoplasmic membrane. The inner S-layer was formed from a surplus of S-layer subunits after disintegration of the cytoplasmic membrane. (e) After fixation of the native sacculus with glutaraldehyde, the peptidoglycan was digested with lysozyme. This procedure did not destroy the original shape of the sacculus. pg, peptidoglycan; os, outer S-layer; is, inner S-layer [(a), (d) and (e) from Sára and Sleytr, 1987b, with permission]. (f) Ultra-thin section of S-layer self-assembly products of *Bacillus stearothermophilus*, PV72. The contrast of the structure has been enhanced by ruthenium red/tannic acid treatment. Bars, 100 nm. (b) and (c) from T. J. Beveridge (unpublished), with permission; (a), (d) and (e) from Sára and Sleytr (1987b), with permission.

fragments and self-assembly products suitable for two- or three-dimensional image reconstructions (Sleytr and Glauert, 1982; Stewart, 1986; Baumeister and Engelhardt, 1987).

Negative staining is achieved by using solutions of heavy metal salts which surround the specimen adsorbed to the support and penetrate to some extent into holes and cavities of the S-layer structure. The staining solution should not give any specific positive staining with the specimen (Haschemayer and Myers, 1972). The degree of penetration of a certain stain may be influenced by charged groups and hydrophobic and hydrophilic regions on the specimen surface. During subsequent air drying the stain solidifies into a smooth glassy film, which under optimal conditions should be completely free of crystallites. The stain acts in a two-fold way: (1) it increases the weak contrast of the biological specimens; and (2) it reduces the radiation damage on the specimen caused by interaction with the electron beam in the microscope.

B. Supports and stains

Negative staining is usually performed on Formvar- or Parlodion-coated 200–400 mesh EM copper grids as support. The plastic films can be stabilized by evaporation of a thin layer of carbon (Haschemayer and Myers, 1972; Nermut, 1973, 1982). Pure carbon films are superior to plastic films because they are thinner, and hence contribute less to the formation of the image. For high resolution electron microscopy of negatively stained isolated S-layer preparations "holey films" can be used. In this case the perforated plastic film (Formvar or Parlodion) is only used as support for a very thin carbon layer produced by deposition of carbon on mica (for details see Nermut, 1973). The hydrophobic carbon layer can be rendered hydrophilic by glow discharge in air for an even spreading of the staining solution on the grid. On the other hand, S-layers with a more hydrophobic surface (e.g. S-layers fixed with glutaraldehyde) may adsorb preferentially to the (hydrophobic) surfaces produced by glow discharge in an alkylamine atmosphere (Kistler et al., 1977; Dubochet et al., 1982).

Different stains (1–2% aqueous solutions) have been used for studying S-layers. The most common stains are:

(1) uranium salts (uranyl acetate, pH 4.2–4.5, uranyl oxalate, pH 5.5–7.0, and uranyl formate, pH 4.5–5.2);

(2) tungsten salts (phosphotungstic acid, pH 7.2, or sodium silicotungstate, pH 5.0–7.5); and

(3) other substances (ammonium molybdate, pH 6.5–7.0, sodium zirconium glycolate, pH 4.0–8.5) (for reviews see Haschemayer and Myers, 1972; Horne, 1975; Nermut, 1982).

For negative staining of intact bacteria (carrying an S-layer), of cell envelope preparations or of S-layer self-assembly products, uranyl acetate or phosphotungstic acid are most commonly applied. Ammonium molybdate appears to be best for sensitive S-layer structures because it is isotonic at the concentrations used for negative staining, and can be applied without any prefixation of the protein (Beveridge, 1979). To prevent denaturation or degradation of the specimen, the crystalline arrays can be fixed with glutaraldehyde before applying the staining solution. For example, S-layers of different *Bacillus stearothermophilus* strains have been shown to lose their regular structure if the protein lattice is not properly fixed before staining (Messner *et al.*, 1984). The lack of a regular structure in self-assembly products of S-layers of various *Bacillus sphaericus* strains may be explained by such denaturation processes (Word *et al.*, 1983). Fixation may also lessen the structural changes which occur during exposure of the specimen to the electron beam.

C. Preparation methods

Negative staining of S-layers, both on intact bacteria or on isolated preparations, is generally performed using the "drop method" (Haschemayer and Myers, 1972; Nermut, 1973, 1982). The whole procedure is performed in a Petri dish on dental wax or on a piece of Parafilm. Staining can be done either simultaneously, by suspending the specimen into the staining solution, or successively. For studying S-layers preferentially the successive technique is used. The method allows both washing steps with buffers (e.g. ammonium acetate, ammonium bicarbonate) and/or fixation steps (e.g. with glutaraldehyde) to take place after application of the specimen onto the grid. If the S-layer protein is fixed, thorough washing of the grid over several drops of water or buffer prior to the transfer to the staining solution is necessary. Then the residual water on the grid is removed by blotting with filter paper, and the specimen (which must not be dry) is stained for 1–2 min on a drop of an appropriate staining solution. Before air drying the excess stain is removed with filter paper.

For high resolution electron microscopy the adsorption of the specimen onto carbon layers as described by Valentine *et al.* (1968) can be recommended. With some specimens this method proves to be superior to conventional successive negative staining methods because concentration of the solutes (originating from buffers) at the moment of drying of the preparation does not occur.

Structural changes of the specimen can be lessened by low dose electron microscopy of negatively stained specimens (Unwin and Henderson, 1975) and/or embedding the specimen in a hydrophilic matrix of substances like

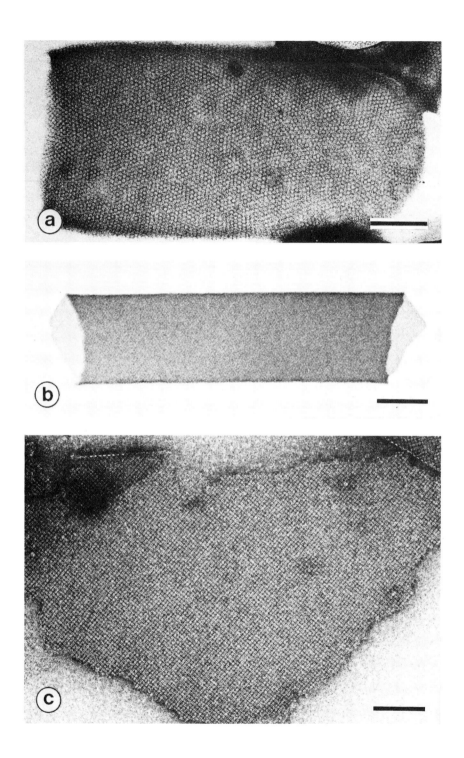

glucose or aurothioglucose (Kühlbrandt, 1982). These matrices do not polymerize, but dry like negative stains. Because the physicochemical properties of these substances resemble water, no significant structural changes are expected when water is substituted by them. Technically, the procedure is performed like negative staining. Aurothioglucose has been used for high resolution three-dimensional reconstructions of eubacterial and archaebacterial S-layers (Rachel *et al.*, 1986; Baumeister and Engelhardt, 1987).

D. Applications and interpretation

Negative staining has been used successfully for demonstrating the presence of S-layers on whole cells and attached to cell wall fragments, and for high resolution studies on S-layer self-assembly products. Furthermore, negative staining represents an ideal method for studying the dynamic process of assembly of isolated S-layer subunits, both in the presence and absence of surfaces suitable for adhesion (Sleytr, 1975, 1981; Hastie and Brinton, 1979b; Jaenicke *et al.*, 1985).

Due to its simplicity, negative staining is most commonly used for monitoring the isolation steps of S-layer subunits or fragments from cell surfaces or cell envelope preparations (Messner and Sleytr, 1988). Cell wall or envelope fragments are obtained by mechanical disruption of the cells (Fig. 5a). In general, chaotropic agents are used for detaching the S-layer protein from the supporting envelope components (Sleytr, 1978; Sleytr and Messner, 1983; Koval and Murray, 1984). Upon dialysis against water or appropriate buffer solutions strain-specific S-layer self-assembly products can frequently be obtained. *In vitro* S-layer self-assembly products may have the shape of open-ended cylinders (Fig. 5b), flat sheets (Fig. 5c) or closed vesicles (for a review see Sleytr, 1981). In the course of negative staining air drying will induce the collapse of the cell wall fragments or self-assembly products onto the support. Depending on the orientation of the superimposed layers, characteristic Moiré patterns will be obtained (Fig. 5a) (for examples see Crowther and Sleytr, 1977; Messner *et al.*, 1986a, 1986b).

Negative staining techniques have also been used to demonstrate that isolated S-layer subunits or S-layer fragments from a variety of Gram-positive and Gram-negative bacteria have the ability to reattach to the cell envelope

---

Fig. 5. Negatively stained preparations of (a) a cell wall fragment of *Clostridium thermohydrosulfuricum*, L111-69, showing the Moiré pattern of the two superimposed p6 S-layer lattices; (b) cylindrical S-layer self-assembly product of *Bacillus stearothermophilus*, NRS 2004/3a, showing a p2 lattice; and (c) sheet-like S-layer self-assembly product of *Desulfotomaculum nigrificans*, NCIB 8706, showing a p4 lattice. Bars in (a) and (c), 200 nm; in (b), 500 nm.

component from which they were originally removed. With isolated S-layers of some *Clostridia* and *Lactobacilli* species both homologous and hetero-logous re-attachment could be observed (Sleytr, 1975, 1976; Masuda and Kawata, 1980, 1981).

In comparison to S-layers on collapsed cell wall or envelope fragments, S-layer self-assembly products reveal a considerably better long-range order. Thus, they are excellent candidates for high resolution electron microscopy. Resolutions down.to 0.83 nm have been reported with aurothioglucose as contrast medium (Rachel *et al.*, 1986), but more usually values of *c.* 2 nm are obtainable. Two- and three-dimensional image reconstruction techniques have been applied for studying S-layers (see below) and have provided new insights into the spatial organization of these crystalline structures (Sleytr and Glauert, 1982; Stewart, 1986; Baumeister and Engelhardt, 1987).

For the interpretation of negatively stained images of S-layers two kinds of artefacts must be considered. (1) Drying artefacts, common to air-dried preparations, can lead to considerable structural changes (Kellenberger and Kistler, 1979) and are described in detail in Section III. For example, in double layers of *Aquaspirillum serpens* VHA S-layers the monolayer in contact with the support reveals a significantly different mass distribution in comparison to the superimposed layer (Dickson *et al.*, 1986). (2) Staining artefacts result from an interaction of the stain with the specimen. Such structural changes, leading to different stain exclusion patterns, can be recognized and possibly circumvented by changing the staining solution to a more appropriate one.

VI.   Image reconstruction procedures

Attempts to determine the fine structure of S-layers at molecular resolution are jeopardized by the extreme sensitivity of these two-dimensional protein crystals to radiation damage (Glaeser, 1975; Baumeister, 1982). Though stains are often used to enhance the contrast, the image signal of the almost invisible specimen will still remain in the noise level due to the indispensable low electron doses. Quantum and structure noise dominate the image formation. Therefore high resolution details in S-layers can only be determined by averaging over many low dose images of similar motifs in order to improve the signal-to-noise ratio. Such averaging is easy for S-layers since orientation and location of the motifs are determined by the two-dimensional lattice.

The oblique, crystalline S-layer of *Bacillus stearothermophilus* strain NRS 2004/3a has been used to demonstrate the capabilities and limitations of image processing methods in the investigation of S-layers. Isolated S-layer subunits of this strain have shown the ability to reassemble into flat sheets and cylinders (Sleytr *et al.*, 1982b; Messner *et al.*, 1986b). The cylinders can be divided into three classes according to their diameters: small monolayer cylinders, medium and large double-layer cylinders. The double layers themselves can be differentiated into five classes according to the orientation of the constituent monolayers with respect to each other (Messner *et al.*, 1986b). The Moiré pattern of one type of double layer is shown in Fig. 6a.

A. Optical diffractometry

Diffraction of light (e.g. laser light) by an ordered arrangement such as an S-layer produces a pattern, while diffraction by a random arrangement does not (Born and Wolf, 1980). The ordered, periodic component of the image is transformed into a regular lattice of intensity maxima—called the reciprocal lattice—in the back focal plane of a lens. Since the signal-to-noise ratio in the diffraction plane attains much higher values than its counterpart in the image plane, all lattice parameters can be calculated more easily from the reciprocal ones. The centre-to-centre spacing d in the real lattice is related to the distance R of the first-order reflection from the origin (undiffracted beam) by $Rd = c/\sin\phi$, where c is a constant ($c = f\lambda$, $f =$ focal length of the lens, $\lambda =$ wavelength of the laser light) and ϕ the interaxial angle. The resolution of structural details is correspondingly defined by the distance of higher order reflections from the origin. Finally, the modulation of the noise spectrum shows the contrast transfer function of the electron microscope. It is sufficient to know that the dark rings should be circular and all visible reflections within this so-called "first-zero crossing". Deviation from the circular symmetry indicates image defects and lens aberrations (for example elliptical rings indicate astigmatism). Figure 6b shows the diffraction pattern of Fig. 6a. Due to the double-layer assembly two reciprocal lattices are present. The periodic signal may be restored by filtering. The optical filter is an opaque filter mask with holes punched at the reciprocal lattice points. Placed in the diffraction plane, rays focused in the intensity maxima may pass and form the periodic component in the image, but rays contributing only to noise are blocked. Since every micrograph requires its own specifically tailored filter mask, filter operations are usually carried out by digital methods.

B. Digital methods

Handling data by numerical methods necessitates digitizing the micrographs with a microscanner. The mathematical description of light diffraction is established by the Fourier transform. The numerically calculated digital diffraction pattern is equivalent to the optical diffraction pattern described previously. The "pseudo-optical" filter is the digital implementation of the optical filter mask. As well as the flexibility of "computerized punching", a further advantage of the digital method is the availability of the numerical

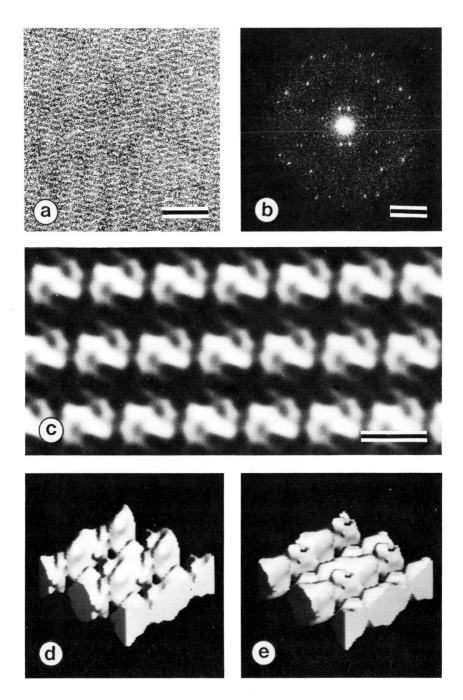

data in the transform. In contrast to the space domain where all data points are real, in the Fourier domain one works with complex values (real and imaginary parts or equivalently amplitude and phase values) for every data point. The amplitude defines the weight, and the phase defines the shift from the origin of a harmonic wave in the image plane. The distance from the origin in the Fourier domain determines the frequency of this wave. A crystal which has a periodic pattern is composed of a set of such harmonic waves. An amplitude/phase pair at a reciprocal lattice point is called a structure factor. In the digital diffraction pattern both the intensities, which are the squared amplitudes, and the phases are available. This information is necessary for crystallographic averaging, where the symmetry properties of the S-layers are used to average within the subunits over symmetry-related protomers. This is possible since the symmetry properties are preserved by the Fourier transform and expressed by identical structure factors of symmetry-related reflections. Prior to the crystallographic averaging a single structure factor for every reciprocal lattice point has to be calculated and then the location of the symmetry operator (e.g. rotation axis) is shifted to the origin, where the symmetry operation is finally applied to the structure. Usually the structure factors have to be calculated from the values on the nearest digital raster points; since these are always superposed by noise, an estimate is determined by applying a least-squares fit procedure on the ideal shape of each peak profile. Since only one amplitude/phase pair (a delta pulse) is calculated for each reflection, this filter procedure is known as the Delta filter (Kübler, 1980; Roberts *et al.*, 1981; Pum and Kübler, 1984). The phase origin is determined by shifting the phases between zero and 360° and monitoring the differences within symmetry-related structure factors. This procedure is called phase origin refinement. The reconstruction of the structure is finally obtained by Fourier synthesis of the structure factors. Figure 6c shows the "one-sided" Delta-filtered reconstruction of a monolayer of Fig. 6a. The space group symmetry (Henry and Lonsdale, 1969) is p2, which means that two-fold rotation axes are the only symmetry operators. The subunit is composed of a dimer with centre-to-centre spacings of 9.4 and 11.6 nm and an interaxial angle of 78°.

Fig. 6. Image reconstruction of the oblique (p2) S-layer of *Bacillus stearothermophilus*, NRS 2004/3a. (a) Moiré pattern of a negatively stained double-layer sheet. (b) Diffraction pattern of (a). The two reciprocal lattices indicate the back-to-back orientation of the two monolayers of the double-layer sheet. (c) "One-sided" Delta-filtered reconstruction of the S-layer of *Bacillus stearothermophilus*. Protein is represented in white and stain in black. (d) Top view of a three-dimensional model of the monolayer of (a). (e) Bottom view of a three-dimensional model of the monolayer. Bars in (a), 100 nm; in (b), $1/15 \text{ nm}^{-1}$; and in (c), 10 nm. (a)–(c) from Messner *et al.* (1986b), with permission.

An alternative method is correlation averaging, introduced by Saxton and Baumeister (1982). The basic idea is to use a template motif consisting of one or several unit cells as a reference, and to locate all other motifs by matched filtering techniques. The motif positions are defined as the local maxima in the cross-correlation function between the image and the reference cell. The averaging is performed by simply shifting the image to the appropriate positions and accumulating by weighted superposition. The averaged cell may be used as a new reference for repeating the procedure in order to eliminate any arbitrary bias in favour of the initially chosen reference. If the averaged cell exhibits rotational symmetry the average may be symmetrized. The symmetry axis is determined by cross-correlating the average with a rotated copy of itself. Once the symmetry axis has been located, the symmetrization is achieved by rotating the average and accumulating by summation.

C.   Three-dimensional image reconstruction

A three-dimensional model of the mass distribution of an S-layer can be obtained by taking a number of micrographs with the specimen tilted at different angles with respect to the viewing direction (the direction of the electron beam) (Unwin and Henderson, 1975; Amos et al., 1982; Baumeister and Engelhardt, 1987). Each of the micrographs gives a projected image in the space domain and a central section through the origin in the Fourier domain, as in the untilted case. The micrographs are first analysed by two-dimensional image processing methods as outlined in the previous sections. A set of structure factors as supplied by the Delta filter is required for each tilted view in the further steps. A hybrid method combining correlation averaging for two-dimensional analysis and Fourier techniques in three-dimensional space was introduced recently (Saxton et al., 1984). Combining all views into a single three-dimensional data set requires the determination of a common phase origin for all sections. This is done by determining a common phase origin for all untilted views first, as described in the previous section. All tilted views are then merged into the data set starting with the smallest tilt angles (e.g. $\pm 10°$, $\pm 15°$, ...) and making use of intrinsic symmetry properties. In order to improve the accuracy of the phase origin determination, an iterative refinement procedure is used. This takes the form of a number of lattice lines which extend perpendicular to the plane of the crystal through the reciprocal lattice points. Since the structure under investigation is only one layer thick the amplitude/phase information on the lattice lines is continuous. The reconstructed three-dimensional model is finally obtained by sampling the amplitudes and phases on the lattice lines at regular intervals according to the thickness of the layer ($1/2t$, $t =$ thickness) and calculating the inverse transform by Fourier synthesis. The three-dimensional structure is usually represented by making a balsa wood model or by displaying the surface profile with three-dimensional computer programs on a television monitor. Figure 6d shows a top view of the three-dimensional reconstruction of the monolayer which is shown in projection in Fig. 6c. Figure 6e shows a bottom view of the same reconstruction.

A severe limitation of three-dimensional electron microscopy is the restriction of the range of tilt angles to less than $\pm 90°$. There always remains a so-called "missing cone" around the axis perpendicular to the crystal where no information is available. This information is required in order to define the shape of the reconstruction uniquely; without this information all slices in the crystal plane have a mean value of zero, which makes the thresholding both difficult and ambiguous. This problem is discussed in more detail elsewhere in this volume (see chapter 16 by H. Engelhardt).

References

Aebi, U., Smith, P. R., Dubochet, J., Henry, C. and Kellenberger, E. (1973). *J. Supramol. Struct.* **1**, 498–522.

Amos, L. A., Henderson, R. and Unwin, P. N. T. (1982). *Prog. Biophys. Molec. Biol.* **39**, 183–231.

Bachhuber, K. and Frösch, D. (1983). *J. Microsc.* **130**, 1–9.

Baumeister, W. (1982). *Ultramicroscopy* **9**, 151–158.

Baumeister, W. and Engelhardt, H. (1987). In *Electron Microscopy of Proteins* (J. R. Harris and R. W. Horne, eds), Vol. 6, pp. 109–154. Academic Press, London.

Baumeister, W., Kübler, O. and Zingsheim, H. P. (1981). *J. Ultrastruct. Res.* **75**, 60–71.

Beveridge, T. J. (1979). *J. Bacteriol.* **139**, 1039–1048.

Beveridge, T. J. (1981). *Int. Rev. Cytol.* **72**, 229–317.

Beveridge, T. J. and Murray, R. G. E. (1974). *J. Bacteriol.* **119**, 1019–1038.

Beveridge, T. J. and Murray, R. G. E. (1975). *J. Bacteriol.* **124**, 1529–1544.

Beveridge, T. J. and Murray, R. G. E. (1976). *Can. J. Microbiol.* **22**, 567–582.

Born, M. and Wolf, E. (1980). In *Principles of Optics*, 6th edn. Pergamon Press, Oxford.

Carlemalm, E., Garavito, R. M. and Villinger, W. (1982). *J. Microsc.* **126**, 123–143.

Caspar, D. L. D. and Klug, A. (1962). *Cold Spring Harbor Symp. Quant. Biol.* **27**, 1–24.

Crowther, R. A. and Sleytr, U. B. (1977). *J. Ultrastruct. Res.* **58**, 41–49.

Deatherage, J. F., Taylor, K. A. and Amos, L. A. (1983). *J. Mol. Biol.* **167**, 823–852.

Dickson, M. R., Downing, K. H., Wu, W. H. and Glaeser, R. M. (1986). *J. Bacteriol.* **167**, 1025–1034.

Dubochet, J., Groom, M. and Mueller-Neuteboom, S. (1982). In *Advances in Optical and Electron Microscopy* (R. Barer and V. E. Cosslett, eds), Vol. 8, pp. 107–135. Academic Press, London.

Dubochet, J., McDowall, A. W., Menge, B., Schmid, E. N. and Lickfeld, K. G. (1983). *J. Bacteriol.* **155**, 381–390.

Glauert, A. M. (1974). In *Practical Methods in Electron Microscopy* (A. M. Glauert, ed.), Vol. 3, pp. 1–207. North-Holland, Amsterdam.

Glaeser, R. M. (1975). In *Physical Aspects of Electron Microscopy and Microbeam Analysis* (B. M. Siegel and D. R. Beaman, eds), pp. 205–239. John Wiley & Sons, New York.

Gross, H. (1987). In *Cryotechniques in Biological Electron Microscopy* (R. A. Steinbrecht and K. Zierold, eds), pp. 205–215. Springer-Verlag, Berlin.

Guckenberger, R. (1985). *Ultramicroscopy* **16**, 357–370.

Harris, W. F. and Scriven, L. E. (1970). *Nature (London)* **228**, 827–829.

Haschemayer, R. H. and Myers, R. J. (1972). In *Principles and Techniques of Electron Microscopy* (M. A. Hayat, ed.), Vol. 2, pp. 101–147. Van Nostrand Reinhold, New York.

Hastie, A. T. and Brinton, Jr, C. C. (1979a). *J. Bacteriol.* **138**, 999–1009.

Hastie, A. T. and Brinton, Jr, C. C. (1979b). *J. Bacteriol.* **138**, 1010–1021.

Hayat, M. A. (ed.) (1971). In *Principles and Techniques of Electron Microscopy*, Vol. 1. Van Nostrand Reinhold, New York.

Hayat, M. A. (1981). In *Fixation for Electron Microscopy*, pp. 120–128. Academic Press, New York.

Henry, N. F. M. and Lonsdale, K. (1969). In *International Tables for X-Ray Crystallography*. The Kynoch Press, Birmingham.

Hobot, J. A., Carlemalm, E., Villinger, W. and Kellenberger, E. (1984). *J. Bacteriol.* **160**, 143–152.

Holt, S. C. and Leadbetter, E. R. (1969). *Bacteriol. Rev.* **33**, 346–378.

Horne, R. W. (1975). In *Advances in Optical and Electron Microscopy* (R. Barer and V. E. Cosslett, eds), Vol. 6, pp. 227–274. Academic Press, London.

Jaenicke, R., Welsch, R., Sára, M. and Sleytr, U. B. (1985). *Biol. Chem. Hoppe-Seyler* **366**, 663–670.

Kellenberger, E. (1987). In *Cryotechniques in Biological Electron Microscopy* (R. A. Steinbrecht and K. Zierold, eds), pp. 35–63. Springer-Verlag, Berlin.

Kellenberger, E. and Kistler, J. (1979). In *Unconventional Electron Microscopy for Molecular Structure Determination* (W. Hoppe and R. Mason, eds), pp. 49–79. F. Vieweg & Sohn, Braunschweig.

Kellenberger, E., Ryter, A. and Séchaud, J. (1958). *J. Biophys. Biochem. Cytol.* **4**, 671–678.

Kistler, J. and Kellenberger, E. (1977). *J. Ultrastruct. Res.* **59**, 70–75.

Kistler, J., Aebi, U. and Kellenberger, E. (1977). *J. Ultrastruct. Res.* **59**, 76–86.

König, H. and Stetter, K. O. (1986). *Syst. Appl. Microbiol.* **7**, 300–309.

Koval, S. F. and Murray, R. G. E. (1984). *Can. J. Biochem. Cell Biol.* **62**, 1181–1189.

Koval, S. F. and Murray, R. G. E. (1986). *Microbiol. Sci.* **2**, 357–361.

Kübler, O. (1980). *J. Microsc. Spectrosc. Electron.* **5**, 561–575.

Kühlbrandt, W. (1982). *Ultramicroscopy* **7**, 221–232.

Lickfeld, K. G. (1976). In *Methods in Microbiology* (J. R. Norris, ed.), Vol. 9, pp. 127–176. Academic Press, London.

Luft, J. H. (1971). *Anat. Rec.* **171**, 347–368.

Luft, J. H. (1973). In *Biological Electron Microscopy* (J. K. Koehler, ed.), pp. 1–34. Springer-Verlag, Berlin.

Masuda, K. and Kawata, T. (1980). *Microbiol. Immunol.* **24**, 299–308.

Masuda, K. and Kawata, T. (1981). *J. Gen. Microbiol.* **124**, 81–90.

McDowall, A. W., Chang, J.-J., Freeman, R., Lepault, J., Walter, C. A. and Dubochet, J. (1983). *J. Microscopy* **131**, 1–9.

Messner, P. and Sleytr, U. B. (1988). In *Bacterial Cell Surface Techniques* (I. C. Hancock and I. R. Poxton, eds), John Wiley & Sons, Chichester, pp. 97–104.

Messner, P., Hollaus, F. and Sleytr, U. B. (1984). *Int. J. Syst. Bacteriol.* **34**, 202–210.

Messner, P., Pum, D., Sára, M., Stetter, K. O. and Sleytr, U. B. (1986a). *J. Bacteriol.* **166**, 1046–1054.

Messner, P., Pum, D. and Sleytr, U. B. (1986b). *J. Ultrastruct. Mol. Struct. Res.* **97**, 73–88.

Mizuhira, V. and Futaesaku, Y. (1971). *Proc. 29th Annu. Meet. Electron Microsc. Soc. Am.*, p. 494.

Muller, L. L. and Jacks, T. J. (1975). *J. Histochem. Cytochem.* **23**, 107–110.

Nermut, M. V. (1973). In *Methodensammlung der Elektronenmikroskopie* (G. Schimmel and W. Vogell, eds), Ch. 3.1.2.3, pp. 1–22. Wissenschaftliche Verlagsgesellschaft, Stuttgart.

Nermut, M. V. (1977). In *Principles and Techniques of Electron Microscopy* (M. A. Hayat, ed.), Vol. 7, pp. 79–117. Van Nostrand Reinhold, New York.

Nermut, M. V. (1982). In *New Developments in Practical Virology* (C. R. Howard, ed.), pp. 1–58. Alan R. Liss, New York.

Nermut, M. V. and Williams, L. D. (1977). *J. Microscopy* **110**, 121–132.

Nermut, M. V. and Williams, L. D. (1980). *J. Microscopy* **118**, 453–461.

Neugebauer, D.-Ch. and Zingsheim, H. P. (1979). *J. Microscopy* **117**, 313–315.

Plattner, H. (1971). In *Methodensammlung der Elektronenmikroskopie* (G. Schimmel and W. Vogell, eds), Ch. 2.2.1, pp. 3–47. Wissenschaftliche Verlagsgesellschaft, Stuttgart.

Plattner, H. and Bachmann, L. (1982). *Int. Rev. Cytol.* **79**, 237–304.

Plattner, H. and Zingsheim, H. P. (1987). In *Elektronenmikroskopische Methodik in der Zell- und Molekularbiologie*, pp. 31–49. G. Fischer Verlag, Stuttgart.

Pum, D. and Kübler, O. (1984). *Proc. 8th Europ. Congr. Electron Microsc.* **3**, 1331–1340.

Rachel, R., Jakubowski, U., Tietz, H., Hegerl, R. and Baumeister, W. (1986). *Ultramicroscopy* **20**, 305–316.

Reid, N. (1974). In *Practical Methods in Electron Microscopy* (A. M. Glauert, ed.), Vol. 3, pp. 213–353. North-Holland, Amsterdam.

Remsen, C. C., Watson, S. W. and Trüper, H. G. (1970). *J. Bacteriol.* **103**, 254–258.

Reynolds, E. S. (1963). *J. Cell Biol.* **17**, 108–212.

Robards, A. W. and Sleytr, U. B. (1985). In *Practical Methods in Electron Microscopy* (A. M. Glauert, ed.), Vol. 10. Elsevier, Amsterdam.

Roberts, K., Shaw, P. J. and Hills, G. J. (1981). *J. Cell Sci.* **53**, 295–321.

Roberts, K., Hills, G. J. and Shaw, P. J. (1982). In *Electron Microscopy of Proteins* (J. R. Harris, ed.), Vol. 3, pp. 1–40. Academic Press, London.

Robinson, D. G., Ehlers, U., Herken, R., Hermann, B., Mayer, F. and Schürmann, F.-W. (1986). In *Präparationsmethodik in der Elektronmikroskopie*, pp. 23–71. Springer-Verlag, Berlin.

Sabatini, D. D., Bensch, K. and Barrnett, R. J. (1963). *J. Cell Biol.* **17**, 19–58.

Sára, M. and Sleytr, U. B. (1987a). *J. Bacteriol.* **169**, 2804–2809.

Sára, M. and Sleytr, U. B. (1987b). *J. Bacteriol.* **169**, 4092–4098.

Saxton, W. O. and Baumeister, W. (1982). *J. Microscopy* **127**, 127–138.

Saxton, W. O., Baumeister, W. and Hahn, M. (1984). *Ultramicroscopy* **13**, 57–70.

Shaw, P. J. and Hills, G. J. (1982). *J. Mol. Biol.* **162**, 459–471.

Shaw, P. J., Hills, G. J., Henwood, J. A., Harris, J. E. and Archer, D. B. (1985). *J. Bacteriol.* **161**, 750–757.

Sitte, H. and Neumann, K. (1983). In *Methodensammlung der Elektronenmikroskopie* (G. Schimmel and W. Vogell, eds), Ch. 1.1.2, pp. 1–248. Wissenschaftliche Verlagsgesellschaft, Stuttgart.

Sitte, H., Edelmann, L. and Neumann, K. (1987). In *Cryotechniques in Biological Electron Microscopy* (R. A. Steinbrecht and K. Zierold, eds), pp. 87–113. Springer-Verlag, Berlin.

Sleytr, U. B. (1975). *Nature (London)* **257**, 400–402.
Sleytr, U. B. (1976). *J. Ultrastruct. Res.* **55**, 360–377.
Sleytr, U. B. (1978). *Int. Rev. Cytol.* **53**, 1–64.
Sleytr, U. B. (1981). In *Cell Biology Monographs* (O. Kiermayer, ed.), Vol. 8, pp. 3–26. Springer-Verlag, Vienna.
Sleytr, U. B. and Friers, G. P. (1978). *Proc. 9th Int. Congr. Electron Microsc.* **2**, 346–347.
Sleytr, U. B. and Glauert, A. M. (1975). *J. Ultrastruct. Res.* **50**, 103–116.
Sleytr, U. B. and Glauert, A. M. (1976). *J. Bacteriol.* **126**, 869–882.
Sleytr, U. B. and Glauert, A. M. (1982). In *Electron Microscopy of Proteins* (J. R. Harris, ed.), Vol. 3, pp. 41–76. Academic Press, London.
Sleytr, U. B. and Messner, P. (1983). *Ann. Rev. Microbiol.* **37**, 311–339.
Sleytr, U. B. and Messner, P. (1988). *J. Bacteriol.* **170** (in press).
Sleytr, U. B. and Robards, A. W. (1977). *J. Microscopy* **111**, 77–100.
Sleytr, U. B. and Sára, M. (1986). *Appl. Microbiol. Biotechnol.* **25**, 83–90.
Sleytr, U. B., Messner, P., Pum, D. and Eder, J. (1982a). *Mikroskopie (Wien)* **39**, 215–232.
Sleytr, U. B., Messner, P., Schiske, P. and Pum, D. (1982b). *Proc. 10th Int. Congr. Electron Microsc.* **3**, 1–8.
Sleytr, U. B., Messner, P., Sára, M. and Pum, D. (1986a). *Syst. Appl. Microbiol.* **7**, 310–313.
Sleytr, U. B., Sára, M., Küpcü, Z. and Messner, P. (1986b). *Arch. Microbiol.* **146**, 19–24.
Smit, J. (1987). In *Bacterial Outer Membranes as Model Systems* (M. Inouye, ed.), pp. 343–376. John Wiley & Sons, New York.
Smith, P. R. and Ivanov, I. E. (1980). *J. Ultrastruct. Res.* **71**, 25–36.
Smith, P. R. and Kistler, J. (1977). *J. Ultrastruct. Res.* **61**, 124–133.
Spurr, A. R. (1969). *J. Ultrastruct. Res.* **26**, 31–43.
Steinbrecht, R. A. and Müller, M. (1987). In *Cryotechniques in Biological Electron Microscopy* (R. A. Steinbrecht and K. Zierold, eds), pp. 149–172. Springer-Verlag, Berlin.
Stewart, M. (1986). In *Ultrastructure Techniques for Microorganisms* (H. C. Aldrich and W. J. Todd, eds), pp. 333–363. Plenum, New York.
Stewart, M. and Murray, R. G. E. (1982). *J. Bacteriol.* **150**, 348–357.
Stewart, M., Beveridge, T. J. and Sprott, G. D. (1985). *J. Mol. Biol.* **183**, 509–515.
Tsuboi, A., Tsukagoshi, N. and Udaka, S. (1982). *J. Bacteriol.* **151**, 1485–1497.
Umrath, W. (1977). *Mikroskopie (Wien)* **33**, 11–29.
Umrath, W. (1983). *Mikroskopie (Wien)* **40**, 9–37.
Unwin, P. N. T. and Henderson, R. (1975). *J. Mol. Biol.* **94**, 425–440.
Valentine, R. C., Shapiro, B. M. and Stadtman, E. R. (1986). *Biochemistry* **7**, 2143–2152.
Watson, S. W. and Remsen, C. C. (1970). *J. Ultrastruct. Res.* **33**, 148–160.
Westphal, C. and Frösch, D. (1984). *J. Ultrastruct. Res.* **88**, 282–286.
Wildhaber, I. and Baumeister, W. (1987). *EMBO J.* **6**, 1475–1480.
Williams, R. C. (1953). *Exp. Cell Res.* **4**, 188–201.
Word, N. S., Yousten, A. A. and Howard, L. (1983). *FEMS Microbiol. Lett.* **17**, 277–282.
Würtz, M., Kistler, J. and Hohn, T. (1976). *J. Mol. Biol.* **101**, 39–56.

3

Analysis of the Structure and Development of Bacterial Membranes (Outer, Cytoplasmic and Intracytoplasmic Membranes)

JOCHEN R. GOLECKI

*Institut für Biologie II, Mikrobiologie, Universität Freiburg,
Federal Republic of Germany*

I. Introduction

Many attempts have been undertaken to elucidate the cytology of bacteria since Leeuwenhoek first observed bacteria through a microscope in 1676. However, due to their small size and the insufficient resolution of the light microscope, only the outer shape of the bacteria could be observed. Some additional information about the organization of the cell wall and internal structures was obtained through the development of specific staining procedures. Weigert was the first, in 1875, to stain killed bacteria with aniline stain.

However, no light microscopical method developed since then is as effective as the electron microscope in investigating the structural organization of the bacterial cell. One example is the frequently used Gram-stain, which was introduced in 1884 by Gram, a contemporary of Robert Koch. Using this simple method it is possible to discriminate, using the light microscope,

METHODS IN MICROBIOLOGY
VOLUME 20 ISBN 0 12 521520 6

between two groups of bacteria—Gram-positive and Gram-negative. However, the different cell wall organization responsible for the different staining behaviour can only be revealed by the use of the electron microscope. Today, many modern electron microscopic preparation techniques are available which allow the ultra-structural organization of bacteria to be demonstrated at levels which closely reflect the patterns of molecular organization.

This chapter deals with ultrastructural findings obtained by application of negative staining, ultra-thin sectioning, freeze-etching and freeze-fracturing. Detailed descriptions of these techniques are given elsewhere in this volume.

II. Electron microscopic examination

A. Negative staining

If bacterial cells are attacked by phages, enzymes or chemicals, the cell envelope becomes leaky and negative staining solution can penetrate into the inner part of the bacteria. As well as the cell wall and the cytoplasmic membrane, intracytoplasmic membranes are also visible, as shown in Fig. 1 for a cell of the photosynthetic bacterium *Rhodobacter capsulatus* infected with RC-1 phages.

B. Ultra-thin sectioning

Internal structures are usually examined in ultra-thin sections or freeze-fractured specimens. In ultra-thin sections the cell walls of bacteria present some typical organization forms. The ultrastructural organizations of Gram-positive and Gram-negative bacteria exhibit some characteristic morphological differences.

The Gram-negative cell wall has a more complex organization than its Gram-positive counterpart. The Gram-positive cell wall (Fig. 2a) is visible in ultra-thin sections as a thick (20–80 nm) single layer of uniform electron-dense material, usually separated from the underlying cytoplasmic membrane by an electron transparent space. The Gram-positive cell wall contains peptido-glycan as a major polymer (up to 90% of the cell wall dry weight) associated with polysaccharides and teichoic acids. The cell wall of Gram-negative bacteria (Figs 2b, 8 and 13) is composed of at least two readily distinguishable layers—the outer membrane and the peptidoglycan layer. Both are thinner than the Gram-positive cell wall. The typical outer membrane of Gram-negative bacteria is characterized by its ultrastructure, i.e. a double-track layer with two electron-dense lines separated by an electron transparent middle zone. The outer membrane of Gram-negative bacteria (width 8–10 nm) is composed of phospholipids, proteins and lipopolysaccharides (Nikaido and

Vaara, 1985). The components of the outer membrane are organized asymmetrically. The lipopolysaccharides are located in the outer leaflet, whereas the phospholipids are mostly located in the inner leaflet of this membrane (Smit et al., 1975). The proteins of the outer membrane are divided into major and minor proteins (for their localization within the membrane, see freeze-fracture results). Among the major proteins are the lipoprotein detected by Braun and Rehn (1969), and the porins which form transmembrane channels (Nikaido and Vaara, 1985). The lipoprotein exists in both free and bound forms; the bound form is covalently bound to the peptidoglycan. The peptidoglycan layer of Gram-negative bacteria is localized below the outer membrane. This thin layer is not always easily detectable because it is usually situated very close to the outer membrane. Where it is separated from the outer membrane (Fig. 2b), it is clearly visible as an electron-dense layer 2–3 nm wide. The peptidoglycan layer of a typical Gram-negative cell wall represents only 1–10% of the total cell wall dry weight.

The cyanobacteria (Figs 2c and 3) possess a cell wall organization similar to that of typical Gram-negative bacteria, but with some characteristic differences. Below the outer membrane, visible as a double-track layer, a peptidoglycan layer is located, which exhibits a thickness (usually 1–10 nm) greater than in Gram-negative bacteria. In the region of septum formation (Fig. 3), the peptidoglycan layer can reach a thickness of up to 200 nm. The septum is formed by an ingrowth of the cytoplasmic membrane and the peptidoglycan layer. The outer membrane is excluded from the septum until the cells begin to separate. Chemical analysis of the cyanobacterial peptidoglycan layer has revealed some similarities to the peptidoglycan of the Gram-positive bacteria (Jürgens et al., 1983, 1985). The ultrastructural and chemical organization of the cyanobacteria may place them in a separate group combining structural elements of the Gram-negative type (outer membrane composed of lipopolysaccharide, lipids, proteins and carotenoids) and the Gram-positive cell wall type (thick and highly cross-linked peptidoglycan with covalently linked polysaccharide) (Jürgens et al., 1983, 1985; Jürgens and Weckesser, 1986).

The cell walls and membranes of the archaebacteria are quite different from those of the other bacteria with respect to their structural organization and chemical composition. The archaebacteria are a heterogeneous group of newly discovered bacteria divided into three major groups: the methanogens, the halophiles and the thermoacidophiles (Woese and Wolfe, 1985). Their cell walls are composed of glycoproteins and a special form of peptidoglycan called pseudomurein. The membranes are composed of special glycerol ethers. A great diversity (Kandler and König, 1985) exists concerning their cell wall ultrastructure: *Methanobacter* contains a thick and homogeneous cell wall resembling the Gram-positive cell wall. *Methanobrevibacterium* presents a

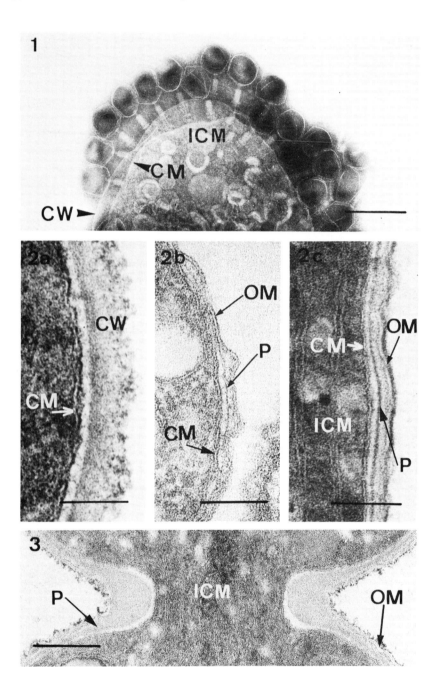

triple-layered cell wall. *Thermoplasma* lacks a cell wall, but its cell membrane contains lipopolysaccharide and glycoprotein. The cell wall of *Sulfolobus* forms a distinct layer of regular subunits outside the cytoplasmic membrane. The continuity of the cell wall layers is interrupted where the flagellae are inserted. Their basal component is formed by a highly ordered and complex structure consisting of two pairs of discs (Coulton and Murray, 1978). Some other pores have been detected in the cell walls of cyanobacteria. For example, in the septum between heterocyst and vegetative cells of filamentous cyanobacteria microplasmodesmata have been discovered in the peptido-glycan layer, both in freeze-fractured and ultra-thin sectioned specimens. Through these plasmatic connections the nitrogen-fixing heterocyst is supplied with nutrients from the neighbouring vegetative cells.

In ultra-thin sections of plasmolysed cells of *Escherichia coli* and *Salmonella typhimurium*, adhesion sites between the cytoplasmic and the outer membrane have been demonstrated. The adhesion sites (200–400 per cell) have been interpreted to be the sites for the export of newly synthesized lipopoly-saccharides into the outer membrane (Mühlradt *et al.*, 1973), and they function as receptor sites for the adsorption of phages (Bayer, 1968).

The cell wall (the outer membrane and peptidoglycan layer) is usually separated from the cytoplasmic membrane by the periplasmic space, which contains periplasmic proteins. These proteins have a variety of different functions—there are proteins for transport, binding proteins for sugars and amino acids, and enzymes for the inactivation of antibiotics and other toxic compounds. With regard to the "periplasmic space", the results of Hobot *et al.* (1984, 1985) should be mentioned. These authors applied Lowicryl resins in combination with cryo-methods of electron microscopy. Their electron micrographs demonstrate the intimate contact of the peptidoglycan layer with both the outer membrane and the cytoplasmic membrane. On the basis of these findings the authors proposed the model of a "periplasmic gel": the space between cytoplasmic and outer membrane is filled by a highly hydrated gel. In

---

Figs 1–3.
Fig. 1. Negative staining preparation of the photosynthetic bacterium *Rhodo-bacter capsulatus* infected with RC-1 phages. The perforation of the cell wall (CW) by phages allows the contrast medium (phosphotungstic acid) to penetrate into the cell interior. Cytoplasmic (CM) and intracytoplasmic (ICM) membranes are visible. Bar, 200 nm.
Fig. 2. Ultra-thin sections demonstrating the typical cell wall organization of (a) Gram-positive *Bacillus subtilis*; (b) Gram-negative *Rhodospirillum rubrum*; (c) the cyanobacterium *Anacystis nidulans*. Note the thick cell wall of *B. subtilis* (a) in comparison to *R. rubrum* (b) and *A. nidulans* (c). The peptidoglycan layer (P) is prominent in *A. nidulans*. Bars represent 100 nm. OM = outer membrane.
Fig. 3. Ultra-thin section of a dividing cyanobacterium *Synechocystis* PCC 6714. The peptidoglycan layer is thickened in the division plane. Bar, 200 nm.

the outer portion the peptidoglycan is sufficiently cross-linked to be visible in the electron microscope and to be isolated as an intact structure (sacculus, see Golecki, this volume). In the inner parts of the gel there is less cross-linking, and near the inside there is none at all. The inner parts of the gel are therefore not discernible on standard micrographs; this area appears as the electron-transparent "periplasmic space".

Next to the cell wall is the inner or cytoplasmic membrane surrounding the protoplast. (In the group of *Mycoplasmas*, which lack a cell wall, the cytoplasmic membrane is the only layer surrounding the cell.) Like the outer and intracytoplasmic membranes, the cytoplasmic membrane also exhibits, in ultra-thin sections, a typical unit membrane structure with a thickness of 7–8 nm (Figs 2, 3, 8, 13 and 15). However, chemical composition and functional properties of the cytoplasmic membrane differ from both other membrane types (Oelze and Drews, 1981). The cytoplasmic membrane of the prokaryotes represents a selective barrier between the environment and the interior of the cell. The cytoplasmic membrane contains special transport proteins and biosynthetic enzymes mediating the synthesis of cell wall and membrane components. In many bacteria, energy-yielding mechanisms like ATP generation and the respiratory electron transport system are incorporated into this membrane. In photosynthetic bacteria, some components of the photosynthetic apparatus are located in the cytoplasmic membrane before they are incorporated in the developing intracytoplasmic membrane system, which is formed by invagination of the cytoplasmic membrane.

C. Freeze-fracturing and freeze-etching

The freeze-fracture method is particularly suited to following differentiation processes in bacterial membrane systems. This method exposes the internal architecture of the membranes on two complementary fracture faces by splitting the membranes along their hydrophobic interior. The outer membrane of Gram-negative bacteria, for example (Figs 4, 6 and 7), is densely covered on its exoplasmic fracture face with numerous small particles 6–10 nm in diameter. The number of particles decreased in Re-mutants of *Salmonella typhimurium* in parallel with the reduction in proteins (Smit et al., 1975). Therefore, a proteinaceous nature was concluded for these particles (Nikaido and Vaara, 1985). However, according to results obtained by Van Alphen et al. (1978, 1979), the particles of the outer membrane, with complementary pits on the opposite plasmic fracture face (Fig. 6), are formed by hemicelles of lipopolysaccharides complexed with proteins forming transmembrane pores. These particles are also present in the outer membranes of the cyanobacteria. They are frequently attached to each other and form triangular structures with a central hole 3 nm in diameter (Golecki, 1977).

The peptidoglycan layer is visible in freeze-fractured samples as a small, smooth edge between the fracture planes of the outer and cytoplasmic membrane. The complementary fracture faces of the cytoplasmic membrane are usually densely covered with intramembrane particles of varying sizes on the plasmic fracture face, and sparsely covered with large particles on the exoplasmic fracture face (Figs 4–7 and 10). The number of intramembrane particles is usually higher on the plasmic than on the exoplasmic fracture face. This may be an indicator of the activity of the membrane; a higher number of particles means a higher membrane activity. Thus far, there has only been one exception where a greater number of intramembrane particles on the exoplasmic face has been found. This is the photosynthetic bacterium *Rhodobacter tenuis* (formerly *Rhodospirillum tenue*) (Fig. 7). When transferred from chemotrophic to phototrophic conditions, instead of forming intracytoplasmic membranes, *Rb. tenuis* increases the area of cytoplasmic membrane by changing the cellular shape. The newly formed photosynthetic units are then placed in the exoplasmic fracture face of the cytoplasmic membrane (Wakim *et al.*, 1978). The number of intramembrane particles on the exoplasmic face increases in *Rb. tenuis* cells by the factor 6.9 to 3711 particles per μm^2 after transfer from chemotrophic conditions to photosynthesis. The number of intramembrane particles on the plasmic fracture face remains almost unchanged.

In other photosynthetic bacteria the cytoplasmic membrane plays an important role in the differentiation process of the photosynthetic apparatus. The participation of the cytoplasmic membrane in the formation of photosynthetically active structural components varies in the different members of the group of photosynthetic bacteria (Golecki and Oelze, 1980).

In the green bacteria *Chlorobium* and *Chloroflexus*, the photosynthetic apparatus is located in the cytoplasmic membrane and the chlorosomes (Fig. 5). The chlorosomes are structurally and functionally linked to the cytoplasmic membrane by a highly ordered baseplate. Changes in the cellular bacteriochlorophyll contents are represented on a structural basis by changes in the number and size of the baseplates and chlorosomes (Golecki and Oelze, 1987).

The purple membrane of *Halobacterium halobium* and other extreme halophiles has been demonstrated to be a constitutive part of the cytoplasmic membrane (Stoeckenius *et al.*, 1979). The purple membrane represents a light energy transducing membrane containing, as the only protein, bacteriorhodopsin. The purple membrane is visible as discrete patches on the fracture faces of the cytoplasmic membrane.

In many other photosynthetic bacteria intracytoplasmic membranes carrying the photosynthetic apparatus are formed by invaginations of the cytoplasmic membrane (Fig. 6). The intracytoplasmic membranes of

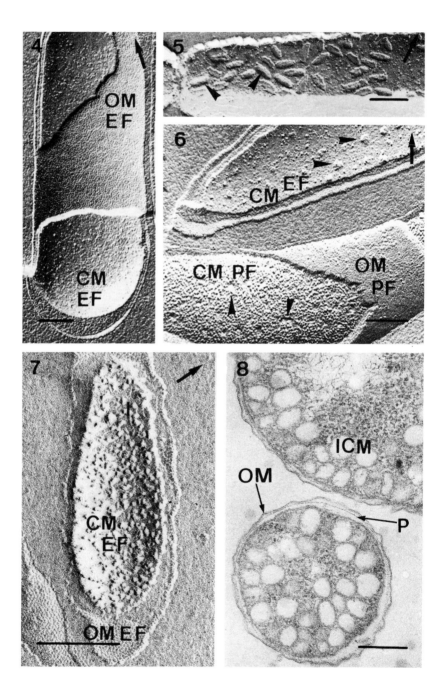

photosynthetic bacteria consist of a network of vesicles (Fig. 8), thylakoid-like flat lamellae (Fig. 9), or tubules (Oelze and Drews, 1981). The intracytoplasmic membranes are connected to each other and/or to the cytoplasmic membrane. In a large number of other microbial organisms such as nitrifying (Murray and Watson, 1965; Remsen et al., 1967), nitrogen fixing and methanogenic bacteria (see Section III), intracytoplasmic membranes with similar ultrastructure but different function have been detected. Azotobacter, a nitrogen-fixing Gram-negative bacterium, presents intracytoplasmic membranes of the vesicular type (Wyss et al., 1961). Nitrogen-fixing cells of Azotobacter vinelandii vary the overall cellular membrane contents only slightly. However, the proportion of the intracytoplasmic to cytoplasmic membrane as a response to the oxygen contents in the medium varies drastically (Post et al., 1982). A special type of intracytoplasmic membrane has been demonstrated in specific strains of E. coli as extraordinary membranes in large quantities (Weigand et al., 1970). For their interpretation as mesosomes, see Section III.

In cyanobacteria the intracytoplasmic membranes, also called thylakoids (Menke, 1961), are flattened membraneous sacs containing the photosynthetic apparatus. In contrast to the thylakoids of plant chloroplasts, the thylakoids of the cyanobacteria are not enclosed by membranes forming cell organelles (Lang and Rae, 1967). They are generally arranged in the peripheral region of the cell in 3–6 layers running parallel to the cell envelope (Fig. 10). The growth of the intracytoplasmic membranes in cyanobacteria seems to be independent from the cytoplasmic membrane (Stanier and Cohen-Bazire, 1977); connections between the two membranes are rarely observed and may

---

Figs 4–7. Freeze-fracture preparations exposing the different exoplasmic (EF) and plasmic (PF) fracture faces of the outer and cytoplasmic membrane. Arrows in the upper right-hand corners indicate the directions of shadowing.

Fig. 4. Typical exoplasmic fracture face of a Gram-negative cell wall exposing the outer membrane densely covered with particles. The cytoplasmic membrane exhibits a lower particle density. Bar, 200 nm.

Fig. 5. Exoplasmic fracture face of the cytoplasmic membrane in the photosynthetic green bacterium Chloroflexus aurantiacus. The typical bag-shaped chlorosomes (arrowheads) are anchored in the cytoplasmic membrane. Bar, 200 nm.

Fig. 6. Two photosynthetic cells of R. rubrum exposing the complementary fracture faces of the cytoplasmic membrane with numerous invaginations (arrowheads) forming intracytoplasmic membranes. Bar, 200 nm.

Fig. 7. Rhodocyclus tenuis, which does not form intracytoplasmic membranes; it incorporates structural equivalents of the photosynthetic apparatus in the exoplasmic leaflet of the cytoplasmic membrane. Bar, 200 nm.

Fig. 8. Ultra-thin section of two cells of R. rubrum exhibiting the Gram-negative cell wall organization and the vesicular type of intracytoplasmic membranes. Bar, 200 nm.

J. R. GOLECKI

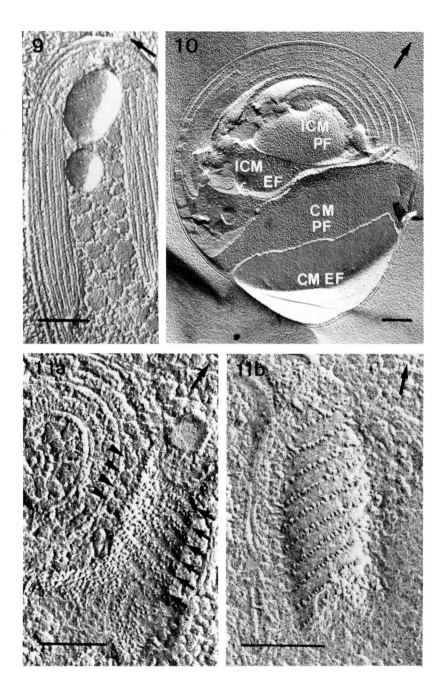

represent only temporary fusions. The thylakoids form an anastomosing network of concentric shells (Fig. 10). Whereas the cytoplasmic membranes of all cyanobacteria examined seem to have the same ultrastructure, the thylakoids show considerable variations in structure and arrangement depending on species, cell type (vegetative cells, heterocysts, akinete) and culture conditions. During cell division (Fig. 3) the thylakoids are separated and distributed into the daughter cells by formation of a septum. This process seems to be controlled by the division of the chromosome, because in nonseptate mutants the thylakoids are grouped into discrete bundles without any sign of cell wall invagination (Ingram and Thurston, 1970).

The complex architecture of the thylakoids in cyanobacteria, displayed particularly well by the freeze-fracture method, exhibits a characteristic arrangement of intramembrane particles (reviewed by Golecki and Drews, 1982). The plasmic fracture face of the thylakoid membrane, densely covered with particles (2.5–12 nm in diameter), is interrupted at regular intervals by parallel grooves free of particles (Fig. 11a). The grooves have a spacing of 35–50 nm. The corresponding exoplasmic fracture face reveals regular rows of 10-nm particles in parallel alignment (Fig. 11b). The distances between the rows of particles are the same as those between the grooves, and are also the same as the spacings of the phycobilisomes outside the thylakoid membranes (Fig. 11a). The close correspondence between the rows of particles and the phycobilisomes indicates that the 10-nm particles on the exoplasmic fracture face represent structural equivalents of the photosystem II apparatus (Giddings and Staehelin, 1979; Golecki, 1979). Due to the higher activity of the photosynthetic membranes, the number of intramembrane particles is always higher in the fracture faces of thylakoids than on cytoplasmic membranes (Fig. 10).

Also, outside the normal cell wall (outside the outer membrane), some

---

Figs 9–11.
Fig. 9. Cross-fractured cell of *Rhodopseudomonas palustris* with two bundles of thylakoid-like lamellar intracytoplasmic membranes. Bar, 200 nm.
Fig. 10. Parallel thylakoids in the cyanobacterium *Plectonema* sp. The complementary fracture faces of the membranes present higher particle densities on the thylakoids (ICM EF, 3822 particles μm^{-2}; ICM PF, 4563 particles μm^{-2}) than on the cytoplasmic membrane (CM EF, 846 particles μm^{-2}; CM PF, 4236 particles μm^{-2}) (from Golecki and Drews, 1982). Bar, 200 nm.
Fig. 11. Complementary (a) plasmic and (b) exoplasmic fracture faces of thylakoids in the cyanobacterium *Nostoc muscorum*. (a) Plasmic fracture face of a thylakoid with several particle-free grooves (arrows) and adjacent phycobilisomes (arrowheads) outside the thylakoids. (b) Exoplasmic fracture face of a thylakoid with parallel rows of particles representing the structural equivalents of photosystem II. The spacing (35–50 nm) of particle rows is identical with the spacing of the grooves and phycobilisomes in (a). Bars, 200 nm.

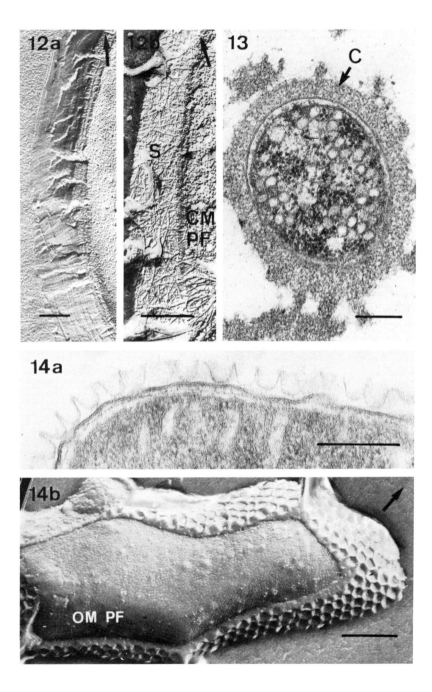

additional layers have been demonstrated in several bacteria and cyano-bacteria using the freeze-fracture method. To protect the heterocysts (Fig. 12a) of cyanobacteria against oxygen, the cell wall is thickened by many additional layers of glycolipids (Winkenbach *et al.*, 1972) with a periodicity of 7–8 nm (Golecki and Drews, 1974).

Some bacteria produce sheaths or capsules outside their cell walls. These external substances are usually polysaccharides. Figure 12b shows the sheath of *Anabaena variabilis*, visible after freeze-etching as a dense network of fibres. In ultra-thin sections (Fig. 13) a highly hydrated capsule is detectable only if a collapse of the capsule during the dehydration for embedding is prevented by pretreatment with capsule-specific antibodies (Bayer and Thurow, 1977; Omar *et al.*, 1983). As shown in Fig. 13, *Rhodobacter capsulatus* is surrounded by two different external layers: the outermost one is only loosely associated and can be removed by washing the cells with saline. The inner layer has a constant thickness and is not removed by saline (Omar *et al.*, 1983); its chemical composition is different from the outer layer.

Bacteria from extreme habitats such as halophilic or thermophilic organisms very often present, external to their cell walls, additional layers with highly ordered structures (see Sleytr *et al.*, this volume). Figures 14a and 14b present the ultrastructure of such an additional outer surface layer from a sulphate-reducing bacterium. The cup-like structures forming the outer surface layer are of proteinaceous nature (Golecki and Bache, unpublished results).

III. Artefacts

The multiplicity of different electron microscopic preparation methods suggests abundant possibilities for the production of artefacts. From the great variety of possible artefacts, only a few characteristic examples will be presented and discussed here.

Mesosomes have been reported as unique membraneous structures, first in Gram-positive (Fig. 15), and later also in some Gram-negative bacteria. The

---

Figs 12–14.
    Fig. 12. Additional layers outside the normal cell wall. (a) Multilayered cell wall of a heterocyst of *Anabaena variabilis* (from Golecki and Drews, 1974). (b) The fibrillar sheath (S) of *A. variabilis* (vegetative cell) as revealed after freeze-etching (from Golecki and Drews, 1974). Bars, 200 nm.
    Fig. 13. Ultra-thin section of *Rhodobacter capsulatus*. The capsule (C) was stabilized with specific antibodies. Bar, 200 nm.
    Fig. 14. Highly ordered surface structures on the cell wall of *Desulfuromonas* sp. (a) Ultra-thin section; (b) freeze-etch preparation. Bar, 200 nm.

mesosomes were defined as sac-like invaginations of the cytoplasmic membrane (Reavely and Burge, 1972). They were thought to be involved in several fundamental processes such as replication and segregation of the chromosomes, and cell division. For the demonstration of mesosomes the bacteria were fixed with glutardialdehyde and/or osmium tetroxide. In freeze-fracture preparations the samples were frozen with cryoprotectants. However, in the recent literature doubts about the classical mesosome concept have been expressed by an increasing number of reports. First, Nanninga (1971) reported that the number and size of mesosomes varied with the preparation technique. Later, the results of several investigations using modern cryofixation methods (Ebersold et al., 1981; Dubochet et al., 1983; Hobot et al., 1985) produced evidence that the classical mesosomes are artefacts of the preparation technique.

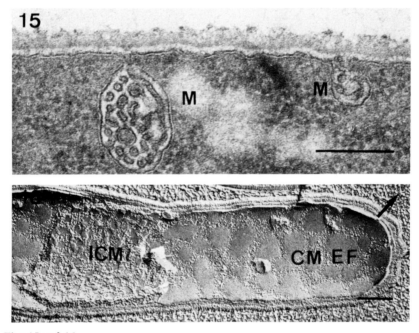

Figs 15 and 16.
   Fig. 15.   Ultra-thin section of *Bacillus subtilis* exhibiting the Gram-positive cell wall organization. Two artificial mesosomes (M) are visible as invaginations of the cytoplasmic membrane. Bar, 200 nm.
   Fig. 16.   Freeze-fracture preparation of *Synechococcus lividus* OH-53s, a thermophilic strain growing at 52°C. Exoplasmic fracture faces of the cytoplasmic and intracytoplasmic membranes can be seen with an artificial separation of the proteins (= particles) and lipids (smooth areas) as a result of freezing the cells from below their phase transition temperature of 30–35°C (from Golecki, 1979). Bar, 200 nm.

Another artificial creation seems to be the demonstration of internal membranes in methanogenic bacteria. As reported for *Methanobacterium thermoautotrophicum* (Aldrich *et al.*, 1987), the artefactual formation of intracytoplasmic membranes is induced by the use of fixatives of low concentrations in combination with phosphate or Tris buffer and by the use of cryoprotectants. According to the results of cryofixation and cryosubstitution experiments, in which internal membranes are absent, the internal membranes of *M. thermoautotrophicum* are indeed artefacts of fixation.

A further artefact is the phase separation of lipids and proteins which takes place in membranes if they are cooled below their specific phase separation temperature. This is particularly liable to happen in thermophilic bacteria if they are cooled from their normal high growth temperature (for example, 52°C for *Synechococcus lividus* OH-53s) to room temperature. This phenomenon is especially visible on freeze-fractured preparations on both fracture faces of the membranes in the form of an aggregation of the intramembrane particles (Fig. 16). The particle aggregation is caused by a phase transition of the lipids from the liquid crystalline into the gel state and the concomitant exclusion of the intramembrane particles (= proteins) from the regions containing the solidified lipids. To avoid the artefactual phase separation it is necessary to hold the bacteria above their phase transition temperature before shock-freezing in liquid propane. A similar artefactual reorientation of membrane components occurs during the isolation of the outer membrane of *Salmonella typhimurium*, resulting in a symmetrical distribution of the lipopolysaccharide molecules after disturbance of the peptidoglycan layer (see Golecki, Chapter 11 of this volume).

IV. Conclusion

The production of artefacts as described above may act as a discouragement against carrying out examinations by electron microscopy. However, the large amount of important information on the ultrastructural organization of bacteria that can be obtained by using the electron microscope should be incentive enough to try. This chapter has shown that reliable results are more likely to be achieved by using several different preparation methods, rather than by choosing only one.

Acknowledgement

I would like to thank Mary Hall for help with the editing of this chapter.

References

Aldrich, H. C., Beimborn, D. B. and Schönheit, P. (1987). *Can. J. Microbiol.* 33, 844–849.
Bayer, M. E. (1968). *J. Virol.* 2, 346–356.
Bayer, M. E. and Thurow, H. (1977). *J. Bacteriol.* 130, 911–936.
Braun, V. and Rehn, K. (1969). *Eur. J. Biochem.* 10, 426–438.
Coulton, J. W. and Murray, R. G. E. (1978). *J. Bacteriol.* 136, 1037–1049.
Dubochet, J., McDowall, A. W., Menge, B., Schmid, E. N. and Lickfeld, K. G. (1983). *J. Bacteriol.* 130, 381–390.
Ebersold, H. R., Cordier, J.-L. and Lüthy, P. (1981). *Arch. Microbiol.* 130, 19–22.
Giddings Jr, T. H. and Staehelin, L. A. (1979). *Biochim. Biophys. Acta* 546, 373–382.
Golecki, J. R. (1977). *Arch. Microbiol.* 114, 35–41.
Golecki, J. R. (1979). *Arch. Microbiol.* 120, 125–133.
Golecki, J. R. and Drews, G. (1974). *Cytobiologie* 8, 213–227.
Golecki, J. R. and Drews, G. (1982). In *The Biology of Cyanobacteria* (N. G. Carr and B. A. Whitton, eds), pp. 125–141. Blackwell Scientific, Oxford.
Golecki, J. R. and Oelze, J. (1980). *J. Bacteriol.* 144, 781–788.
Golecki, J. R. and Oelze, J. (1987). *Arch. Microbiol.* 148, 236–241.
Hobot, J. A., Carlemalm, E. and Kellenberger, E. (1984). *J. Bacteriol.* 160, 143–152.
Hobot, J. A., Villiger, W., Escaig, J., Maeder, M., Ryter, A. and Kellenberger, E. (1985). *J. Bacteriol.* 162, 960–971.
Ingram, L. O. and Thurston, E. L. (1970). *Protoplasma* 71, 55–75.
Jürgens, U. J. and Weckesser, J. (1986). *J. Bacteriol.* 168, 568–573.
Jürgens, U. J., Drews, G. and Weckesser, J. (1983). *J. Bacteriol.* 154, 471–478.
Jürgens, U. J., Golecki, J. R. and Weckesser, J. (1985). *Arch. Microbiol.* 142, 168–174.
Kandler, O. and König, H. (1985). In *The Bacteria*, Vol. 8: "Archaebacteria" (C. R. Woese and R. S. Wolfe, eds), pp. 413–457. Academic Press, London and New York.
Lang, N. J. and Rae, P. M. M. (1967). *Protoplasma* 64, 67–74.
Menke, W. (1961). *Z. Naturforsch.* 16b, 543–545.
Mühlradt, P. F., Menzel, J., Golecki, J. R. and Speth, V. (1973). *Eur. J. Biochem.* 35, 471–481.
Murray, R. G. E. and Watson, S. W. (1965). *J. Bacteriol.* 89, 1594–1609.
Nanninga, N. (1971). *J. Cell Biol.* 48, 219–224.
Nikaido, H. and Vaara, M. (1985). *Microbiol. Rev.* 49, 1–32.
Oelze, J. and Drews, G. (1981). In *Organization of Prokaryotic Cell Membranes* (B. K. Ghosh, ed.), Vol. II, pp. 131–195. CRC Press, Boca Raton, Florida.
Omar, A. S., Flammann, H. T., Golecki, J. R. and Weckesser, J. (1983). *Arch. Microbiol.* 134, 114–117.
Post, E., Golecki, J. R. and Oelze, J. (1982). *Arch. Microbiol.* 133, 75–82.
Reavely, D. A. and Burge, R. E. (1972). In *Advances in Microbial Physiology* (A. H. Rose and D. W. Tempest, eds), Vol. 7, pp. 1–81. Academic Press, New York.
Remsen, C. C., Valois, F. W. and Watson, S. W. (1967). *J. Bacteriol.* 94, 422–433.
Smit, J., Kamio, Y. and Nikaido, H. (1975). *J. Bacteriol.* 124, 942–958.
Stanier, R. Y. and Cohen-Bazire, G. (1977). *Ann. Rev. Microbiol.* 31, 225–274.
Stoeckenius, W., Lozier, R. H. and Bogomolni, R. A. (1979). *Biochim. Biophys. Acta* 505, 215–278.
Van Alphen, L., Verkleij, A., Leunissen-Bijvelt, J. and Lugtenberg, B. (1978). *J. Bacteriol.* 134, 1089–1098.
Van Alphen, L., Van Alphen, W., Verkleij, A. and Lugtenberg, B. (1979). *Biochim. Biophys. Acta* 556, 233–243.

Wakim, B., Golecki, J. R. and Oelze, J. (1978). *FEMS Microbiol. Lett.* **4**, 199–201.
Weigand, R. A., Shively, J. M. and Greenawalt, J. W. (1970). *J. Bacteriol.* **102**, 240–249.
Winkenbach, F., Wolk, C. P. and Jost, M. (1972). *Planta* **107**, 69–80.
Woese, C. R. and Wolfe, R. S. (1985). *The Bacteria*, Vol. 8: "Archaebacteria". Academic Press, London and New York.
Wyss, O., Marilyn, P., Neumann, G. and Socolofsky, M. (1961). *J. Biophys. Biochem. Cytol.* **10**, 555–565.

4

Analysis of Refractile (R) Bodies

JORGE LALUCAT

Departamento de Biología, Facultad de Ciencias, Universidad de las Islas Baleares, 07071 Palma de Mallorca, Spain

I. Introduction

Refractile (R) bodies are bacterial inclusion bodies consisting of a convoluted proteinaceous ribbon, found naturally in *Caedibacter* (Preer and Preer, 1984) (an obligate endosymbiont of *Paramecium*) and in some strains of the genus *Pseudomonas* (*P. taeniospiralis*, Lalucat *et al.*, 1982; *P. avenae*, Wells and Horne, 1983; and *Pseudomonas* EPS-5028, Fusté *et al.*, 1986). R bodies are plasmid coded in *Caedibacter taeniospiralis* and the genes responsible for their synthesis have been cloned and are expressed in *Escherichia coli* P678-54 (pBQ65) (Quackenbush and Burbach, 1983). Their function is still unclear, but in some cases they seem to be involved in the killing of sensitive paramecia. R bodies can be seen under the light microscope in bright phase-contrast or dark-field in cells of *Caedibacter* or *P. avenae*, having the appearance of hollow cylindrical structures approximately 0.5 μm in diameter and length. The R bodies of *P. taeniospiralis* and *Pseudomonas* EPS-5028 are smaller (0.2–0.3 μm in diameter) and easily confused in *P. taeniospiralis* under light microscopy with poly-β-hydroxybutyric acid granula. Electron microscopy is the only way to distinguish the R bodies in these bacteria.

R bodies are actually long proteinaceous ribbons (6–30 μm long, 0.2–0.5 μm wide and 11–13 nm thick) rolled up inside the bacterial cell. After heat treatment or pH changes the R bodies can unroll and appear like long twisted filaments. There is considerable variation in the length and diameter of the R

METHODS IN MICROBIOLOGY
VOLUME 20   ISBN 0 12 521520 6

TABLE I
Types and characteristics of R bodies

Producing bacteria	R body					Reference and figure
	Type	Length (μm)	Width (μm)	Acute ends	Unrolling	
Endosymbionts						
C. taeniospiralis	51	Up to 20	0.4	Both	Inside	Preer and Preer (1984); Figs 1 and 13
C. varicaedens ⎫						
C. pseudomutans ⎬	7	Up to 20	0.4	Inner	Outside	Preer and Preer (1984)
C. paraconjugatus ⎭						
C. macronucleatum	Cm		0.8	Inner	Inside	Estève (1978)
Free-living						
P. taeniospiralis	Pt	Up to 7	0.25	Inner	Inside	Lalucat et al. (1979); Figs 2–6
P. avenae	Pa	Up to 30	0.8	Both	Outside	Wells and Horne (1983); Fig. 12
Pseudomonas EPS 5028			0.2			Fusté et al. (1986)

bodies of the same strain, probably indicating different stages in the synthesis. All the R bodies share the same basic morphology and dimensions. Several characteristics of the different types of R bodies are indicated in Table I.

II. Methodology

A.   Ratio of cells containing R bodies

R bodies are found in only a fraction of the cells in a bacterial population, always being present in less than 50% of the cells. Growth conditions can be modified in order to obtain the highest ratio of cells with R bodies; this is necessary to facilitate isolation, physiological and morphological studies.

In *Caedibacter* the highest proportion can be obtained by culturing the host paramecia at a doubling time of 24 hours; the proportion is lower when growing at a faster rate (Sonneborn, 1970), usually being about 5% of the bacteria in most populations. In *Escherichia coli* P678-54 (pBQ65) the ratio is less than 3%.

Irrespective of the growth phase, the substrate greatly influences the proportion of cells with R bodies in *Pseudomonas taeniospiralis*. Sucrose induces less than 1% of cells with R bodies, whereas gluconate induces up to 45%. In the other bacteria the best conditions for induction have not been well established.

B. Preparation of purified R bodies

For most studies on R bodies it is necessary to purify them to some extent or to homogeneity (Fig. 8). Different methods are used for each bacterial species, but the general steps are the same. In all cases, it is recommended to start with cultures having a high ratio of R body-containing cells.

R bodies are very stable proteins, resisting many physical and chemical treatments, thus facilitating their isolation. Sonication (as indicated in Table II) or freezing merely result in the unrolling of the R bodies.

Table II shows the steps which have to be followed for the isolation of the different types of R bodies once the bacterial cells have been harvested. Methodologies for obtaining high populations of *Caedibacter* are described by Sonneborn (1970). Purified R bodies can be stored at 4°C or frozen at −20°C for long periods of time without damage.

C. Electron microscopy of R bodies

Electron microscopy is indispensable for research on R bodies. It has been used to establish the different morphological types, assess the purity of R body suspensions, follow their synthesis, and to determine the effects of chemical agents and proteolytic enzymes on their structure. It has also been useful for comparative purposes with the immunogold negative staining technique.

The accelerating voltage used in the microscope is usually 75 or 80 kV. A few observations have been made by Wells and Horne (1983) on *Pseudomonas avenae* R bodies with a high voltage electron microscope working at 500 kV, revealing a triple-layer structure forming the lamella which could not be resolved at conventional accelerating voltages.

1. Negative staining

Negative staining has been used preferentially for studying isolated R bodies and also for detecting R bodies inside cells. In no case have subunits of the protein constituents been detected in extended or rolled-up R bodies, indicating a low molecular weight of the subunits.

There are essentially two procedures: the direct mounting of liquid suspensions onto filmed grids, and the floating technique.

(a) Direct method

One droplet of the sample is deposited directly onto a copper grid (200 mesh) coated with a film of Formvar or Pioloform (0.3% or 2% in chloroform, respectively) and stabilized with a thin layer of evaporated carbon. The cells or

TABLE II

Procedures to isolate R bodies from various bacteria

	Caedibacter (Sonneborn, 1970)	*E. coli* P678-54 (pBQ65) (Kanabrocki et al., 1986)	*P. taeniospiralis* (Lalucat et al., 1979)	*P. avenae* (Wells et al., 1983)
Suspend cells in:	Phosphate buffer 10 mM, pH 7.0	5 ml 25% sucrose in Tris-HCl 50 mM, pH 8.0	5 g wet weight in 100 ml phosphate buffer 50 mM, pH 7	Distilled water
Cell lysis	SDS 0.1% 18 h room temperature with stirring *or* Ultrasound 20 kc at intensity 1 for 4 min (LS75 Branson)	Lysozyme 2.5 mg ml^{-1} 15 min 4°C DNase, RNase 2 mg ml^{-1} 30 min 4°C 15 ml 2% sodium lauroyl sarcosinate in 50 mM Tris-HCl, pH 8	SDS 0.1% 18 h room temperature, stirring DNase	Equal volume SDS 2%, 16 h 4°C
Purification	As *P. avenae*	31 000g 20 min, resuspend in buffer	10 000g 30 min, resuspend in buffer (20 g original wet weight/100 ml)	27 000g 30 min, resuspend in 10 ml 50 mM Tris-HCl, pH 8.5; 63g 5 min and supernatant 27 000g 30 min, resuspend in 50 mM Tris-HCl, pH 8.5
Sonication			2 min ml^{-1} at 60 W 40 000g 30 min 4°C, resuspend in 7 ml phosphate buffer	2 min ml^{-1}
Gradient centrifugation		Three rounds in 10–55% (w/v) linear sucrose gradients in 25 ml, 1500g 45 min	Linear 55–80% (w/v) at 200 000g 4°C 33 h	Percoll 1:1 in buffer, 20 min 30 000g
Collect fraction with R bodies	White band	White band	White band at 70–75% sucrose	White band
	Centrifuge 31 000g 15 min, resuspend in buffer	Centrifuge 40 000g 15 min, resuspend in buffer	Centrifuge 40 000g 15 min, resuspend in buffer

the R bodies are allowed to adsorb for 1–2 minutes; the excess liquid is then removed using a piece of filter paper. The specimens are dried at room temperature. One droplet of the stain is deposited on the grid and removed with filter paper after 1–1.5 minutes. Uranyl acetate (1–4% w/v in water, pH 4.8), methylamine tungstate (2% w/v in Tris-HCl, pH 8.5), ammonium molybdate (3% w/v in water, pH 6.8) or phosphotungstic acid (1% w/v in water, pH 7.0) gave good results. In our hands, phosphotungstic acid is the best stain, penetrating in the cells and allowing the best recognition of the R bodies (Figs 1–4, 7 and 8).

This method is very easy to do and has been used for finding the ratio of cells with R bodies in *P. taeniospiralis*, giving the same results as obtained with metal shadowing.

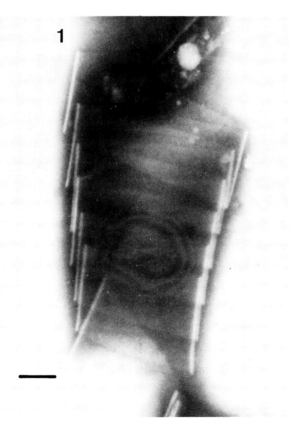

Fig. 1. Partially unwound R body of *Caedibacter taeniospiralis* (type 51). Negative stain by the direct method with phosphotungstic acid. Bar, 0.1 μm.

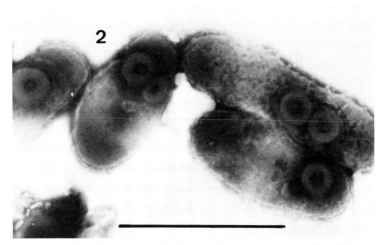

Fig. 2. Autotrophically grown cells of *Pseudomonas taeniospiralis* with R bodies negatively stained by the direct method with phosphotungstic acid. Bar, 1 μm.

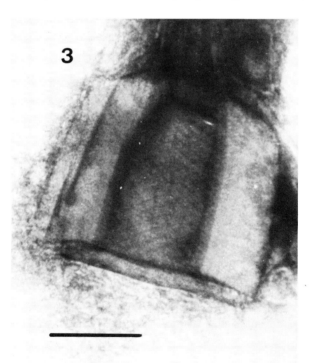

Fig. 3. Negatively stained R body of *P. taeniospiralis* in side view, looking like an optical section. Bar, 0.1 μm.

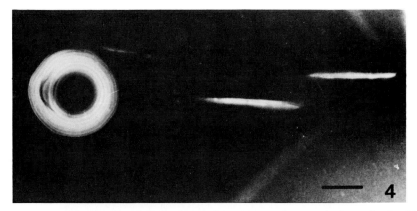

Fig. 4. Partially unwound R body of *P. taeniospiralis* negatively stained with phosphotungstic acid by the direct method. Bar, 1 μm.

(b) Floating technique

Follow the technique of Valentine as described elsewhere in this volume. It gives a very good contrast on isolated R bodies but not on whole cells. Whole cells are better preserved than in the direct method, and the stain does not penetrate so well into the cytoplasm. The same stains as in the direct method can be used.

2. Metal shadowing

Metal shadowing has been used for detecting R bodies inside whole cells and for studying isolated R bodies in the extended form, allowing the

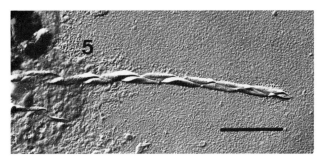

Fig. 5. Extended R body of *P. taeniospiralis* metal shadowed with carbon–platinum at an angle of 30°. Bar, 1 μm.

differentiation of two surfaces, one with a more granular texture in the inner side and the other, at the outer surface, relatively smooth (Fig. 5).

Samples adsorbed onto Formvar–carbon grids are allowed to dry and are then shadowed with palladium–platinum evaporated from a tungsten wire or from carbon. An angle of 10° to the specimen surface allows a good differentiation of the granular and smooth surfaces, and an angle of 30–40°

Fig. 6.   Cell of *P. taeniospiralis* metal shadowed with carbon–platinum at an angle of 30°. The R body protrudes from the flattened cell. Bar, 1 μm.

makes it possible to count the cells very easily. The R bodies protrude from the flattened cells because they are more stable (Fig. 6).

3. Thin sections

Ultra-thin sections of whole cells show that the R bodies are rolled-up double membrane systems with two electron-dense zones. The average thickness is 11–16 nm (Figs 9 and 10). Most of the R bodies sectioned contain a fibrous core material or spherical bodies, which could be associated with their assembly (Lalucat and Mayer, 1978; Wells and Horne, 1983).

For thin sections the cells grown on solid media can be taken from the agar surface with a cork borer and processed following the usual methods. The cells cultured in liquid media can be fixed overnight by adding an aliquot of 25% (w/v) stock solution of glutaraldehyde to the culture media (2.5% w/v, final concentration) or embedded in small agar cubes before fixation. The methods described for *P. taeniospiralis* (Lalucat and Mayer, 1978), *P. avenae* (Wells *et al.*,

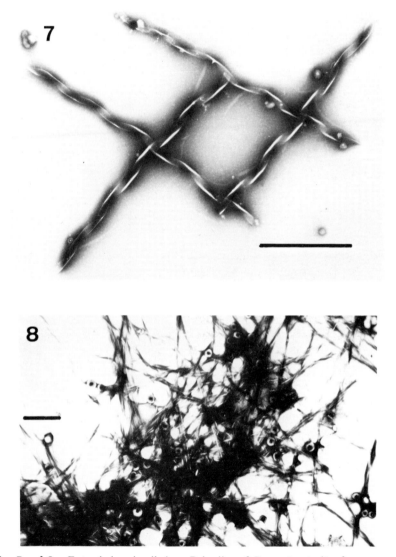

Figs 7 and 8. Extended and rolled-up R bodies of *P. taeniospiralis* after a sucrose gradient centrifugation. The fraction contains only R bodies. Bars, 1 μm.

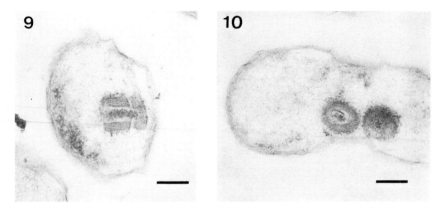

Figs 9 and 10. Thin sections of cells of *P. taeniospiralis* in side and face on view showing the layers of the ribbon. Bar, 0.2 μm.

1983) and *E. coli* P678-54 (pBQ65) (Quackenbush and Burbach, 1983) are very similar and useful for the three types of R bodies:

1. Fixation: 3% glutaraldehyde in 50 mM phosphate buffer, pH 6.8, 90 min, room temperature;
   2–2.5% glutaraldehyde in 50 mM cacodylate buffer, pH 7.0;
   2% paraformaldehyde/5% glutaraldehyde in 100 mM phosphate buffer, pH 7.4.

2. Postfixation: 1% osmium tetroxide in 50 mM phosphate buffer, pH 6.8, 2 h 4°C;
   1% osmic acid in 50 mM phosphate buffer, pH 7.0, 1 h;
   2% osmium tetroxide.

3. Dehydration: series of acetone or ethanol.

4. Embedding: Spurr's low viscosity medium or Epon via toluene.

5. Poststaining: uranyl acetate and lead citrate.

6. Sectioning: 70–80 nm.

4. Immunogold negative stain

This method has been used for comparing the antigenic cross-reactions between types of R bodies. It clearly demonstrates homologous antigens in *P. taeniospiralis* and *P. avenae* R bodies, but none with *C. taeniospiralis* R bodies (Lalucat *et al.*, 1986; Gibson *et al.*, 1987).

Antisera against R bodies from *C. taeniospiralis* and *P. taeniospiralis* have been obtained by (1) weekly intravenous injections into rabbits of 200 μl of R bodies suspended in 1.1 mg ml^{-1} phosphate-buffered saline, pH 7.4 (8 g litre^{-1} NaCl, 0.2 g litre^{-1} KCl, 1.15 g litre^{-1} Na$_2$HPO$_4$, 0.2 g litre^{-1} KH$_2$PO$_4$)

(Kanabrocki *et al.*, 1986); or (2) purified R bodies in phosphate buffer (50 mM, pH 7.0) at 2 mg ml^{-1} are mixed 1:1 with Freund's coadjuvant and 1 ml is injected intramuscularly twice over a 2-week period. Two weeks later serum is collected and stored at $-20°C$ (Lalucat *et al.*, 1986).

The immunogold labelling procedure used is based on that of Beesley *et al.* (1984). The R body suspension is dried on gold grids with carbon-coated Parlodion film as in the direct method for negative staining. Antibody is diluted 10 times in 0.5 M Tris-HCl buffer, pH 7.4, containing 0.1% gelatin, 1% Tween 20 and 1% bovine serum albumin, and a 50-µl drop is deposited on a hydrophobic surface (Petri dish). The grid is floated on the drop for 1 h at room temperature, washed with distilled water and labelled for 1 h at room

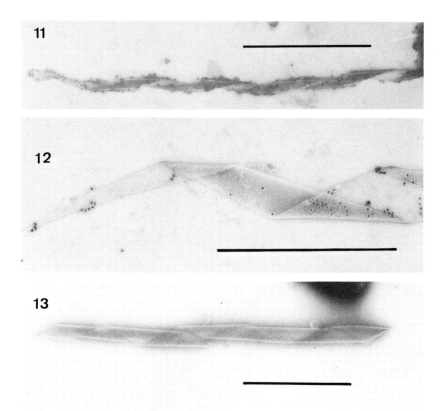

Figs 11–13. Immunogold negative staining of isolated R bodies of *P. taeniospiralis* (11), *P. avenae* (12) and *Caedibacter taeniospiralis* (13) with antisera obtained against purified R bodies of *P. taeniospiralis*, showing the cross-reaction between both *Pseudomonas* R bodies, but not against *Caedibacter*. Bar, 1 µm.

temperature with GAR G15 EM grade goat anti-rabbit IgG linked to 15 nm gold colloid (Jansen Pharmaceutical) diluted 20 times with the same buffer. The grid is washed with distilled water and negatively stained with 2% uranyl acetate in water (Figs 11–13).

References

Beesley, J. E., Day, S. E. J., Betts, M. P. and Thorley, C. M. (1984). *J. Gen. Microbiol.* **130**, 1481–1487.
Estève, J.-C. (1978). *Protistologica* **14**, 201–207.
Fusté, M. C., Simon-Pujol, M. D., Marqués, A. M., Guinea, J. and Congregado, F. (1986). *J. Gen. Microbiol.* **132**, 2801–2805.
Gibson, I., Bedingfield, G., Dobbs, H. and Shackleton, J. (1987). *Micron Microsc. Acta* **18**, 71–75.
Kanabrocki, J. A., Lalucat, J., Cox, B. J. and Quackenbush, R. L. (1986). *J. Bacteriol.* **168**, 1019–1022.
Lalucat, J. and Mayer, F. (1978). *Z. Allg. Mikrobiol.* **18**, 517–521.
Lalucat, J., Meyer, O., Mayer, F., Parés, R. and Schlegel, H. G. (1979). *Arch. Microbiol.* **121**, 9–15.
Lalucat, J., Parés, R. and Schlegel, H. G. (1982). *Int. J. System. Bacteriol.* **32**, 332–338.
Lalucat, J., Wells, B. and Gibson, I. (1986). *Micron Microsc. Acta* **17**, 243–245.
Preer, J. R. and Preer, L. B. (1984). In *Bergey's Manual of Systematic Bacteriology* (N. R. Krieg and J. G. Holt, eds), Vol. 1, pp. 795–811. Williams and Wilkins, Baltimore.
Quackenbush, R. L. and Burbach, J. A. (1983). *Proc. Natl Acad. Sci. USA* **80**, 250–254.
Sonneborn, T. M. (1970). In *Methods in Cell Physiology*, Vol. 4, pp. 241–339. Academic Press, New York.
Wells, B. and Horne, R. W. (1983). *Micron Microsc. Acta* **14**, 329–344.
Wells, B., Horne, R. W., Lund, B. M. and King, N. R. (1983). *Micron Microsc. Acta* **1**, 11–28.

5

Electron Probe Microanalysis of Cryosections from Cell Suspensions

KARL ZIEROLD

Max-Planck-Institut für Systemphysiologie, Rheinlanddamm 201,
4600 Dortmund 1, Federal Republic of Germany

I. Introduction

Many processes in biological cells, for example ion transport through membranes or cytosis of substances, are connected with the uptake or loss of specific elements. The element content in single cells, in particular in bacteria, is usually studied by means of flame photometry or by using radioactive tracers. It is the aim of this chapter to show that electron probe microanalysis of cryosections from cell suspensions can be used as a supplementary method with three additional options: (1) usually, all elements with the atomic number higher than 10 can be measured simultaneously in one and the same cell; (2) as single cell sections are measured it becomes possible to discover intercellular variations; and (3) in some cells, depending on the dry mass distribution, elements can be localized in intracellular compartments.

In the following section on methodology the physical and technical basis for electron probe microanalysis of biological specimens is outlined. Preparation and analysis of cryosections are then described. Results obtained from cell

METHODS IN MICROBIOLOGY
VOLUME 20 ISBN 0-12-521520-6

suspensions of *Escherichia coli* and *Saccharomyces cerevisiae* illustrate the capability of this analytical method. Finally, methodological and microbiological aspects of electron probe microanalysis of cryosectioned cells are discussed.

II. Methodology

A. Physical and technical basis

The electron microscope is not only a tool to produce enlarged images of cells and their structural constituents; the interaction of the electron beam with matter can also be used to analyse the elemental composition and the distribution of mass within cells. For example, the intensity of the elastically scattered electrons in a sufficiently thin specimen is proportional to the mass thickness (mass density × thickness of the specimen). Accelerated electrons penetrating the specimen lose energy, the loss depending on the atomic composition of the specimen. This effect is utilized by electron energy loss spectroscopy, which is described in more detail in a separate chapter by R. Bauer in this volume. The most important interaction of accelerated electrons with matter for analytical purposes is the generation of X-rays: the impinging electron beam ejects an inner shell electron from the atom. The remaining free electron site is reoccupied by an outer shell electron dropping down to the inner shell. The energy difference between the two electron shells is emitted as an X-ray photon. The energy (E) of these X-rays depends on the atomic number (Z) of the atoms in the specimen according to Moseley's law, $E \sim Z^2$. Thus, measurements of the energy distribution of the X-rays generated in the electron-irradiated specimen provide information on the elemental composition. For technical reasons generally all elements with the atomic number higher than 10 can be measured simultaneously. This method is called energy dispersive X-ray microanalysis. In order to achieve a high spatial resolution, this analytical method is preferably combined with scanning transmission electron microscopy (STEM). In this technique a thin electron beam scans the area of interest of the specimen. The electron scattering is used for imaging and mass measurements, and from the X-rays generated the distribution of elements can be determined. For more information on the physical and technical basis of electron probe microanalysis, the reader is referred to Hren *et al.* (1979) and Goldstein *et al.* (1981).

However, the application of electron probe microanalysis to biological problems requires that the element distribution in the specimen in the electron microscope is the same as it was in the functional state of life to be investigated. Electron probe microanalysis of biological specimens has been reviewed, for example, by Hayat (1980), Moreton (1981), Hutchinson and

Somlyo (1981), Hall and Gupta (1983), Morgan (1985), and Hall (1986). Conventional preparation techniques for electron microscopy of biological cells based on chemical fixation, heavy metal staining, dehydration by alcohol, and resin embedding, cause redistribution and loss of diffusible and soluble elements. Therefore, they are inappropriate for analytical purposes such as electron probe microanalysis to localize diffusible substances, in particular ions, in biological cells (Zierold and Schäfer, 1983; Morgan, 1985; Roos and Barnard, 1985; Hall, 1986). The method of choice is a cryopreparation technique, which starts by quick freezing of the cells in their functional state. Then, ultra-thin (100 nm thick) cryosections are prepared from this frozen specimen. After freeze-drying, these sections are studied by X-ray micro-analysis in STEM.

B. Preparation of cryosections

The preparation of cryosections and the analysis in STEM, as outlined schematically in Fig. 1 and described below, are treated in more detail by Zierold (1982, 1986a). Cryofixation means the solidification of an aqueous biological sample by quick freezing without any chemical pretreatment. The biological sample is cooled within approximately 10 ms to a temperature below 170 K and kept below this temperature in order to avoid any recrystallization of water. For this purpose three different techniques have been established: (1) plunging the specimen into a cold liquid, preferably liquid propane or liquid ethane; (2) propane jet cryofixation, where a stream of liquid propane is directed onto the specimen; and (3) metal mirror cryofixation. In the last technique the specimen is pressed quickly onto a cold polished metal surface. None of these techniques can be accepted as being ideally suited for all kinds of specimens and biological problems. The above mentioned techniques only allow small samples or superficial specimen layers up to a thickness of approximately 20 μm to be frozen without visible ice crystal damage. For a review on cryofixation techniques the reader is referred to Plattner and Bachmann (1982), Robards and Sleytr (1985), Menco (1986), and to the chapter by M. Müller in this volume.

For cryoultramicrotomy the cell suspension has to be frozen as a small droplet in order to achieve a sufficiently small cutting width below 100 μm in diameter. This can be achieved by freezing a droplet of the cell suspension between two 0.1 mm thick copper sheets (Fig. 2). One of these sheets has a smoothly polished surface with a cavity in its centre; the other sheet has a rough surface. After freezing, the copper sheets are detached from each other in the cold chamber of the cryoultramicrotome. In most cases the frozen droplet remains on the rough copper surface, which is mounted on the object holder of the cryoultramicrotome for sectioning.

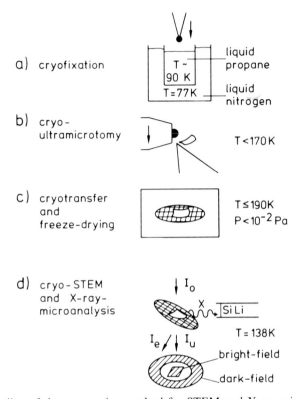

a) cryofixation $T \sim$ 90 K $T = 77K$ liquid propane   liquid nitrogen

b) cryo-ultramicrotomy $T < 170K$

c) cryotransfer and freeze-drying $T \leq 190K$ $P < 10^{-2} Pa$

d) cryo-STEM and X-ray-microanalysis I_0 X   SiLi $T = 138K$ I_e / I_u bright-field   dark-field

Fig. 1. Outline of the preparation method for STEM and X-ray microanalysis of freeze-dried ultra-thin cryosections. (a) Cryofixation is achieved by quick plunging the specimen into liquid propane cooled by liquid nitrogen. (b) Approximately 100 nm thick cryosections are prepared in the cryoultramicrotome by means of a dry glass knife. (c) The cryosections are transferred to the STEM in a transportable cryochamber in a cold nitrogen gas atmosphere at a temperature of 120 K. The sections are freeze-dried within this transfer chamber after evacuation and increasing the temperature. (d) The sections are placed in the cryostage of the STEM. The electron beam, I_0, interacts with the specimen. Unscattered (I_u) and inelastically scattered (I_e) electrons are recorded by the central bright-field detector, while the elastically scattered electrons reach the annular dark-field detector. A fraction of the excited X-rays is detected by the Si(Li) crystal of the X-ray microanalyser.

Cryosections are cut by a dry glass knife at a temperature below 160 K. Sections with a thickness of approximately 100 nm appear as irregularly shaped transparent sheets at the knife edge. The sections are picked up by an eyelash probe and are placed onto a Pioloform®-coated and carbon-reinforced copper grid. The sections are pressed onto the grid by a cold metal rod with a polished surface. The grid is mounted onto a cryospecimen holder

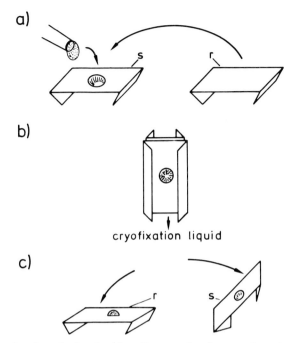

Fig. 2.   Cryofixation of a droplet of a cell suspension for cryoultramicrotomy. (a) The droplet is transferred to a smoothly polished copper sheet (s) with a central cavity by means of a glass pipette. A flat copper sheet of the same size with a rough surface (r) is placed on the sheet with the droplet. (b) This arrangement is plunged into liquid propane by means of forceps. In the cryoultramicrotome the copper sheets are detached from each other at low temperature. In most cases the droplet remains on the sheet with the rough surface (r).

and transferred to the STEM by a transportable cryochamber filled with cold nitrogen gas. In this chamber the grid bearing the sections is kept at a temperature of 120 K. This cryochamber is attached to the STEM and evacuated before freeze-drying and transfer of the grid into the cold stage of the electron microscope is performed. After increasing the temperature to 190 K at a pressure below 10^{-2} Pa, the sections are freeze-dried in the cryotransfer chamber.

C.   Microanalysis and data evaluation

In the STEM the grid is placed in a cold stage at a temperature of 138 K, as drawn schematically in Fig. 1. The grid is tilted by 30° towards the energy dispersive X-ray analyser, which is placed 12 mm from the specimen. The microanalyser consists of a SiLi crystal cooled by liquid nitrogen, which transforms the impinging X-ray quanta into electrical pulses according to their energy. These pulses are sorted by a multichannel analyser, thus yielding the X-ray spectrum of the specimen area irradiated by the primary electron beam. The SiLi crystal is shielded from the microscope column by a 7 μm thick beryllium window in order to prevent scattered electrons from reaching the detector. These electrons would interfere with the detection of X-rays. The primary electrons are accelerated to 100 keV; the beam current is 4×10^{-9} A. X-ray microanalysis is carried out either in the spot mode, or by scanning the electron beam over an area ranging from 70 nm × 110 nm to 7 μm × 11 μm. The counting time for one spectrum is 100 s.

The cryosections are imaged by means of a central bright-field detector and an annular-shaped dark-field detector, preferably by ratio contrast (bright-field/dark-field signal). The electron optical contrast obtained from unstained freeze-dried cryosections in STEM is mainly due to elastically scattered electrons, which are recorded by the dark-field detector.

The typical X-ray spectrum consists of a broad background radiation caused by deceleration of the primary electrons in the specimen, and of peaks at defined energies caused by excitation of electrons bound with characteristic energies in the electron shells of the atoms. The background radiation is made up of three contributions: (1) X-rays from the support film; (2) X-rays from electrons scattered into the column of the electron microscope, the so-called extraneous radiation; and (3) X-rays from the specimen itself. This background contribution from the cryosection can be used as a measure of the organic dry mass of the specimen present. It is beyond the scope of this chapter to describe the evaluation of the specimen-dependent background radiation (see e.g. Statham, 1979; Roomans and Kuypers, 1980; Zierold, 1986b). According to Hall and Gupta (1979, 1982), the basic idea for quantitative evaluation of X-ray spectra of cryosections is that the specimen-dependent peak-to-background ratio (p/b) is proportional to the dry weight concentration, C_{d}, of the element emitting the characteristic X-rays resulting in the peak "p":

$$C_{\mathrm{d}} = k \times p/b \qquad (1)$$

The constant, k, depends on the particular element; the characteristic X-ray peak, p, originates from and depends on the mean atomic number composition of the irradiated part of the specimen. Equation (1) allows us to determine dry weight concentrations of elements by comparison with X-ray spectra from cryosections of standard specimens. For example, salt solutions of known concentration were mixed either with dextran or another organic substance. These mixtures were frozen in liquid propane for cryoultramicrotomy. The cryosections were transferred and analysed by the same procedure as described above. They have proved to be feasible standard specimens for quantitative evaluation of X-ray spectra (Roomans and Sevéus, 1977; Zierold, 1986b). The QUANTEM FLS program by Link Systems, based on the peak-to-background ratio method, was used for quantitative evaluation (Hall and Gupta, 1979, 1982; Statham, 1979). The detection limit was found to be about 10 mmol (kg dry weight)$^{-1}$ for all elements with atomic number (Z) higher than 12, and 20 and 30 mmol (kg dry weight)$^{-1}$, respectively, for magnesium ($Z = 12$) and sodium ($Z = 11$). The spatial resolution is limited by the thickness of the cryosections, due to electron scattering in the specimen. In 100 nm thick freeze-dried cryosections the lateral analytical resolution is below 50 nm. Thus, less than 1000 atoms concentrated in a sufficiently small volume of 100 nm edge width provide a detectable X-ray signal.

In sufficiently thin specimens the annular dark-field intensity is proportional to the mass thickness (see e.g. Carlemalm et al., 1985; Reichelt and Engel, 1986). Therefore, the annular darkfield intensity can be used to determine the mass thickness, which is equivalent to the dry weight content in cells and their major compartments (Zierold, 1986b). Calibration measurements have been carried out with freeze-dried cryosections of the same thickness prepared from mixtures consisting of dextran and water in varying concentrations. Dextran is assumed to represent the organic dry mass in biological specimens with respect to the elastic electron scattering conditions. For sufficiently constant thickness the dry weight portion, d, is given by

$$d = d_0(i - 1) \qquad (2)$$

where d_0 = calibration constant; i = relative dark-field intensity, defined as the ratio of the dark-field signal of the cryosection placed on the support film and the dark-field signal of the bare support film. The corresponding water portion is then

$$w = 1 - d \qquad (3)$$

By combining the results of X-ray microanalysis, which give the dry weight concentrations of elements, with the measurements of the relative dark-field intensity, the wet weight element concentration, C_w, and the aqueous element concentration, C_a, can be determined:

$$C_w = C_d \times d \qquad (4)$$
$$C_a = C_w/(1 - d) = C_d \times d/(1 - d) \qquad (5)$$

In particular, in physiological studies concerning the distribution of electrolyte ions, the wet weight concentration and the aqueous concentration of ions are often more significant than the bare dry weight concentration obtained from X-ray microanalysis of freeze-dried cryosections.

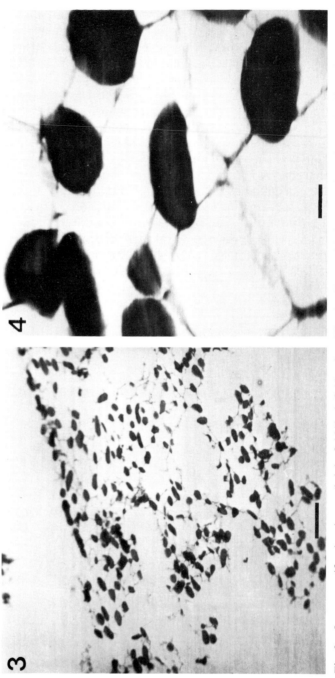

Fig. 3. Low magnification image (bright-field/dark-field) of a freeze-dried cryosection of *E. coli* cells. Bar, 2.5 µm.

Fig. 4. Freeze-dried cryosection of *E. coli* cells, ratio contrast (bright-field/dark-field). Intracellular structures are not revealed. The fibrillar network between the cells represents the segregation pattern of the extracellular organic material after freeze-drying. Bar, 250 nm.

III.   Results from cryosectioned cell suspensions

Cells were prepared and analysed by energy-dispersive X-ray microanalysis in STEM, as described above. Figures 3 and 4 show freeze-dried cryosections from a cryofixed suspension of *Escherichia coli* bacteria, ATCC 11775, harvested in LB-medium. No intracellular structures were revealed from the cryosectioned unstained cells. Figures 5 and 6 show typical intracellular and extracellular X-ray spectra. However, the extracellular spectrum is only qualitatively useful, because the high water content of the extracellular space results in random aggregation of organic matter and ions. Reliable quantitative data can only be obtained from compartments with a sufficiently homogeneous organic dry matrix. The segregation compartments of the remaining dry matrix caused by ice crystal growth have to be significantly smaller than the scanning area for quantitative evaluation (Zierold, 1986a). Quantitative data obtained from cryosectioned *E. coli* cells are compiled in Table I. This table shows that electron probe microanalysis of freeze-dried cryosections enables the study of element content in individual cells. The

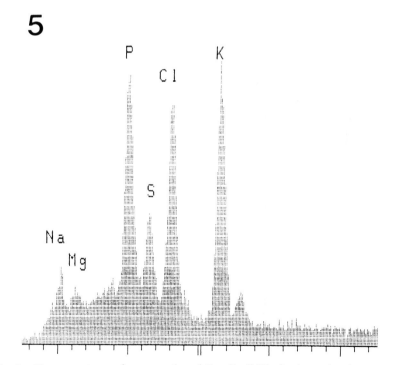

Fig. 5.   X-ray spectrum of a cryosectioned *E. coli* cell. The horizontal scale represents the X-ray energy, the vertical scale the number of X-ray counts.

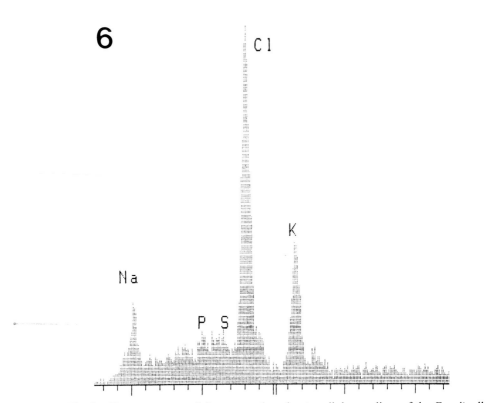

Fig. 6. X-ray spectrum of the cryosectioned extracellular medium of the *E. coli* cell suspension.

TABLE I

Dry weight portions and element concentrations (mmol (kg dry weight)$^{-1}$) measured in a freeze-dried cryosection of *Escherichia coli* cell suspension. The intracellular values in the single cell were calculated from the X-ray spectrum in Fig. 5, the extracellular values from the X-ray spectrum in Fig. 6. The intracellular mean values are based on measurements of 13 cells.

	Intracellular single cell	Extracellular	Intracellular mean ± S.D.
d	0.31	—	0.33 ± 0.03
Na	182	1589	169 ± 48
Mg	83	200	73 ± 16
P	407	210	400 ± 55
S	167	193	162 ± 14
Cl	290	2616	287 ± 60
K	340	993	313 ± 50
Ca	—	—	—

intercellular variation becomes evident from the standard deviation calculated from the measurements of many cells. The corresponding wet weight and liquid concentrations can be calculated by applying equations (4) and (5) to the data in Table I.

In many cell types electron probe microanalysis in STEM makes it possible to determine element concentrations in different intracellular compartments. Figure 7 shows a cryosection of *Saccharomyces cerevisiae* yeast cells and

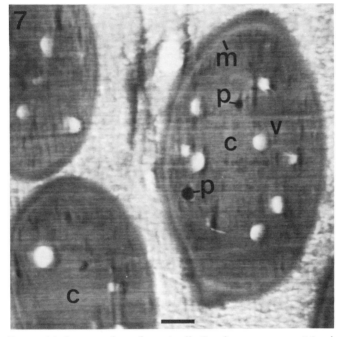

Fig. 7. Freeze-dried cryosection of yeast cells *Saccharomyces cerevisiae*, imaged by ratio contrast (bright-field/dark-field). c = cytoplasm, m = region of the cell membrane, p = electron dense particle, v = vacuole; bar, 500 nm.

corresponding X-ray spectra from different intracellular compartments (Figs 8–10). Before cryofixation, the yeast cells were suspended in a buffer medium of sucrose and Tris acetate at pH 8.2. In particular, electron dense particles containing high amounts of phosphorus, magnesium, and calcium in comparison to the surrounding cytoplasm were found. The region close to the cell membrane contains approximately the same element pattern as the cytoplasm, but in lower concentration (Fig. 10). The quantitative evaluation of the X-ray spectra in Figs 8–10 is summarized in Table II.

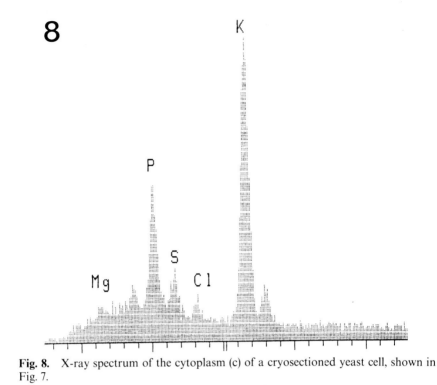

Fig. 8. X-ray spectrum of the cytoplasm (c) of a cryosectioned yeast cell, shown in Fig. 7.

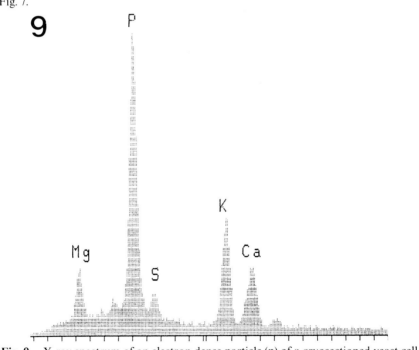

Fig. 9. X-ray spectrum of an electron dense particle (p) of a cryosectioned yeast cell shown in Fig. 7.

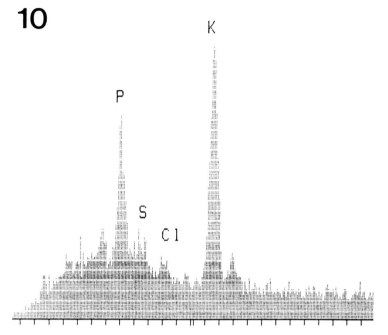

Fig. 10. X-ray spectrum of the region of the cell membrane of a cryosectioned yeast cell shown in Fig. 7.

TABLE II

Dry weight portions and element distribution measured in the freeze-dried cryosectioned yeast cell (*Saccharomyces cerevisiae*) (Fig. 7). The element concentrations (mmol (kg dry weight)$^{-1}$) were calculated by evaluation of the X-ray spectra in Fig. 8 (cytoplasm), Fig. 9 (electron-dense granule), and Fig. 10 (cell membrane region).

	Cytoplasm	Electron dense granule	Cell membrane region
d	0.23	0.34	0.18
Na	—	—	—
Mg	31	386	24
P	258	1039	164
S	87	99	39
Cl	31	—	19
K	460	329	223
Ca	—	156	—

IV. Discussion

A. Methodological aspects

Due to their small size, most single cells and microorganisms are ideally suited objects for cryofixation. They can be frozen without or with only minimal ice crystal damage. The freezing method described earlier (Fig. 2) guarantees that no water evaporates from the extracellular medium of the cell suspension immediately before cryofixation, as the copper sheets protect the droplet from the contact with air. Thus, undesired osmotic alterations which could affect the ion content in the cells are avoided. If the elemental composition of the cells is expected to be affected by copper ions, the copper sheets can be coated by a thin layer of gold. Gold has been found to be physiologically inactive and it has a good thermal conductivity, a prerequisite for cryofixation. Cryoprotectants must not be used in cryofixation for microanalytical purposes, as they could influence the ion composition of the cells. For the same reason sections have to be cut dry without any trough liquid. Dry cryosectioning causes unidirectional deformation of the section perpendicular to the cutting direction (Zierold, 1986a). This can be seen by the elliptical shape of the originally round yeast cells (Fig. 7). Hitherto, this effect could not be reduced, neither by variation of the cutting temperature between 120 K and 180 K nor by adjusting the cutting speed between $0.2 \, mm \, s^{-1}$ and $10 \, mm \, s^{-1}$. Scratch marks occasionally observed on sections cut by a glass knife with an inhomogeneous edge can be avoided by the use of a diamond knife. In other respects diamond knives have not yet proved to be superior to glass knives in cryoultramicrotomy. The cryosections are transferred from the knife to the electron microscopical grid by an eyelash probe as described in Christensen (1971)–still the best known method. However, this step remains a weak link in cryoultramicrotomy, as it requires patience and some experience in handling. The often observed electrical charging of cryosections can be reduced by applying an electrical discharge device.

It is obvious that electron microscopy of frozen-hydrated cryosections requires the use of a cryotransfer system. Furthermore, cryotransfer has also turned out to be advantageous for STEM and microanalysis of freeze-dried cryosections which are extremely hygroscopic. Freeze-dried cryosections attract dust and water particles from the air. Storage under air and transfer of these sections through air impairs the recognition of structural details and may also disturb the distribution of ions. A closed cryotransfer system prevents any damage that could be caused by contact with a humid atmosphere.

Frozen-hydrated cryosections are considered to maintain best the structure and the composition inherent to cells in the living state. Thus, excellent micrographs of cryosections from vitrified bacteria in the frozen-hydrated state have been obtained, for example, by Dubochet *et al.* (1983). Unfortunately, 100 nm thick frozen-hydrated cryosections are extremely sensitive to radiation damage, as can be seen by the "bubbling" that occurs with increasing electron irradiation, resulting in mass loss (Zierold, 1985, 1986a). Electron microprobe analysis requires considerably higher electron doses than are necessary for bare imaging. For example, scanning the 100 kV electron beam of 1 nA for 100 s over an area of 2 μm^2 in size–corresponding to an electron dose of 3×10^5 e nm^{-2}—would etch away a 100 nm thick ice layer. For ultra-thin frozen-hydrated sections even such rather moderate electron microprobe working conditions are beyond the tolerance limit. In 1 μm thick sections the etch depth of 100 nm can still be accepted as negligible (Gupta and Hall, 1981; Hall and Gupta, 1982; Hall, 1986), but such thick specimens are not feasible for the analysis of small cells such as *E. coli* bacteria, which would be thinner than the whole section. Thus, ultra-thin frozen-hydrated cryosections are useless for electron microprobe analysis of most microbiological objects.

After freeze-drying, cryosections are stable in the electron beam even at high electron doses up to 10^8 e nm^{-2}, as required for high resolution X-ray microanalysis. Freeze-drying has turned out to be a necessary preparation step. Incompletely dried sections, however, show phenomena of mass loss similar to those described above, indicating that the radiolysis of water is the main problem in electron microscopy and microprobe analysis of frozen-hydrated specimens (Talmon, 1984). Caused by the sublimation of water, structures of organic material shrink more or less, and redistribution of ions in originally very aqueous compartments has to be taken into account. In most intracellular compartments with a dry mass portion of at least 10%, the displacement of ions is negligible, as they are assumed not to move further than to the next organic structures. In compartments with sufficient organic mass portion these diffusion paths can be assumed to be shorter than the spatial electron microprobe resolution. But in very aqueous compartments, such as the extracellular space, ice crystal growth during cryofixation or recrystallization causes a segregation pattern of the remaining dry mass, which appears as a fibrillar network after freeze-drying (Figs 3, 4). Diffusible ions aggregate randomly at these structures. Therefore, quantitative evaluation of X-ray spectra according to equation (1) yields reliable data only for compartments with sufficiently dense and homogeneous dry mass, whereas X-ray spectra of extracellular structures represent random element distributions with only qualitative significance. The segregation pattern of the remaining dry mass may limit the lateral analytical resolution. In well cryoprepared cells with invisible fine segregation pattern the lateral resolution of electron microprobe analysis is determined mainly by the section thickness. In 100 nm thick freeze-dried cryosections the lateral analytical resolution was found to be in the range of 30–50 nm. The quantitative evaluation of the X-ray spectra by use of the peak-to-background ratio (Hall and Gupta, 1979, 1982) yields element concentrations in terms of the number of atoms per dry mass. The detection limit of 10 mmol (kg dry weight)$^{-1}$ for elements with an atomic number higher than 12 is independent of the size of the irradiated area. Chang *et al.* (1986) report measurements of calcium concentrations in *E. coli* of less than 3 mmol (kg dry weight)$^{-1}$, presumably due to an improved data evaluation program.

The determination of the dry mass portion of defined compartments in ultra-thin cryosections should be possible by measurements of the specimen-dependent background of the X-ray spectrum. But the specimen background is found to depend on variations in extraneous X-ray background, making this method less precise than the relative dark-field intensity method described above. The relative dark-field intensity measurements yield reliable dry weight portions in the range between 10% and 50%. The prerequisite for the linear relationship between dark-field intensity and dry mass is that the mean free path of elastic electron scattering—this is the mean distance between two elastic scattering events in the specimen—is much larger than the specimen thickness. For example, in a freeze-dried cryosection consisting of 50% dry mass and 50% vacuum (formerly water), this mean free path is 420 nm. Thus, the relative dark-field intensity method is limited to sufficiently thin cryosections. As an upper limit, a thickness of 200 nm should not be exceeded. The relative dark-field intensity method for the determination of dry weight portions depends critically on there being a constant section thickness. Variations in section thickness would affect directly the accuracy of mass determination, as the dark-field intensity is proportional to the product of mass density and section thickness. Previous experience indicates a standard deviation of approximately 10% for the measured dry weight portions.

B. Microbiological aspects

Applications of electron probe microanalysis to microbiological objects are rare. Scherrer and Gerhardt (1972) measured calcium in spores of *Bacillus cereus* and *B. megaterium* by means of a wavelength-dispersive X-ray spectrometer. They prepared whole cells on either quartz or silicon oxide layers. After chemical fixation and embedding, they found in dry cut plastic sections of the spore that calcium was located in a central part of the spore. Roomans and Boekestein (1978) measured the distribution of sodium, potassium, and phosphorus in freeze-dried cells of *Neurospora crassa* by X-ray microanalysis in SEM. Noll and Zierold (1981, 1984) studied iron and manganese accumulation planktonic microorganisms, e.g. *Trachelomonas hispida*, by energy-dispersive X-ray microanalysis in SEM. As these heavy metals were bound in the hard outer shell of the cells, chemical fixation of the whole cells followed by air drying and carbon coating was sufficient for specimen preparation. Similar element analyses on *Gallionella* stalks were reported by Ridgway *et al.* (1981). The influence of the preparation method on the element content of bacterial cells was studied by Sigee *et al.* (1985). These authors compared X-ray spectra obtained from whole cells, cell fragments, and plastic sections of the bacterium *Pseudomonas tabaci*. They showed that glutaraldehyde fixation and dehydration by ethanol change the elemental composition considerably.

To the author's knowledge Roomans and Sevéus (1976) carried out the first investigations on cryosections of microorganisms by electron probe microanalysis. They performed ion uptake experiments with rubidium and caesium in *Saccharomyces cerevisiae* yeast cells and reported quantitative data on ion concentrations in the cytoplasm, the nucleus, and the vacuole evaluated from energy-dispersive X-ray spectra. They found an intracellular potassium wet weight concentration of 160 mmol kg^{-1}, which is in the range of $150–200 \text{ mmol kg}^{-1}$ cell water, reported by Rothstein (1974). After correcting for differences in the estimation of the dry weight portion, the cytoplasmic potassium content agrees roughly with that reported in this chapter (Table II). Stewart *et al.* (1980, 1981) published X-ray distribution maps for magnesium, silicon, phosphorus, sulphur, calcium, manganese, and iron of cryosectioned spores of *B. cereus* and *B. coagulans*. They found different element distributions in the protoplast and in the surface coat.

Cryosections from *E. coli* cells were studied by means of X-ray microanalysis and electron energy loss spectroscopy by Chang *et al.* (1986). The authors reported dry weight concentrations of sodium, magnesium, phosphorus, sulphur, chlorine, potassium, and calcium found in the cells in the logarithmic growth phase and in dividing cells. Except for sulphur and chlorine, the element concentrations measured in the log-phase were distinctly higher than those reported in this chapter (Table I). Chang and co-workers observed considerable element loss after washing the cells in water, except for magnesium and calcium, which even increased. Additionally, they found in the cells in the log-phase 26 mmol kg^{-1} calcium in the cell envelope, and only 1.5 mmol kg^{-1} calcium in the cytoplasm. Dividing cells had about 33 mmol kg^{-1} calcium in both the cytoplasm and the cell envelope. In the studies on *E. coli* cells discussed in this chapter (Figs 3–6, Table I) no calcium was found in either the central part or in the envelope region. This means that the calcium concentration was below $10 \text{ mmol (kg dry weight)}^{-1}$. However, these cells were not identified with respect to their growth state. Nevertheless, the results obtained (Table I) indicate the capability of electron probe microanalysis of cryosectioned suspensions of bacteria in ion transport studies.

Data on potassium and magnesium concentrations measured under

different conditions and with different methods in *E. coli* cells are compiled in
Table III. Most data available on ion transport in bacteria are based on flame
photometry and experiments with radioactive isotopes. For example, Schultz
and Solomon (1961) measured in the early logarithmic growth phase of *E. coli*
an intracellular potassium concentration up to $250\,\text{mmol litre}^{-1}$ cell water,
which decreased with cell age below $50\,\text{mmol litre}^{-1}$ after 45 h. The
intracellular sodium increased within the same period of time from
$50\,\text{mmol litre}^{-1}$ cell water to above $100\,\text{mmol litre}^{-1}$ cell water. These data are

TABLE III

Intracellular potassium and magnesium concentrations (mmol (kg dry weight)$^{-1}$ $\pm$
standard deviation, where indicated) as measured in *E. coli* cells by different authors.
The wide range of data reflects the different experimental conditions as indicated. Dry
weight portions were taken into account according to the data given in the particular
publication.

K	Mg	Authors
167–837		Schultz and Solomon (1961), depending on cell age
457–1020		Epstein and Schultz (1965), depending on extracellular osmolarity
175–790	79–195	Lusk *et al.* (1968), depending on extracellular magnesium concentration and growth phase
308	117	Kung *et al.* (1976)
1100	329	Moncany and Kellenberger (1981)
300–600		Meury and Kepes (1981), depending on the metabolic activity of the cells
576 $\pm$ 181	205 $\pm$ 53	Chang *et al.* (1986), logarithmic growth phase
169 $\pm$ 68	201 $\pm$ 66	Chang *et al.* (1986), dividing washed cells
313 $\pm$ 50	73 $\pm$ 16	Zierold, see Table I, undefined growth state

based on the estimate of the dry weight portion of 0.23, which is less than the
value of 0.27 determined by Winkler and Wilson (1966) and lower than the
value of 0.33 measured by the relative dark-field intensity method in STEM
(Table I). Epstein and Schultz (1965) found that the intracellular potassium
concentration increases with the extracellular osmolarity in growing *E. coli*
cells, whereas the intracellular sodium content is independent of the
extracellular osmolarity and depends largely on the sodium concentration of
the extracellular medium. Influx of potassium due to increase of the medium
osmolarity was observed by Meury *et al.* (1985). The intracellular potassium
level in *E. coli* depends also on the metabolic activity of the cells. In the
presence of a carbon source, such as sucrose, in the medium the metabolism is
activated, and the potassium concentration reaches about $200\,\text{mmol litre}^{-1}$,
whereas in the absence of a carbon source the intracellular potassium level is

below 100 mmol litre^{-1} (Meury and Kepes, 1981). The potassium concentration of 313 ± 50 mmol (kg dry weight)$^{-1}$, as given in Table I, is in good agreement with these data and with the value measured by use of an X-ray fluorescence spectrometer by Kung *et al.* (1976), who found 308 mmol (kg dry weight)$^{-1}$. The intracellular magnesium concentrations in *E. coli* published in the literature are higher than those found in this study on cryosections − 73 ± 16 mmol (kg dry weight)$^{-1}$ according to Table I. For example, Kung *et al.* (1976) reported 115 mmol (kg dry weight)$^{-1}$, Moncany and Kellenberger (1981) reported 329 mmol (kg dry weight)$^{-1}$, and Chang *et al.* (1986) reported approximately 200 mmol (kg dry weight)$^{-1}$. These differences could be due to the different magnesium content in the extracellular medium. Lusk *et al.* (1968) reported magnesium concentrations between 79 and 195 mmol (kg protein)$^{-1}$, depending on the extracellular concentration and on the growth phase of the cells.

This preliminary overview of the literature of element determinations in microorganisms, in particular *E. coli*, shows only rough agreement and some discrepancies, probably due to different experimental conditions such as the growth phase of the cells and the composition of the extracellular medium. It remains to be proved by future studies whether electron probe microanalysis of cryosectioned cells would help to elucidate ion transport and storage of elements in bacteria and other microorganisms.

V. Summary

Single cells in suspension, such as bacteria or other microorganisms, are the specimens of choice for cryofixation. Ultra-thin (100 nm thick) cryosections are prepared from these specimens, cryotransferred to a scanning transmission electron microscope, and, after freeze-drying, they are studied by energy-dispersive X-ray microanalysis. This method allows the measurement of all elements with an atomic number, Z, higher than 12 with a detection limit of 10 mmol (kg dry weight)$^{-1}$. For magnesium ($Z = 12$) and sodium ($Z = 11$) the detection limits are 20 and 30 mmol (kg dry weight)$^{-1}$, respectively. The lateral analytical resolution is less than 50 nm. Thus, less than 1000 atoms can be detected if they are concentrated within a volume of 100 nm edge width. The local dry weight portion or water portion is determined by measuring the relative dark-field intensity in STEM. Thus, dry weight element concentrations can be converted to biologically often more significant wet weight concentrations, or element concentrations per mass of cell water.

This method could be particularly useful for studies of ion transport and element accumulation in cells. In comparison to other established methods in this field such as flame photometry or application of radioactive isotopes, electron probe microanalysis enables the measurement of the element content in individual cells. The localization of elements in defined intracellular compartments depends on inhomogeneities in the organic dry mass distribution, which is responsible for the contrast available from unstained cryosections in STEM. Although the densely packed organic matrix present throughout the cells of most bacteria does not permit the determination of element distributions in subcellular structures, this method allows us to account for intercellular variations.

Acknowledgements

I would like to thank Miss S. Dongard for her careful technical assistance and Mrs B. Menge and Professor Dr K. G. Lickfeld (University of Essen) for providing *E. coli* cells for electron probe microanalysis.

References

Carlemalm, E., Colliex, C. and Kellenberger, E. (1985). *Adv. Electr.* **63**, 269–334.
Chang, Ch.-F., Shuman, H. and Somlyo, A. P. (1986). *J. Bacteriol.* **167**, 935–939.
Christensen, A. K. (1971). *J. Cell Biol.* **51**, 772–804.
Dubochet, J., McDowall, A. W., Menge, B., Schmid, E. N. and Lickfeld, K. G. (1983). *J. Bacteriol.* **155**, 381–390.
Epstein, W. and Schultz, S. G. (1965). *J. Gen. Physiol.* **49**, 221–234.
Goldstein, J. I., Newbury, D. E., Echlin, P., Joy, D. C., Fiori, Ch. and Lifshin, E. (1981). *Scanning Electron Microscopy and X-ray Microanalysis.* Plenum Press, New York and London.
Gupta, B. L. and Hall, T. A. (1981). *Tissue Cell* **13**, 623–643.
Hall, T. A. (1986). *Micron Microsc. Acta* **17**, 91–100.
Hall, T. A. and Gupta, B. L. (1979). In *Introduction to Analytical Electron Microscopy* (J. J. Hren, J. I. Goldstein and D. C. Joy, eds), pp. 169–197. Plenum Press, New York and London.
Hall, T. A. and Gupta, B. L. (1982). *J. Microsc.* **126**, 333–345.
Hall, T. A. and Gupta, B. L. (1983). *Quart. Rev. Biophys.* **16**, 279–339.
Hayat, M. A. (ed.) (1980). *X-ray Microanalysis in Biology.* Baltimore University Park Press, Baltimore.
Hren, J. J., Goldstein, J. I. and Joy, D. C. (eds) (1979). *Introduction to Analytical Electron Microscopy.* Plenum Press, New York and London.
Hutchinson, Th. E. and Somlyo, A. P. (eds) (1981). *Microprobe Analysis of Biological Systems.* Academic Press, New York, London, Toronto, Sydney and San Francisco.
Kung, F.-C., Raymond, J. and Glaser, D. A. (1976). *J. Bacteriol.* **126**, 1089–1095.
Lusk, J. E., Williams, R. J. P. and Kennedy, E. P. (1968). *J. Biol. Chem.* **243**, 2618–2624.
Menco, B. Ph. M. (1986). *J. Electron Microsc. Tech.* **4**, 177–240.
Meury, J. and Kepes, A. (1981). *Eur. J. Biochem.* **119**, 165–170.
Meury, J., Robin, A. and Monnier-Champeix, P. (1985). *Eur. J. Biochem.* **151**, 613–619.

Moncany, M. L. J. and Kellenberger, E. (1981). *Experientia* **37**, 846–847.
Moreton, R. B. (1981). *Biol. Rev.* **56**, 409–461.
Morgan, A. J. (1985). *X-ray Microanalysis in Electron Microscopy for Biologists*, Microscopy Handbooks 05, Royal Microscopical Society. Oxford University Press, Oxford.
Noll, M. and Zierold, K. (1981). *Arch. Hydrobiol.* **91**, 242–253.
Noll, M. and Zierold, K. (1984). *Verh. Dtsch. Ges. Ökol.* **XII**, 417–424.
Plattner, H. and Bachmann, L. (1982). *Int. Rev. Cytol.* **79**, 237–304.
Reichelt, R. and Engel, A. (1986). *Ultramicroscopy* **19**, 43–56.
Ridgway, H. F., Means, E. G. and Olson, B. H. (1981). *Appl. Environ. Microbiol.* **41**, 288–297.
Robards, A. W. and Sleytr, U. B. (1985). *Low Temperature Methods in Biological Electron Microscopy*. Practical Methods in Electron Microscopy, Vol. 10 (A. M. Glauert, ed.). Elsevier, Amsterdam, New York and Oxford.
Roomans, G. M. and Boekestein, A. (1978). *Protoplasma* **95**, 385–392.
Roomans, G. M. and Kuypers, G. A. J. (1980). *Ultramicroscopy* **5**, 81–83.
Roomans, G. M. and Sevéus, L. A. (1976). *J. Cell Sci.* **21**, 119–127.
Roomans, G. M. and Sevéus, L. A. (1977). *J. Submicr. Cytol.* **9**, 31–35.
Roos, N. and Barnard, T. (1985). *Ultramicroscopy* **17**, 335–344.
Rothstein, A. (1974). *J. Gen. Physiol.* **64**, 608–621.
Scherrer, R. and Gerhardt, Ph. (1972). *J. Bacteriol.* **112**, 559–568.
Schultz, S. G. and Solomon, A. K. (1961). *J. Gen. Physiol.* **45**, 355–369.
Sigee, D. C., El-Masry, M. H. and Al-Rabaee, R. H. (1985). *Scan. Electron. Microsc./ 1985/III*, 1151–1163.
Statham, P. J. (1979). *Mikrochim. Acta* **8**, 229–242.
Stewart, M., Somlyo, A. P., Somlyo, A. V., Shuman, H., Lindsay, J. A. and Murrell, W. G. (1980). *J. Bacteriol.* **143**, 481–491.
Stewart, M., Somlyo, A. P., Somlyo, A. V., Shuman, H., Lindsay, J. A. and Murrell, W. G. (1981). *J. Bacteriol.* **147**, 670–674.
Talmon, Y. (1984). *Ultramicroscopy* **14**, 305–316.
Winkler, H. H. and Wilson, T. H. (1966). *J. Biol. Chem.* **241**, 2200–2211.
Zierold, K. (1982). *Ultramicroscopy* **10**, 45–54.
Zierold, K. (1985). *J. Microsc.* **140**, 65–71.
Zierold, K. (1986a). In *The Science of Biological Specimen Preparation 1985* (M. Müller, R. P. Becker, A. Boyde and J. J. Wolosewick, eds), pp. 119–127. Scanning Electron Microscopy Inc. AMF O'Hare, Chicago.
Zierold, K. (1986b). *Scan. Electron Microsc./1986/II*, 713–724.
Zierold, K. and Schäfer, D. (1983). *Acta Histochem.* (Suppl.) **XXVIII**, 63–72.

6

Electron Spectroscopic Imaging: An Advanced Technique for Imaging and Analysis in Transmission Electron Microscopy

RICHARD BAUER

Carl Zeiss Oberkochen, Application for Transmission Electron Microscopy, D-7082 Oberkochen, Federal Republic of Germany

I. Introduction

The imaging and examination of microorganisms make very high demands on the instrumentation. Modern conventional transmission electron microscopes (CTEM) have now reached the stage where they meet successfully all standard requirements. For research purposes, however, new procedures are constantly being developed to refine and extend the information available from the microscope and from the specimen. This means new methods of preparation are required, and with them new methods of imaging and analysis.

CTEMs produce good results with thin stained specimens. With very thin

METHODS IN MICROBIOLOGY
VOLUME 20 ISBN 0-12-521520-6

sections, especially if specimens are unstained, the image contrast is often too weak. Results are also unsatisfactory with thicker sections. Not only is there a reduction in contrast, but the resolution is also impaired.

Electron Spectroscopic Imaging (ESI) is a new technique for improving imaging and analysis in the transmission electron microscope, which uses an electron energy loss spectrometer permanently integrated in the electron optical column. The introduction of spectroscopy into the electron microscopic imaging process is the most significant landmark in the development of the electron microscope since it was invented, comparable with the introduction of filters and spectrometers in light optics.

The spectrometer in the electron microscope enhances the contrast when imaging thin and conventionally prepared specimens. With thick specimens both contrast and resolution are improved. Furthermore, depth of focus remains unaffected. With ultra-thin and unstained specimens electron spectroscopic imaging produces outstanding contrast.

Apart from enhanced imaging, the imaging spectrometer presents new possibilities for preparation: for staining in immunology, for the examination of freeze-dried specimens, and particularly for the analysis of ultra-microscopic structures.

II. Principles

A. Methodology

Electrons can be described either as loaded particles (e^-) with a mass (m), or as a wave phenomenon with a wavelength (λ). When the electrons are accelerated in an electrostatic field, they carry the energy to which they were accelerated, say 80 keV, or the corresponding wavelength of 0.0042 nm (Reimer, 1984). When accelerated electrons meet a specimen, the incident electrons and the atoms of the specimen interact. This interaction is called scattering. It is the basis for the image produced in the electron microscope and the analysis which can be made thereof. ESI is a special technique for the imaging and analysis of transmitted electrons.

With a thin specimen, because of the loose structure of the material (diameter of atom $\sim 10^{-8}$ cm, diameter of atomic nucleus $\sim 10^{-12}$ cm), a large proportion of the incident electrons can pass through the specimen without being scattered. The scatter probability, indicated by the cross-section (σ), depends on the atomic number (Z) and the specimen thickness (Carlemalm et al., 1985). When unscattered electrons pass through the specimen they produce no measurable energy and their trajectory does not alter. They do not give any information about the specimen but they have a significant influence on the brightness of the image. The electrons scattered at the specimen atom are distinguished as follows.

1. Elastically scattered electrons

If an accelerated electron passes through the shell of a specimen atom without meeting a shell electron, it is attracted by the positively charged atomic nucleus and deflected (Fig. 1). Because of the great differences in mass between atomic nucleus and electron (atomic nucleus, 1.7×10^{-24} g; electron, 9×10^{-28} g) the incident electron is scattered elastically. Its trajectory is changed, but not its energy. Elastically scattered electrons remain monoenergetic in so far as the accelerated electrons are monoenergetic. When electrons scatter elastically, they do so at relatively large angles between about 0 and 100 mrad. Electrons with large scatter angles are filtered out of the image by the objective aperture diaphragm. Depending on the aperture diameter, this diaphragm only allows the passage of electrons with scatter angles of approximately 5–20 mrad.

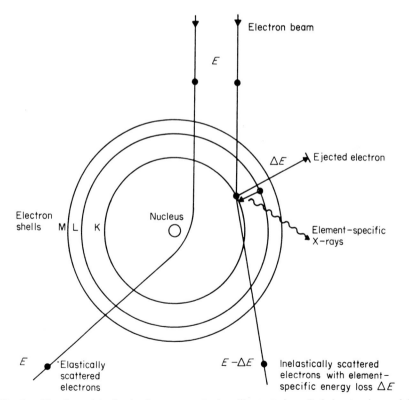

Fig. 1. Elastic and inelastic electron scattering illustrated on Bohr's atomic model.

This discrimination produces the contrast in the image. The elastic scattering increases with rising atomic numbers. Thus it is the heavy elements like uranium (92), lead (82), osmium (76) and tungsten (74) which are preferred for staining specimens for the electron microscope.

2. Inelastically scattered electrons

When an accelerated electron penetrates a specimen atom and strikes a shell electron, various interactions occur. One of them is inelastic scatter. In the inelastic scatter process the incident electron is decelerated, thereby transferring energy to the specimen electron and losing energy itself. This can be measured as an energy loss (ΔE). Energy losses commonly measure from 10–100 eV. So inelastically scattered electrons are polyenergetic, or according to the wave theory can be considered to be polychromatic. The exact energy distribution is described by the inelastic cross-section (σ_{inel}) and represented as an energy loss spectrum. In addition to the energy loss, the inelastic impact also results in a slight change of direction, normally up to 1 mrad. As a result of this limited change of trajectory, almost all the inelastically scattered electrons pass through the objective aperture diaphragm and thus into the image. The effect of these polyenergetic polychromatic electrons is that the image is affected by the chromatic aberration of the objective lens with chromatic aberration constant (C_c). This impairs both the contrast and the resolution.

Inelastic scattering exceptionally results in the accelerated electron transferring so much energy to the shell electron that it is boosted to a higher energy level (excitation) or is ejected from the atomic shell altogether (ionization). The energies necessary to result in excitation or ionization are characteristic for each individual atom and the corresponding shell. If a shell electron is ejected from the nuclear shell (K), the energy loss is characterized by the atom and the shell, e.g. Mg_K. The shells further out from the nucleus consist of several energy levels: the L shell for instance has 3, the M shell 5, the N shell 7 and the O shell 7. Energy losses arising from the excitation or ionization of the outer shells are characterized by the atom, the shell and the respective energy level, for instance Fe_{L1}. Energy levels very close together which cannot be separately resolved clearly are combined in one designation, e.g. $Fe_{L2,3}$.

Thus the energy loss electrons provide the complete information on the specimen atoms hit and at the same time information on the elemental composition of the specimen and the arrangement of the atoms within it.

3. Multiple-scattered electrons

When an accelerated electron passes through a specimen, it is possible that it is scattered twice or even several times. The probability of multiple scatter increases in proportion to the specimen thickness and rising atomic number. All combinations can occur: elastic plus elastic; inelastic plus elastic or multiple inelastic. Multiple-scattered electrons lose energy with each inelastic impact. Thus the spread of energy distribution and the chromatic aberration of the objective lens both increase with increasing specimen thickness. Multiple-scattered electrons also show greater changes in direction, so reducing the proportion of electrons getting through the objective aperture diaphragm and reducing the intensity of the image.

The information content of element-characteristic single scatter electrons is greatly reduced when multiple scatter increases. Multiple-scattered electrons provide no information value for energy loss analysis. For one thing, they enhance the background brightness, and secondly, the more multiple scatter there is the less reliable the analysis for the element-characteristic single scatter. This leads to a considerable reduction of the signal-to-noise ratio and a decrease in detectability.

B.   The electron energy loss spectrum

The energy loss spectrum represents the intensity distribution of the scattered electrons as a function of the energy loss. With thin specimens it provides both qualitative and quantitative information on the atomic and molecular structure of a specimen. The spectrum is divided into the following regions (Fig. 2).

1.   Region around 0 eV energy loss (zero loss peak)

The maximum intensity in the spectrum of a thin specimen always lies at 0 eV, because at this point the unscattered electrons and those scattered elastically at very small angles constitute the major proportion. The median width of the zero loss peak is the criterion for the resolution of the spectrometer and the energy spread of the incident electrons.

2.   Region of approximately 10–30 eV energy loss (plasmon peak)

The plasmon region incorporates about 90% of the inelastic scatter processes. It results from the overlapping of various effects. In crystals inelastic scatter occurs through the interaction of the accelerated electrons and the electrons of the lattice system. When the latter is bombarded by electrons, it is excited into oscillating. The electron system oscillates relative to the lattice of atoms. The quantized excitation states of the oscillating electron systems are called plasmons.

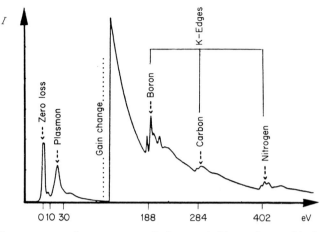

Fig. 2. Electron energy loss spectrum of a boron nitride specimen with the zero loss peak, the plasmon range and the element-specific absorption edges.

The plasmon region is also the region of the collective energy losses in the conduction and valency bands. It is particularly useful for materials research. Energy losses in the valency bands provide information about molecular bonding. Energy losses in the conduction bands permit the study of relative dispersions in solid bodies, e.g. determination of the optical constants of solid matter, such as refractive index and dielectric constant.

The analytical potential of this spectral range is limited by a number of factors. These include the overlap with absorption edges of the outer shells, damage from bombardment to the molecular structures producing the plasmons, and the limited energy-resolving power of the spectrometer. For biological work the range is chiefly useful for imaging in ratio-contrast (Z-contrast; Carlemalm *et al.*, 1985).

3. Region of energy losses > 50 eV (inner shell losses)

This is the most important range for element analysis. It contains all element-characteristic energy losses which have taken place as a result of excitation or ionization of specimen atoms. The shell electrons, however, cannot only take on a fixed value of excitation or ionization energy, but also kinetic energy of a certain bandwidth. For this reason these element-specific energy losses do not appear as sharp peaks on the intensity spectrum. They are indicated rather by a steep rise in intensity which then declines over several hundred eV. This is why the ranges for element-specific energy losses are called absorption edges. They generally show, after the steep rise and within the following 100–200 eV energy loss, fine structure which provides information on the molecular composition of the specimen and the crystallographic structure.

4. The background

The element-characteristic absorption edges are superimposed on a background intensity. This arises as a result of the remaining intensity of lower absorption edges and plasmon losses, and of non-specific energy losses in the field of the specimen electrons and multiple scatter. As the background grows fainter exponentially with rising energy loss, and the signal-to-background ratio is therefore constantly changing, it must be taken into account when making quantitative evaluation.

C. Instrumentation

Various types of spectrometer are used for the spatial separation of electrons of different energy levels. They are divided into magnetic prism spectrometers and imaging spectrometers.

Magnetic prism spectrometers are always an accessory of the electron optical column and are attached beneath the observation screen. They consist of a dual focusing magnetic prism, the energy-selecting slit and an electron detector for measurement of the intensity. Electrons passing through an entrance diaphragm into the spectrometer are diverted into an orbit in the homogenous magnetic field. The radii of the orbits are proportional to the energy of the electrons. Electrons of equal energy are focused in one point in the energy dispersive plane. Consequently the transmitted electrons produce an intensity distribution determined by their energy. If this intensity distribution is shifted by changing the prism's magnetic field strength with the slit, the electron detector measures the electron intensity passing through the slit and applies it as a spectrum corresponding to the energy.

Magnetic prism spectrometers can only be used for recording spectra. To produce images they have to be used in conjunction with scanning equipment (STEM). In this case specimens are scanned either linewise or pointwise and the intensities, which are measured at a fixed energy value, are displayed graphically on a television screen. Owing to the scanning procedure, the local resolution of the image is limited. The recording time for a micrograph is determined by the intensity, the scanning speed and the number of lines or points to be recorded, and is generally very long (Egerton, 1986).

In contrast to magnetic prism spectrometers, imaging energy spectrometers are not an accessory but are permanently integrated into the electron-optical system. They are also described as filters. There are various types of electron energy loss spectrometer, such as the prism–mirror–prism spectrometer, also

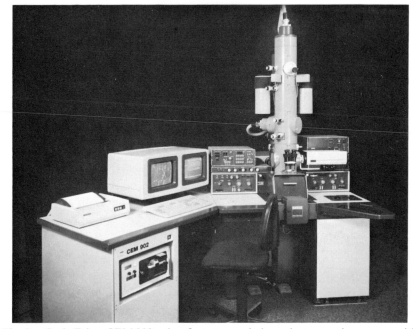

Fig. 3. Carl Zeiss CEM 902, the first transmission electron microscope with integrated imaging electron energy loss spectrometer and integrated computer system for image processing, spectra evaluation and instrument control.

known as the Castaing/Henry filter, and purely electromagnetic spectrometers like the Omega or Alpha filters. As at the time of writing only the Castaing/Henry filter is commercially available, for information on the other spectrometers the reader is referred to the relevant literature (Egerton, 1986).

The prism–mirror–prism spectrometer was first described by R. Castaing and L. Henry in 1962 (Castaing and Henry, 1962). F. P. Ottensmeyer improved it and its function in the electron-optical system and introduced it as a standard component for biological research (Ottensmeyer and Andrew, 1980). The first and currently the only commercially available instrument to incorporate the imaging electron energy loss spectrometer is the Carl Zeiss EM 902 (Fig. 3).

1. Electron-optical configuration and operation

The Castaing/Henry/Ottensmeyer spectrometer is inserted between two projector lens systems (Fig. 4a). The first projector lens system varies the magnification and also images the rear focal plane of the objective (the diffraction plane) in the entrance cross-over of the spectrometer. The magnetic field of the prism is excited to the exact degree that monoenergetic electrons, accelerated to, say, 80 keV, are deflected in the first section of the double prism by exactly 90° to the electron-optical axis. An electrostatic mirror at a more negative potential than the electron gun reflects the electron beam. After changing direction the electron beam again enters the homogenous magnetic field of the spectrometer. There it is bent again by exactly 90° to the electron-optical axis and focused in the energy dispersive plane.

Electrons which have lost energy in the object are deflected by the magnetic field in proportion to their energy loss and do not return to the electron-optical axis. They are spatially separated from each other according to their energy losses in the energy dispersive plane and focused into an intensity spectrum or band (Fig. 4b). This intensity spectrum contains not only the analytical information, but also the total information content of the image, because the rear focal plane of the objective where the diffraction pattern is produced is imaged in the plane of the entrance cross-over.

In the intensity spectrum of a thin specimen the maximum intensity is produced at an energy loss of 0 eV. The image of the spectrum clearly shows the diffraction image and the boundary of the objective aperture diaphragm. Above the 0 eV energy loss the objective aperture diaphragm blurs into a band on account of the progressive intensity change and only the outer edge is distinguishable. Each point on the spectrum represents the diffraction diagram of the corresponding electrons. If the slit is now used to select a spectral range with a specific energy window, the diffraction patterns of the electrons are automatically matched against the selected energy losses. All other diffraction patterns are blocked out and are not imaged.

But the imaging spectrometer does not only image the spectrum from the energy dispersive plane. The second projector lens system enables the user to switch over to the real image of the object in the final image plane of the spectrometer, analogous to switching between diffraction and image in the CTEM. In this case there is then produced in the final image plane the real image of the diffraction patterns selected with the slit. The slit is centred on the electron-optical axis and, like an optical filter, determines the energy width of the imaging electrons. The energy window with this spectrometer is typically < 20 eV.

To select the electrons desired for imaging, the spectrum is shifted relative to the slit. Unlike the magnetic prism spectrometer, where the shift is made by altering the magnetic field, here it is made by increasing the accelerating voltage. The spectrometer is calibrated and the magnetic field set so that, say, 80 keV electrons remain exactly on the electron-optical axis after passing the spectrometer and pass through the slit. If it is desired to image with specific energy loss electrons, the voltage is boosted to exactly the desired energy loss level. But only those 80 kV electrons set by calibration pass through the slit;

122   R. BAUER

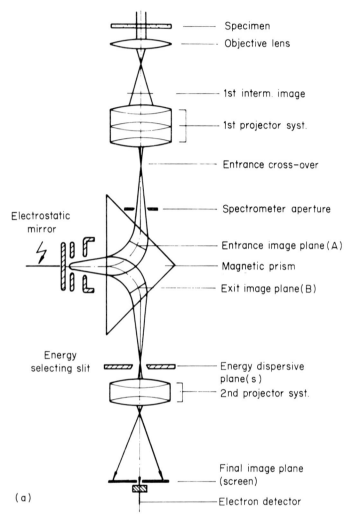

Specimen
Objective lens

1st interm. image

1st projector syst.

Entrance cross-over

Spectrometer aperture

Electrostatic
mirror

Entrance image plane (A)

Magnetic prism

Exit image plane (B)

Energy
selecting slit

Energy dispersive
plane(s)

2nd projector syst.

Final image plane
(screen)

(a)

Electron detector

Fig. 4. (a) Beam path and optical configuration of the imaging electron energy loss spectrometer (Zeiss EM 902).

that is, only those electrons which have lost that energy to which the voltage was increased. In this way the spectrometer and the whole subsequent electron-optical system always work under constant conditions.

2. Imaging methods

Owing to its special prism–mirror–prism electron-optical system, the imaging spectrometer not only produces a point-by-point image of the electron source

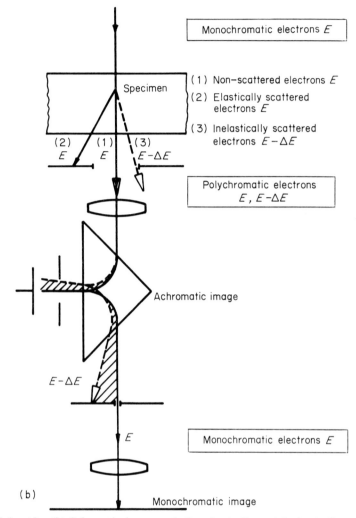

Monochromatic electrons E

Specimen

(1) Non-scattered electrons E

(2) Elastically scattered electrons E

(3) Inelastically scattered electrons $E - \Delta E$

(2) E (1) E (3) $E - \Delta E$

Polychromatic electrons E, $E - \Delta E$

Achromatic image

$E - \Delta E$

E

Monochromatic electrons E

(b)

Monochromatic image

Fig. 4 (cont.). (b) Schematic beam path for elastically and inelastically scattered electrons.

in the energy dispersive plane, but it also produces superb quality images of extended object areas. There are a number of different imaging options.

(a) Global imaging

If the slit is removed from the beam path or the slit width is set very wide, the images which were dispersed according to energy losses by the spectrometer are reassembled by the second projector lens system and projected as a "global" image on the final image screen. When the spectrum is imaged on the final image screen, it represents one section of the total spectrum by parallel detection. Global imaging can be compared to bright-field imaging in CTEM. No further improvement in contrast is possible.

(b) Elastic imaging

For elastic imaging only electrons scattered elastically at low angles and unscattered electrons pass through the slit, i.e. electrons with 0 eV energy loss. Inelastic electrons are eliminated by the slit and thereby from the imaging process. Electrons scattered elastically at large angles are already eliminated by the objective aperture diaphragm. This mode of imaging is applicable for bright-field, dark-field and diffraction imaging. In all cases the elimination of the inelastically scattered electrons results in distinctly improved contrast.

(c) Electron spectroscopic imaging (ESI)

For electron spectroscopic imaging inelastically scattered electrons are selected from the spectrum for imaging, producing both object- and element-specific contrast.

(d) Electron energy loss spectra (EELS)

To record the energy loss spectrum the spectrum from the energy dispersive plane is imaged in the final image plane. An electron detector mounted below the final image screen measures the electron intensity and records it as a factor of the energy loss in a memory or in an XY recorder. The spectrum is shifted relative to the detector opening by increasing the high voltage under digital control. This is a serial defection method.

III.   Applications

A.   Imaging of conventional specimens

For CTEM specimens are usually prepared very thin (less than 100 nm) and stained with heavy metal salts such as uranyl acetate, neutralized phosphotungstic acid or lead citrate. Specimens prepared like this are ideal for CTEM. The images produced have good resolution and contrast. This applies equally to sections and negatively stained suspensions on thin carbon foils.

   If an electron microscope with an electron spectrometer is used to image conventionally prepared specimens, the contrast is always better, but not significantly so. This is because the heavy metal staining results in a higher proportion of elastically scattered electrons. These are eliminated by the objective aperture diaphragm, and this produces a high contrast. Because the specimen is so thin, the proportion of inelastically scattered electrons is insignificant and does not severely impair the contrast.

The outstanding advantages of the imaging electron spectrometer are better exploited at the point where CTEM can no longer produce satisfactory results—that is when imaging thick and ultra-thin unstained specimens.

B. Imaging of thick specimens

In CTEM, where acceleration voltages are about 100 kV, the examination of thick biological specimens is limited less by their transmissivity than by the contrast obtainable and the resolution of fine structures. Once specimens are more than 100–200 nm thick, sharpness and contrast are drastically reduced. This is because the greater the specimen thickness, the higher the frequency of multiple inelastic scattering (Pearce-Percy and Cowley, 1976; Carlemalm et al., 1985). These electrons extend the energy bandwidth δE of the imaging electrons, which has a significant influence on the contrast and the resolution.

If the electrons which have passed through a specimen are recorded with an electron energy loss spectrometer, one obtains the intensity distribution of the imaging electrons as a factor of energy loss (Fig. 5). The spectrum contains only the electrons which pass through the objective aperture. With a specimen thickness of 70 nm, the normal thickness for CTEM examinations, the energy loss spectrum shows a peak intensity of 0 eV. This is produced by the unscattered electrons and those scattered elastically at low angles, in other

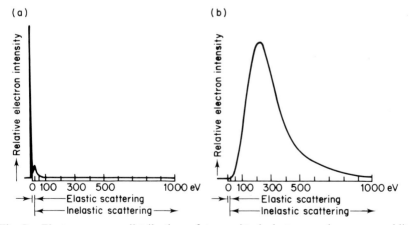

Fig. 5. Electron energy distribution of transmitted electrons using pure araldite specimens with different thicknesses. Thickness: (a) 70 nm; (b) 1 μm.

words monoenergetic electrons with an energy width δE of about 1 eV. With a 1 μm thick section, on the other hand, more than 99% of the imaging electrons are inelastically scattered and are thus polyenergetic. The energy width δE in this case extends to about 300 eV (Peachey, 1986).

The contrast when imaging amorphous specimens is produced partly by the scatter contrast and partly by the phase contrast. The scatter contrast occurs as a result of the filtering out of those electrons scattered at larger angles than the objective aperture (selective angle). The electrons so eliminated are primarily those elastically scattered and multiple-scattered at larger angles, whereas inelastically scattered polyenergetic electrons pass the objective aperture because of their small scattering angles.

A TEM with imaging energy loss spectrometer permits both angle and energy selectivity. Via the spectrometer slit one can select, say, only the zero-loss electrons for imaging. This dual criterion selection process applied to the same areas of the structure results in markedly improved contrast.

Phase contrast in the bright-field mode is generated by the interference of the elastically scattered electron wave with the incident wave of an electron which result in Fresnel fringes. Phase contrast will only occur with coherent mono-energetic electrons of similar wavelength (Scherzer, 1949; Reimer, 1984). Thick specimens, however, show an extended spectrum of inelastically scattered electrons of differing wavelengths. If the whole spectrum of transmitted electrons is applied for imaging, as is the case in CTEM, only a small proportion of them react phasically with each other, so the image does not show any phase contrast.

If all the inelastically scattered electrons are eliminated with a TEM with imaging spectrometer, leaving only the zero-loss electrons for imaging, the intensity is reduced in proportion to the number eliminated, but the phase contrast leads to a marked contrast enhancement in the image. Another factor which influences the contrast and resolution of fine structures is the chromatic aberration of the objective lens, which affects the energy width. When a specimen is focused with the objective, the focusing is only effective for electrons of a specific energy, say the primary energy. Electrons which have lost energy in the specimen are focused in other planes depending on the energy loss sustained. Electrons which have lost energy in relation to the focused primary electrons create blur in the image. A measure of the blur is the radius of the chromatic aberration disc (Δc):

$$\Delta c = C_c \times \alpha \times \delta E/E$$

where C_c is the chromatic aberration constant of the objective, α the objective aperture, δE the energy bandwidth and E the primary energy. Since C_c is normally optimized for an objective and α is determined by the application, Δc can only be influenced by the ratio $\delta E/E$. In CTEM δE can only be kept down by using thin specimens. To achieve a significant increase in the primary energy E, very high voltage electron microscopes must be used.

As well as the chromatic aberration of the objective lens, the change of magnification and rotation of the projector lenses is a contributory factor to blur (Bihr and Egle, 1979; Reimer, 1984):

$$\delta r = C_r \times r \times \delta E/E$$

where δr is the resulting magnification and rotation aberration of all the projector lenses, C_r is the resulting chromatic aberration of the projector lenses, and r is the distance from the optical axis.

The change of magnification and rotation of the projector lenses in modern CTEMs is largely compensated by a computerized correction process. Only with very thick sections and low magnification does it affect sharpness at all.

1. ESI with thick specimens

If ESI is used to examine thick sections, the polyenergetic inelastically scattered electrons are excluded by the spectrometer slit. The imaging takes place only with the electrons which have passed through the spectrometer slit. Typical energy widths are $\delta E < 20$ eV, giving chromatic aberration discs Δc of at most 2.5 nm diameter. Thus with ESI one can obtain micrographs of the same energy width, that is to say with the same image quality, regardless of the object thickness. If one compares the two images of the same object area of a 0.75 µm thick section, Fig. 6a taken with a CTEM at 200 kV and Fig. 6b taken with ESI at only 80 kV, the differences in resolution and contrast are unmistakable. To obtain images of comparable quality with the CTEM by increasing the voltage, acceleration voltages of around 1 MV would be necessary.

2. Depth of focus

When examining thick sections, especially for three-dimensional reconstruction and stereo observation, the depth of focus T is very important (Reimer, 1984):

$$T = \delta s/\alpha$$

where α is the objective aperture, δs is the point which, outside the focusing plane at the aperture angle in question, can only be a disc with a certain size (Fig. 7). When δs is projected with magnification M in the final image plan, $\delta s \cdot M$ cannot be any larger than the resolution of the film grain, otherwise the image is blurred.

If one also considers an existing blurring of the chromatic aberration of the objective with its chromatic aberration disc Δc, then this will reduce the useful depth of focus. With an energy width of about 40 eV at 3000 × magnification, the chromatic aberration disc Δc in the final image plane will already be as

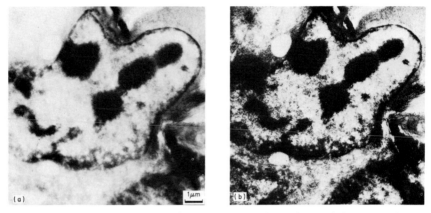

Fig. 6. Comparison between (a) a CTEM micrograph (200 kV) and (b) an ESI (80 kV, $\delta E = 20$ eV) of an ascites tumour cell invading the elastic reticulum of mouse peritoneum. Specimen thickness, 0.75 μm. Specimen courtesy of M. Marko, Albany.

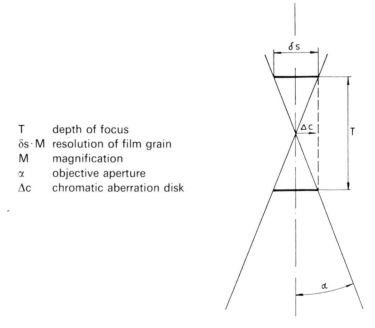

T   depth of focus
$\delta s \cdot M$ resolution of film grain
M   magnification
α objective aperture
Δc chromatic aberration disk

$T < \delta s / \alpha$

Fig. 7. Schematic of the depth of focus.

large as the film grain, about 30 μm (α = 6 mrad). Object details lying outside
the focusing plane cause additional lack of sharpness in the image.

With a section of 1 μm thickness (Fig. 5b) over 90% of all transmitted
electrons have an energy loss of more than 40 eV. In CTEM these inelastically
scattered electrons impair the contrast and the resolution of images and
produce background noise.

For ESI only electrons with an energy width of less than 20 eV are admitted
for imaging. The chromatic aberration of the objective and the changes of
magnification and rotation of the projector lenses are minimized. As a result,
the depth of focus can be fully exploited.

Stereo images of thick biological sections taken by the ESI method can
make structures look as though they were embedded in glass.

3. Contrast-tuning

Objects of varying density show varying intensity distributions (Fig. 5). This
also applies to local mass-thickness variations. Figure 8 shows energy loss
spectra for the areas labelled 1, 2 and 3. Area 1 is thinner as a result of cutting
artefacts and therefore has a lower mass-thickness. The energy loss spectrum
shows a high intensity at 0 eV energy loss and at lower energy losses too. Area
2 represents the matrix for average mass-thickness. Area 3 is of average
thickness, but owing to heavy metal staining has a high density. The total
intensity of 3's spectrum is distinctly lower, because with high density objects
on the whole fewer electrons are transmitted, and as a result of multiple
scattering more electrons are scattered at large angles and are thus eliminated
by the objective aperture diaphragm. Furthermore, the intensity maximum
shifts towards the energy loss electrons. The high brightness range of this
image is caused primarily by artefacts (cutting artefacts in 1, heavy metal
staining in 3). As a result of the high brightness range, interesting object
structures are lost, because the standard film materials can only take a limited
brightness range. Excessive intensities lead to overexposure; if intensities are
too low, the film blackens either too faintly or not at all.

The energy loss spectra in Fig. 8 were taken under the same illumination
conditions and with the same enhancement techniques. So they also convey
information about the differences in brightness of areas 1, 2 and 3 as a factor of
energy loss. If the spectrometer is used to select electrons with specific energy
losses for the purposes of imaging, the intensities of areas 1, 2 and 3 converge
with increasing energy loss. As a result, the brightness range of the image
decreases and more specimen detail is revealed on the film. When all the
spectra have the same intensity, artefacts are no longer visible and the matrix
structure is clearly evident (Fig. 9). Here, too, the chromatic aberration, and
with it the image quality, are determined by the energy width δE. As with
elastic imaging, it is less than 20 eV.

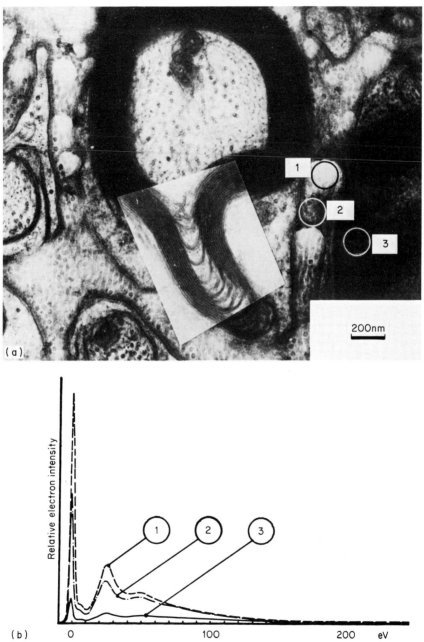

Fig. 8. (a) Nerve tissue of rat with three areas of varying mass-thickness (1, 2, 3). Specimen thickness, 0.5 μm. Specimen courtesy of R. Marx, Marburg. (b) Electron energy loss spectra of the three areas (1, 2, 3) taken under the same illumination and enhancement conditions.

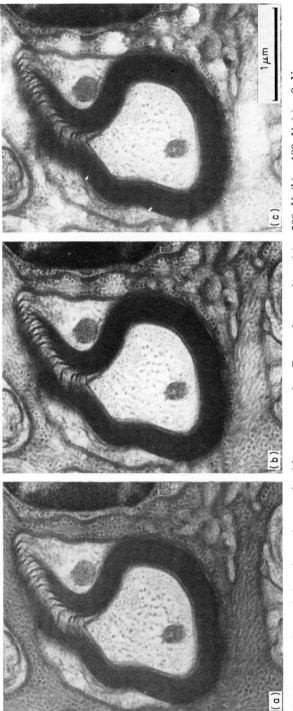

Fig. 9. Intensity variations in a micrograph with contrast-tuning. Energy loss settings: (a) $= 200\,eV$; (b) $= 100\,eV$; (c) $= 0\,eV$.

Electron spectroscopic imaging of biological specimens produces superb results in respect of both contrast and resolution, irrespective of specimen thickness up to about 2 μm. It permits optimum exploitation of the depth of focus, which has particular significance for stereo observation and 3-D reconstructions. Where there are big differences in mass-thickness, the information content of images can be enhanced by contrast-tuning. This applies, for instance, to the imaging of freeze-dried cell cultures, immunolabelled thick sections, or with negative-stained thick specimens on substrates. Furthermore, cutting artefacts such as knife marks can be eliminated from the image.

Thus electron spectroscopic imaging offers a wide range of new possibilities with excellent results.

C. Imaging of frozen-hydrated specimens

One of the main problems when examining frozen-hydrated specimens is the low contrast of biological structures. It makes localization in the section much more difficult. Here, too, ESI can suitably be applied to improve image contrast.

In CTEM contrast in images of amorphous specimens is produced as a result of the scatter contrast and the phase contrast. The scatter contrast with amorphous specimens is determined by the filtering out of the elastically scattered electrons by the objective aperture diaphragm (selective angle). The elastic scattering is strongly dependent on the atomic number (Fig. 10); with elements of low atomic number it is slight and it increases with rising atomic number. Biological specimens consist largely of elements with atomic numbers ≦ 20, so little elastic scatter takes place and heavy metals such as U (92) and Pb (82) are commonly used in their preparation to increase the contrast. For cryopreparations, however, heavy metal staining is undesirable, because if possible specimens should be left in the natural state. This means that with CTEM only low contrasts are obtainable with frozen-hydrated specimens.

If one applies not only angle selectivity but energy selectivity to the scattered electrons, the inelastic electrons can also be eliminated and higher contrast obtained. According to Lenz (1954) and Egerton (1976) the relationship between the differential inelastic scatter cross-section (σ_{in}) and the differential elastic scatter cross-section (σ_{el}) is given by

$$\frac{\sigma_{in}}{\sigma_{el}} = \frac{20}{Z}$$

Thus, with structures consisting of elements of atomic number less than 20

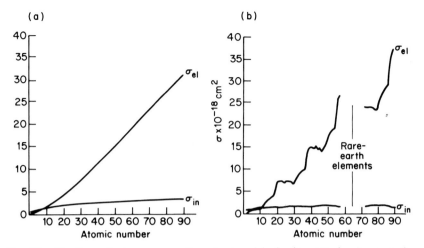

Fig. 10. Elastic scattering cross-sections (σ_{el}) and inelastic scattering cross-sections (σ_{in}) as a function of Z for 100 kV calculated according to (a) Langmore *et al.* (1973) and (b) Eusemann *et al.* (1982).

higher contrasts are obtainable by eliminating the inelastically scattered electrons than by eliminating the elastically scattered ones. Thus the double selection process produces more than twice the contrast. The energy loss spectrum of a frozen-hydrated section about 150 nm thick emphasizes this effect by the high proportion of polyenergetic inelastically scattered electrons (Fig. 11). Thus similar imaging conditions pertain for relatively thin frozen-hydrated specimens as for much thicker, conventionally embedded unstained specimens. The high proportion of inelastically scattered electrons with a large energy bandwidth δE also inhibits phase contrast. In addition, fine structures cannot be seen sharply as a result of chromatic aberration of the objective lens.

If the high proportion of inelastically scattered electrons is eliminated with the imaging electron energy loss spectrometer, a significant increase in contrast is obtained, as shown in Fig. 12. The greater brightness range in the filtered image is caused by the increased scatter contrast, while the phase contrast makes the fine structures visible. Furthermore, the fine structures are imaged clearly because in the filtered image the chromatic aberration of the objective lens has no effect.

It is true that by filtering out the inelastically scattered electrons the specimen suffers greater exposure to radiation. However, by using various technical aids such as the TV image intensifier camera and the low-dose unit together with high sensitivity film material, frozen-hydrated specimens can produce superbly sharp images with good contrast without showing visible signs of radiation damage.

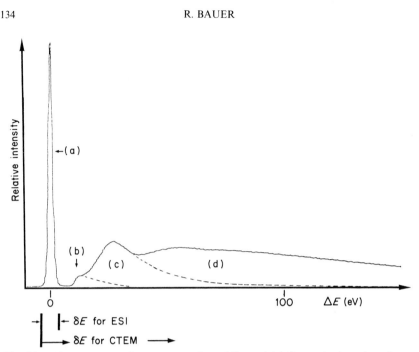

Fig. 11. Electron energy loss spectrum for a 150 nm thick frozen-hydrated section on carbon support film, as shown in Fig. 11. (a) Zero loss electrons. (b) Hydrogen core loss and oxygen low loss electrons. (c) Carbon low loss electrons. (d) Multiple inelastically scattered electrons.

D. Dark-field imaging

The dark-field technique permits the imaging of ultra-fine biological structures such as DNA with high contrast. For dark-field imaging in CTEM all electrons scattered at large angles are admitted. To achieve this a double prism system above the specimen deflects the electron beam so that the primary electrons are eliminated by the objective aperture. Another technique uses a condensor ring diaphragm and an associated objective aperture diaphragm to separate the primary electrons.

In both cases only electrons scattered at high angles are admitted for imaging. The distribution of these electrons includes not only monoenergetic elastically scattered ones, but, even with extremely thin specimens, a considerable proportion of inelastically scattered ones with varying energy losses. We can consider these as primarily bi-scattered electrons, i.e. once inelastically and once elastically. Because of the relatively high proportion of inelastically scattered electrons and their energy width, dark-field imaging is subject to chromatic conditions comparable to those existing in the imaging of thick specimens in bright-field. The elimination of the inelastic electrons by

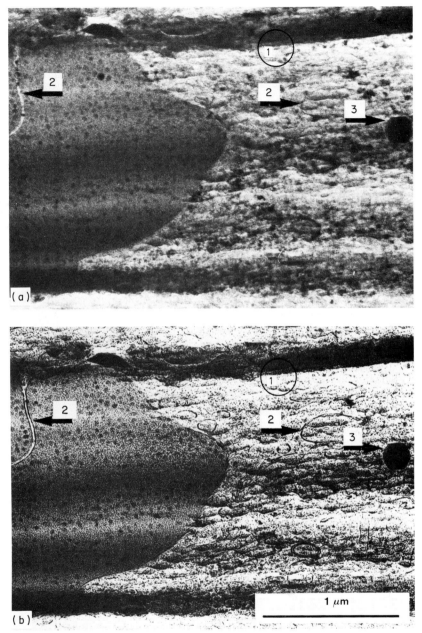

Fig. 12. Contrast of a frozen-hydrated specimen of mouse kidney. Specimen thickness approximately 150 nm. Specimen courtesy of L. Edelmann, Hamburg. (a) Global image $\hat{=}$ CTEM. (b) Zero loss image with ESI. (1) Shows the difference in scattering contrast. (2) Shows the effect of phase contrast. (3) Shows blurring due to chromatic aberration.

means of the spectrometer slit brings a marked increase in contrast and improvement in resolution (Frösch *et al.*, 1987).

ESI is an excellent alternative for dark-field imaging. ESI permits the application of various different staining techniques (Revet *et al.*, 1986).

E. Imaging of very thin specimens

Thin specimens are produced and examined in order to get pictures of the finest structures with the best possible resolution, or to avoid overlapping with other similar or different structures. In such cases the staining sometimes has to be sacrificed because it can overlay the finest specimen structures with a structure of its own and result in artefacts which are difficult to interpret (Müller, 1985).

In the CTEM images of thin unstained specimens like this do not give satisfactory contrast even with low acceleration voltages and minimum objective aperture diaphragms. The reason for this is that with thin specimens more than 90% of all the transmitted electrons pass through the object unscattered (Carlemalm *et al.*, 1985). The proportion of elastically scattered electrons eliminated by the objective aperture diaphragm is very small.

If the imaging electron spectrometer is used for thin specimens, the spectrometer slit can eliminate the inelastically scattered electrons as well. This improves the contrast considerably, but if specimens are unstained the result is still unsatisfactory. However, if selective inelastically scattered electrons are used for imaging the improvement in contrast is dramatic.

While using elastically scattered and unscattered electrons results in a bright-field image, the inelastically scattered electrons produce a high contrast dark-field image. The more strongly scattering specimen details show up brightly against the more weakly scattering background. In contrast to conventional dark-field imaging, it is not the multiple-scattered electrons or those scattered at wide angles which are used, but those scattered inelastically at low angles which pass through the objective aperture diaphragm.

The energy loss spectrum of the inelastically scattered electrons from which those wanted for imaging are selected is influenced chiefly by the specimen thickness and the chemical composition. Each element displays a specific spectral pattern described by the scatter cross-section (σ). If an object structure includes a number of elements, the cross-sections combine proportionally. The varying density of the specimen also affects the spectrum background, so the scatter behaviour of the structures is clearly distinguishable in contrast to their surroundings.

The dominating scatter behaviour in biological specimens results mainly from the scatter due to carbon, because embedding resins and substrate films normally contain more than 60% carbon. Added to this is the carbon content of biological structures. The scatter due to carbon produces a background brightness in the inelastic image, which decreases along with the cross-section as the energy loss is adjusted upwards. The background brightness is reduced to a minimum just before the carbon absorption edge (at 284 eV). Against this low background structures containing elements other than carbon show up in outstanding contrast, because at this point the signal-to-background ratio for all other elements is relatively high (Fig. 13). Thus the phosphorus content, for instance in membranes, ribosomes or cell nuclei, is sufficient to show biological structures with good contrast. Above the carbon absorption edge the carbon characteristic signal is added to the background. Because of the high carbon content of the matrix specimen structures lose contrast severely. For this reason the most suitable spectral range for the non-element specific imaging of thin specimens is between 150 and 260 eV.

For imaging with inelastically scattered electrons it is also possible to select areas from the energy loss spectrum in which the absorption edges of specific elements lie. In these images the selected element appears with greatly enhanced contrast against the background (Fig. 14).

To localize a specific element with absolute precision at least one more photograph should be taken below the absorption edge in order to define the background. Then by subtracting the background images from the picture taken above the absorption edge, one obtains a high resolution element distribution image.

F. Imaging of immunolabelled specimens

The examination of immunolabelled tissue and isolated cells is one of the standard methods in electron microscopy, but the preparation and more especially the localization of markers and their attribution to biological structures can present great problems. Here, too, ESI improves results and offers a number of new, interesting possibilities.

Colloidal gold of different particle sizes, ferritin or horseradish peroxidase are generally used for the immunolabelling of both thin and thick sections.

Filtering out the energy loss electrons clearly enhances the contrast of thick sections. ESI also offers the opportunity of parallel examination in an optical microscope of one and the same section with fluorescence-labelled antibodies.

Contrast-tuning is of particular advantage for isolated cells (Fig. 15).

Colloidal gold is often used for cryofixed, cryosubstituted and freeze-embedded specimens. Such specimens must be stained to detect and localize their biological structure. Labellings can be washed out during staining. ESI does not improve the contrast for the imaging of gold markers, which are easily detected in a CTEM because of their great mass, but enhances the contrast of biological structures for better localization. Thin specimens need

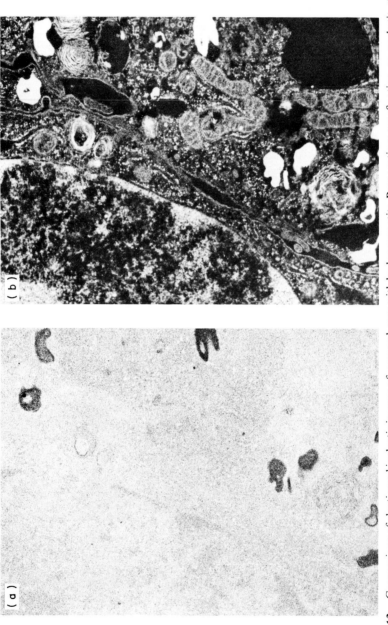

Fig. 13. Comparison of elastic and inelastic images of macrophages with hydroxyapatite. Preparation: cryofixation, cryosubstitution, embedding in araldite/Epon, bright-field, no heavy metal staining. (a) Elastic image. (b) Electron spectroscopic image with energy loss electrons ($\Delta E = 240$ eV).

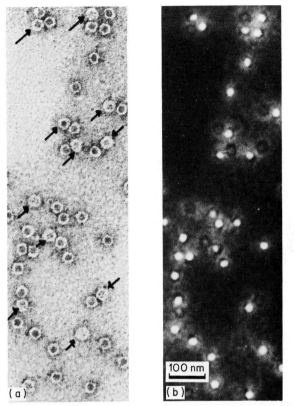

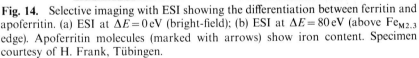

Fig. 14. Selective imaging with ESI showing the differentiation between ferritin and apoferritin. (a) ESI at $\Delta E = 0\,\text{eV}$ (bright-field); (b) ESI at $\Delta E = 80\,\text{eV}$ (above $Fe_{M2,3}$ edge). Apoferritin molecules (marked with arrows) show iron content. Specimen courtesy of H. Frank, Tübingen.

no staining at all if inelastically scattered electrons are selected for imaging (Fig. 13).

The contrast of ferritin is generally low, and it is now rarely used. ESI produces high-contrast images not only of ferritin, but also of the tissue. The user defines by the energy loss setting whether the tissue and the labelling or the labelling alone is to be imaged with high contrast (Fig. 16). This selection may be important for the stereological analysis of labelled substances. The selectivity of element-specific imaging also offers new opportunities for double and multiple labellings. New labellings can be developed besides the known markers, which are then detected by high-resolution elemental distribution, provided the concentration lies within the method's range of detection. How wide the full range of new possibilities will be in the future is at present unimaginable.

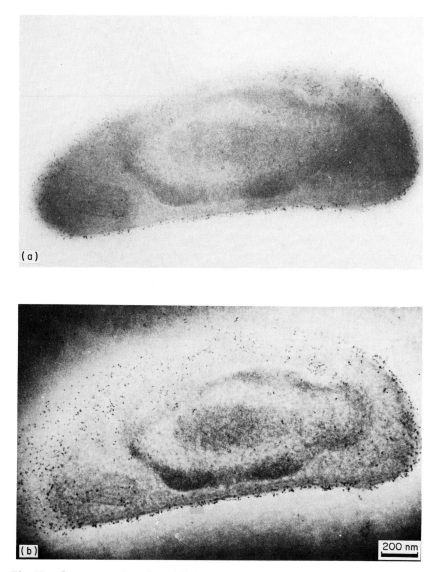

Fig. 15. Contrast-tuning of a gold labelled semi-thick sectioned parasite (*Leishmania mexicana*). (a) ESI at $\Delta E = 0$ eV (comparable to CTEM); (b) ESI at $\Delta E = 100$ eV (contrast-tuning). Specimen courtesy of Y. Stierhof, Tübingen.

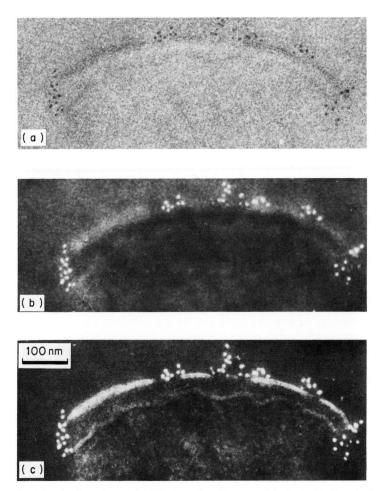

Fig. 16. Selective imaging with ESI of an *E. coli* mutant labelled with ferritin for the OM antigen and uranium stained. (a) ESI at $\Delta E = 0$ eV (bright-field); (b) ESI at $\Delta E = 80$ eV (selective for Fe); (c) ESI at $\Delta E = 114$ eV (selective for U and Fe).

G.   Analysis with electron energy losses

Electron energy loss spectroscopy (EELS) and electron spectroscopic imaging (ESI) provide excellent techniques for analysing thin biological specimens. Theoretically all the elements of the periodic table from hydrogen ($Z = 1$) to uranium ($Z = 92$) can be identified both qualitatively and quantitatively with high detectability, but results are especially good for the light elements such as boron, carbon, nitrogen and oxygen. With alternative methods of analysis, such as energy dispersive X-ray analysis, these light elements are comparatively difficult to detect. The literature on EELS and ESI gives the following detectability limits:

Minimum detectable mass: about 10^{-21} g
Minimum number of detectable atoms:
   Phosphorus: 30–50 (Harauz *et al.*, 1984; Colliex *et al.*, 1984)
   Calcium: 3 (Shuman *et al.*, 1986)
   Uranium: 20 (Krivanek, 1982)
Minimum detectable concentration: about 1 vol %

These spectacular values, however, are only possible with ideally prepared test specimens. But with standard specimens, too, preparation is very important. There are three main points to consider:

(1)   Specimen thickness should be about 30–50 nm. Where concentrations are high, up to about 70 nm thickness is possible.

(2)   The specimens should be prepared in such a way as to avoid washing out the elements of interest.

(3)   The results are more difficult to interpret if other elements are introduced into the specimen during preparation.

EELS and ESI can be applied quite easily to detect pathological increases in concentration, as for instance with secretional conditions. Other applications include element detection in labelled and stained specimens, precipitates and mineralizations such as calcification. The elements naturally occurring in biological specimens are often present in extremely small concentrations close to the limit of detection. The detection of ions is particularly difficult because they are easily washed out or dislocated during specimen preparation.

1. Electron energy loss spectra

Energy loss spectra document the constituent elements of the object area analysed (Fig. 2). The position of the absorption edges provides information about the elements present, while the fine structures in the absorption edges indicate the pattern of molecular bonding (Fig. 17).

The area of the specimen to be analysed is blocked out of the image by the energy spectrometer by means of a diaphragm, the minimum diameter being 8 nm. Instead of having the specimen image on the screen, the user can switch over so that it is the spectrum from the energy dispersive plane which appears (Fig. 4a). If the voltage is altered and the spectrum is scanned by the electron detector below the screen, it measures the intensity and records it as a factor of the energy loss. Recording can be done either graphically or entered in a computer system for further treatment and either qualitative or quantitative

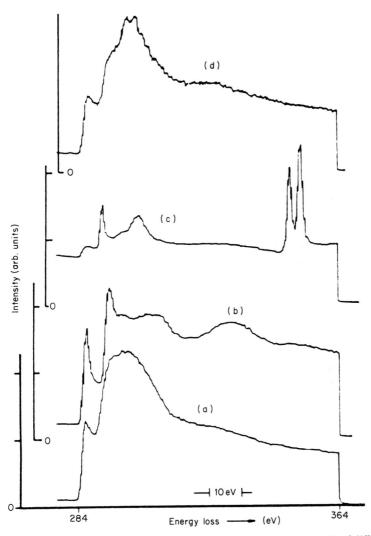

Fig. 17. Fine structures in the carbon absorption edge varying as a result of different specimen preparations. (a) Amorphous carbon foil, 10 nm thick; (b) graphitized carbon black, 30 nm thick; (c) $CaCO_3$ suspension on hole grid, 70 nm thick; (d) KOH suspension on amorphous carbon foil.

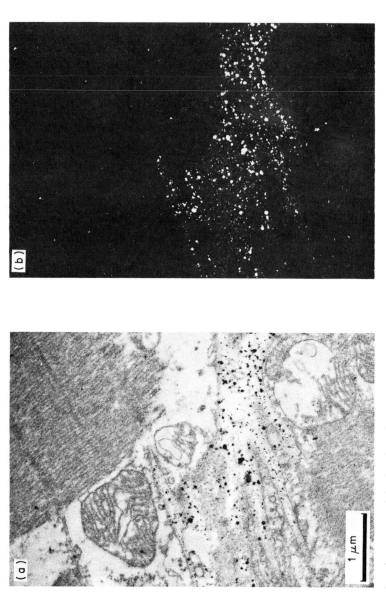

Fig. 18. Lanthanum precipitation in a heart muscle. (a) General view; (b) high resolution lanthanum distribution produced by photographic image subtraction. Specimen courtesy of Ph. Schnabel, Göttingen.

analysis. Results are evaluated by programs which allow for various corrections.

2. High-resolution elemental distribution images

An elemental distribution image shows the local distribution and variations in concentration of a selected element. Spatial resolution is theoretically about 1 nm. It is limited by the local blur of the inelastically scattered electrons.

The elemental distribution image is the result of subtracting images from one another. One image, taken as an ESI above the desired absorption edge, contains the element-specific image information and the unspecific background. One, or better two, other images contain only the unspecific background information. If background images are subtracted from the first, the result is a highly resolved elemental distribution image, which only reveals those areas and structures containing the selected element (Fig. 18). The subtraction process can be done photographically or with an image processing system which analyses and evaluates images and spectra digitally. It can also be used to automate the microscope operation. Entering of the images and the subsequent processing take only a few seconds. Photographic evaluation nevertheless gives the best local resolution in conjunction with large film formats.

Concluding remarks

The integration of an imaging electron energy loss spectrometer into the electron microscope column is a real advance in transmission electron microscopy. The double selection of scattered electrons according to their scattering angle and to their different energy losses improves image contrast and resolution for all imaging methods in transmission electron microscopy. Together with its analytical capabilities like elemental imaging and electron energy loss spectroscopy, it opens a wealth of new and better information of a specimen. It also opens a field of new applications.

Acknowledgments

I gratefully acknowledge the very helpful assistance of Mr W. Egle, Mr E. Zellmann, Dr W. Probst, Dr U. Hezel and Mr J. Bihr in various instances. I also gratefully acknowledge Dr D. Kurz for his support and Miss Judy Norwell (Perth, Great Britain) for translation of the manuscript.

References

Bihr, J. and Egle, W. (1979). *Die Bedeutung des Abbildungssystems eines TEMs für moderne medizinische Anwendung*, Vort. 19.Tag.EM, Tübingen.
Carlemalm, E., Colliex, C. and Kellenberger, E. (1985). In *Advances in Electronics and Electron Physics*, Vol. 63, pp. 269–334. Academic Press, New York.
Castaing, R. and Henry, L. (1962). *C.R. Acad. Sci. Paris B* **255**, 76–78.
Egerton, R. F. (1976). *Phys. stat. sol. (a)* **37**, 663.
Egerton, R. F. (1986). In *Electron Energy-Loss Spectroscopy in the Electron Microscope*. Plenum Press, New York.
Eusemann, R., Rose, H. and Dubochet, J. (1982). *J. Microsc. (Oxford)* **128**, 232–240.
Frösch, P., Westphal, Ch. and Bauer, R. (1987). *J. Microscopy*, **147**, 313–321.
Harauz, G. and Ottensmeyer, F. P. (1984). *Science* **226**, 936–940.
Krivanek, O. L. (1982). *EM Proc. Hamburg* **1**, 167–171.
Langmore, J. P., Wall, J. and Isaacson, M. S. (1973). *Optik* **38**, 335–340.
Lenz, F. (1954), *Z. Naturforschung* **A9**, 185.
Müller, M. (1985). *Zeiss-Inform. MEM Oberkochen* **4**, 26–30.
Ottensmeyer, F. P. and Andrew, J. W. (1980). *J. Ultrastruct. Res.* **72**, 336–345.
Peachey, L. D. (1986). *EMSA Proc.* **44**, 88–91.
Pearce-Percy, H. T. and Cowley, J. M. (1976). *Optik* **44**, 273–288.
Reimer, L. (1984). In *Transmission Electron Microscopy: Physics of Image Formation and Microanalysis*. Springer-Verlag, Berlin.
Revet, B., Delain, E. and Bauer, R. (1986). *EM Proc. Kyoto* **3**, 2409–2410.
Scherzer, O. (1949). *J. Appl. Phys.* **20**, 20–22.
Shuman, H. and Somlyo, A. P. (1986). *Ultramicroscopy* **15**, 110–120.

7

Immunoelectron Microscopy of Surface Antigens (Polysaccharides) of Gram-negative Bacteria using Pre- and Post-embedding Techniques*

GEORG ACKER

Fachgruppe Biologie, Elektronenmikroskopie, Universität Bayreuth, Bayreuth, Federal Republic of Germany

I. Introduction

Gram-negative bacteria are surrounded by a cell envelope which usually consists of a cytoplasmic (or inner) membrane, a peptidoglycan (or murein) layer and an outer membrane. The latter two layers form the cell wall of these

* Dedicated to Professor H. Sitte on his 60th birthday.

147

bacteria. In addition to these layers, the outer membrane of many organisms is covered by capsular layers or by weakly adhering slime polymers, both structural components usually being acidic polysaccharides. In recent reviews the molecular organization and functioning of the outer membrane has been extensively discussed, and models of structure for the outer membrane of bacteria in smooth and rough form have been presented (Lugtenberg and van Alphen, 1983; Vaara and Nikaido, 1984; Freer, 1985). Although the outer membrane is one of the most intensively investigated membranes, very little is known about its detailed structure. A few ultrastructural aspects should be mentioned which seem relevant to this chapter.

The thin sectioning technique, applied to the cell envelope of conventionally fixed and embedded Gram-negative cells, has revealed the well-known layered structure; the outer membrane appears as a trilaminar structure, connected to the cytoplasmic membrane via zones of adhesion. The so-called periplasmic space was considered to be located between these two membranes (Beveridge, 1981; Lugtenberg and Van Alphen, 1983). Recently, however, Kellenberger and associates (Hobot et al., 1984, 1985), applying the progressive lowering of temperature embedding technique with the resin Lowicryl K4M, or using the freeze-substitution procedure, observed no differentiation of the cell envelope into well-defined layers. The cell envelope appeared to be of constant width and was divided into a cytoplasmic and an outer membrane, with a so-called periplasmic gel in between; no "empty" periplasmic space was recognizable, which is in agreement with the observations of Dubochet et al. (1983) on frozen-hydrated sections of fixed and unfixed *Escherichia coli* cells. It is noteworthy that frozen sections of fixed bacteria showed distinct division of the cell envelope into layers (see also Armbruster and Kellenberger, 1986).

Despite the relatively detailed knowledge that we have about the chemistry of important polysaccharide-containing surface antigens such as lipopoly-saccharides (Galanos et al., 1977; Lugtenberg and van Alphen, 1983), enterobacterial common antigen (Mayer and Schmidt, 1979; Kuhn et al., 1984; Rick et al., 1985) and capsular polysaccharides (Ørskov et al., 1977; Troy, 1979; Jann and Jann, 1983), information about their topological relationships to each other, and to the outer membrane, remains rather scarce. This fact can be explained by the observation that the capsular polymers, for example, usually form highly hydrated layers at the cell surface; a strong hydrophilicity also exists on the outer membrane surface of non-encapsulated strains. These hydrophilic layers present preparative problems because they usually collapse during conventional dehydration procedures for microscopic examination; thus the detailed morphology of polysaccharide-containing components outside the outer membrane, also termed bacterial glycocalyx (Costerton et al., 1981), remained obscure during the early years of electron microscopy.

However, capsular structures treated with specific antibodies before dehydration can be preserved to some extent from collapse, an approach applied first using light microscopy for diagnostic purposes (Quellungs-reaktion) and later using electron microscopy (for literature and discussion, see Bayer and Thurow, 1977).

Investigation of cell surface antigens has been greatly assisted by the introduction of ferritin-conjugated antibodies (Singer, 1959). Cells are treated with ferritin conjugate before embedding (the pre-embedding method), and thereafter prepared for the thin sectioning technique. Labelled whole cells can also be examined in the electron microscope (the whole-mount technique). Neither of these two techniques is suitable for examination of masked surface epitopes or intracellular antigens, both inaccessible to antibodies and/or label in morphologically intact cells. In our experiments these difficulties were overcome to some extent by the application of two labelling techniques on thin sections (post-embedding or post-sectioning labelling), namely the progressive lowering of temperature technique using Lowicryl K4M resin and Tokuyasu's cryosection methods. Since ferritin conjugates often bind non-specifically to thin sections (Roth, 1982), we used colloidal gold-conjugated antibodies in thin-section labelling experiments.

Many reviews have recently been published on the preparation of electron-dense markers, in particular colloidal gold, and their conjugation with a variety of biological molecules suitable for both direct and indirect as well as for double and multiple labelling (for discussion and secondary literature, see De Mey and Moeremans, 1986; Smit and Todd, 1986; Plattner and Zingsheim, 1987).

In this chapter the emphasis is mainly on the application of pre- and post-embedding immunoelectron microscopy to the study of three different surface antigens, namely:

(1) capsular polysaccharides, which form well-developed capsular layers;

(2) O-specific chains of lipopolysaccharides, which form a microcapsule on the outer membrane surface; and finally

(3) the enterobacterial common antigen, a further cell surface antigen, which is located in the outer membrane and is present in all enterobacteria.

Monoclonal and polyclonal antibodies have been used. Although the discussion is based mainly on experience obtained with these three surface antigens, it is hoped that some observations will prove useful in planning experiments with other antigens.

II. Methodology

A. Bacterial strains and growth conditions

The smooth and rough forms (Ye 75 S and Ye 75 R) of *Yersinia enterocolitica* strain 75 were used. The LPS of the rough mutant Ye 75 R completely lacks O-specific chains (Acker *et al.*, 1980). The following strains and mutants derived from wild-type *E. coli* 56b have been previously described: D 280 (O8: K27: H −); F 492 (O8: K27$^-$: H $^-$); F 470 (R: K27$^-$: H $^-$); F 782 (R: K27: H $^-$); F 1283 (R: K27$^-$: H$^-$) (Acker *et al.*, 1982). *E. coli* D 1737 O16: K 1, originally a gift from F. Ørskov (Copenhagen) (Ørskov *et al.*, 1979), and LE 392, a derivative of the laboratory strain *E. coli* K 12, were previously used for characterizing the specificity of the monoclonal antibody 735 D4 for K 1 polysaccharide; plasmid pKT 274 is a hybrid cosmid carrying the cloned genes for production of K 1 capsule (Echarti *et al.*, 1983). Plasmid pGB 33, a gamma–delta insertion mutant derivative of pKT 274, specifies the synthesis of K 1 polysaccharide but not the transport of this polymer to the cell surface (Frosch *et al.*, 1985).

When not indicated otherwise, the bacteria were grown in Tryptic soy broth (TSB, Difco) and harvested in the early logarithmic growth phase at an optical density of 0.15–0.3 at 540 nm.

B. Polyclonal antisera

1. ECA-specific antiserum

The ECA specificity of the sera obtained by immunization of New Zealand White rabbits with the ECA-positive R-mutant *E. coli* F 470, followed by absorption with the genetically closely related ECA-negative mutant *E. coli* F 1283, was documented by the passive haemagglutination and indirect immunoferritin techniques as described by Acker *et al.* (1981).

2. E. coli *K27 antiserum*

OK antiserum was obtained by rabbit immunization with formalin-treated cells of *E. coli* D 280. The O8: K27 antiserum was absorbed with the K $^-$ mutant F 492 to produce the anti-K27 serum (Acker *et al.*, 1982).

3. Ye 75 S- and Ye 75 R-antisera

These rabbit antisera were kindly provided by Professor Knapp (Erlangen, Federal Republic of Germany).

4. Anti-human C3

This was purchased from Behringwerke (Marburg, Federal Republic of Germany).

C.   Monoclonal antibodies

Mouse anti-ECA MAb (MAb 898; immunoglobulin G 2a class) has been recently described and its ECA specificity documented (Peters *et al.*, 1985; Rick *et al.*, 1985).

Mouse anti-K1 polysaccharide antibody (MAb 735 D4; IgG class) has also been recently described (Frosch *et al.*, 1985).

D.   Antibody conjugates

Ferritin-conjugated goat anti-rabbit IgG (heavy and light chains) were obtained from Cappel Laboratories Inc. (Chochranville, PA, USA); gold-labelled antibodies (goat anti-rabbit and goat anti-mouse IgG) were purchased from Bayer-Diagnostic (Munich, Federal Republic of Germany).

E.   Exposure of cells to antibodies for indirect immunolabelling of surface antigens and for stabilization of bacterial capsules

Cells of the bacterial strains under investigation were sedimented at a low *g*-value to protect the fragile surface structures. The pellets served for the preparation of cell suspensions in either fresh TSB medium or in an adequate buffer (optical density of 0.15–0.3 at 540 nm). Two parts of the suspension were mixed with one part of the corresponding antiserum (heated for 30 min at 56°C) or with polyclonal or monoclonal IgG fractions and incubated at the growth temperature of the respective culture or at 10°C for 30 min. The cells were then carefully washed twice to remove the antibodies added in excess and the unbound serum components; then the samples were embedded in Epon according to standard procedures or in Lowicryl K4M resin (see below), or used for indirect immunolabelling with ferritin- or colloidal gold-labelled secondary anti-species antibodies. The labelled cells were used for the thin-section technique or for whole-mount electron microscopy (for further details, see Acker *et al.*, 1981).

F.   Electron microscopy

1. Whole-mount electron microscopy

Pellets of the pre-treated sample (see Section II.E) were suspended in distilled water and adsorbed onto carbon film according to the standard procedure for negative staining, or onto carbon film that had previously been deposited in a vacuum evaporator on the surface of freshly cleaved mica sheets. The carbon film (c. 3 mm × 3 mm) is completely floated off from a small portion of mica substrate onto the surface of the sample solution, following the procedure of Valentine et al. (1968). This procedure can be recommended for the preparation of fragile surface structures. It also leads to a more even distribution of the sample on the support film than do standard negative staining procedures. Ferritin-labelled cells can usually be examined without further staining.

2. *Lowicryl K4M thin sections of cells prepared according to the progressive lowering of temperature embedding technique*

The progressive lowering of temperature embedding technique using the resin Lowicryl K4M (Chemische Werke Lowi, Waldkraiburg, Federal Republic of Germany) was carried out as described, with some modifications (Hobot et al., 1984; for technical details and useful discussion of this procedure, see Armbruster and Kellenberger, 1986; Ashford et al., 1986; and EMBO practical course manual, Basel, 1984). Cells were usually fixed directly in their growth medium or after immunolabelling (see Section II.E) with up to 1% (v/v) glutaraldehyde (final concentration). Formaldehyde alone (up to 4%) or in different combinations with glutaraldehyde (preferably 2% formaldehyde/ 0.1% glutaraldehyde) were also used, but no essential differences could be observed with respect to the ultrastructural preservation or labelling density. The fixed cells were enrobed in agar (1% w/v) or gelatin (5–10% w/v) and dehydrated in ethanol or, as has been recently described (Bayer et al., 1985, 1986), in dimethyl formamide (DMF) as follows: 25%, 0°C, 30 min; 50%, −20°C, 60 min; 70%, −35°C, 60 min; 96%, −35°C, 60 min; 100%, −35°C, twice for 60 min each time.

In comparison with the infiltration schedule for tissues, many bacteria, especially smooth and encapsulated strains, required prolonged infiltration times to improve sectioning behaviour. In these cases infiltration of the samples with increasing concentrations of Lowicryl K4M in 100% (v/v) ethanol or DMF at −35°C was as follows: resin–solvent (1:3), 2–3 h; resin–solvent (1:1), overnight; resin–solvent (3:1), 4–5 h; pure resin, 4–5 h; pure resin, overnight. The next day the samples were transferred in Beem capsules and polymerized by indirect UV irradiation (360 nm for 24 h at −35°C and 2–4 days at room temperature).

Thin sections were mounted on Formvar-carbon-coated nickel or copper grids and labelled at room temperature, preferably on the same day. The grids were floated, sections down, on a drop of lysine (0.1 M in phosphate-buffered saline (PBS)) for 30 min, to block free aldehyde groups, then transferred, without rinsing, on a drop of gelatin (2% in PBS, 10 min) or, alternatively, on a drop of milk solution (0.4% up to 1% milk powder in PBS, freshly prepared and centrifuged at 10 000g before use). Occasionally, the latter treatment leads to a lower non-specific background than does the gelatin treatment. After brief rinsing on a drop of PBS, the grids were transferred for 2 h on a drop of primary antibody (diluted in 2% BSA–PBS, centrifuged for 1 min at 10 000g and kept in a moist chamber). Aliquots of poly- and monoclonal antibodies of high quality were stored at $-20°$C (or $-70°$C for longer times). The optimal working dilution for each antibody batch was determined by testing series of dilutions of both primary and secondary antibodies. After primary antibody labelling, the sections were rinsed with PBS (5 × on puddles of PBS for a total of 15 min), transferred for up to 1 h to a drop of secondary gold-labelled anti-species antibody (also diluted in 2% BSA–PBS), then rinsed with PBS (see above) and with distilled water (3 × on puddles for a total of 5 min).

For double labelling experiments, thin sections were mounted on uncoated grids and labelled according to the procedure described by Bendayan (1982), except that the protein A–gold complex was replaced by corresponding secondary gold-labelled antibodies, and the labelling procedure was as described above. After labelling the "face A" of the sections for the K1 antigen, the grid was dried and the second antigen (ECA) was labelled on the "face B" of the section. The labelled and dried sections were stabilized by coating them with a Formvar film or with a thin carbon layer in a vacuum evaporator.

Working with an unknown antigen labelling system, we routinely observed first the distribution pattern and the degree of labelling in the electron microscope after this final rinsing step with distilled water. *Important:* for contrasting with different concentrations of uranyl acetate solutions (up to 5 min), and/or lead citrate (15 to 20 s), we used dried samples only, because in many of our experiments we observed a loss and/or redistribution of considerable amounts of gold label when the sections were contrasted immediately after the final water rinsing step. It is advisable to compare the label patterns obtained with both uncontrasted and contrasted sections.

3. Tokuyasu's cryosection method

Aldehyde-fixed cells (see Section II.F.2) were infused with different sucrose concentrations (1.1 or 2.3 M) in order to prevent ice crystal formation and to improve sectioning behaviour (Tokuyasu, 1973). On the modified end of a specimen holder having a slight central depression, and a total diameter of 0.3 mm, a small droplet of a dense pellet was placed and immediately frozen in melting nitrogen. Alternatively, the infused cells were resuspended in 10% melted gelatin in PBS. Solidified slabs (thickness c. 0.5 mm) prepared from the obtained pellets were cut into triangular pieces (side length up to 3 mm), placed in 1.1 M sucrose–PBS for 1 h, or overnight in 2.3 M sucrose–PBS. Pieces were placed with a droplet of sucrose in the slit of the specimen holder. After removing the excess sucrose with a filter paper from the tip but not from the base of the specimen, and after adequate orientation of the sample for the cutting process, the specimens were frozen in melting nitrogen.

Sections were obtained with a Reichert ULTRACUT/FC4D system at temperatures ranging from −90 to −120°C, depending on the sucrose concentration of the samples. Sections were picked up with a droplet of 2.3 M sucrose in a loop of platinum wire before the sucrose had solidified, and transferred to|Formvar-carbon-coated nickel or copper grids. The picked-up grid was floated, sections down, on a gelatin layer (2% gelatin–PBS in a Petri dish). The sections were immunolabelled as described for Lowicryl K4M sections (see Section II.F.2), except that the periods of incubation with primary and secondary antibodies were reduced to 30 min for each reagent.

The labelled, water-rinsed sections were contrasted for 5 min with 2% uranyl acetate–oxalate (pH 7) to stabilize membrane structure (Tokuyasu, 1980), then briefly washed in three droplets of distilled water (1 min total). Finally, the grids were placed on a solution mixture containing (0.3%) uranyl acetate–methyl cellulose (1.5%) with a viscosity of 25 centipoise (Methocel, Fluka, Buchs, Switzerland) for 10 min after touching two drops of the same mixture. The looped and dried grids were examined in the electron microscope. This methyl cellulose stain treatment protects cell structure, especially membranes, from collapse (Tokuyasu, 1978, 1980; Griffiths et al., 1983, 1984).

In some experiments Tokuyasu's positive contrasting procedure was applied; after final water rinse, the grids were transferred on a mixture containing 2% polyethylene glycol 1540 (Carbowax), 0.1% methyl cellulose (1500 centipoise) and 0.01–0.1% uranyl acetate for up to 10 min. Subsequently they were looped, dried and examined.

As an alternative to the two contrasting procedures mentioned, some samples were negatively stained with different heavy-metal stains (Tokuyasu, 1973; Griffiths, 1983). Labelled and simply air-dried cryosections were also examined; after indirect labelling and water washing (see above), water excess was removed and the sections were simply air-dried without further contrasting or embedding in methyl cellulose. The label intensity and the distribution patterns of label observed on such sections, and on uranyl stained and/or methyl cellulose embedded sections (see above), were compared.

Both the Lowicryl K4M and the frozen sections were obtained using glass or diamond knives. They were examined in a Zeiss EM 109 electron microscope at 80 kV.

III.   Results and discussion

A.   Application of the whole-mount procedure for testing the specificity of mono- and polyclonal antibodies

It is known that the availability of specific, high titre antibodies is an important prerequisite of any immunoelectron microscopic investigation. Therefore the specificity of the mono- and polyclonal antibodies, already documented by serological methods, was also examined by the indirect immunoferritin or immunogold techniques. Figure 1 shows a cell of the ECA-positive strain *Escherichia coli* F470 after incubation with the monoclonal mouse anti-ECA antibody MAb898 and subsequently with gold-labelled secondary antibodies. The gold particles show the distribution of ECA on the cell surface. The specificity of labelling for ECA was demonstrated with the ECA-negative mutant *E. coli* F1283 after the same treatment (Acker *et al.*, 1986); most of the cells were not labelled at all with gold. Only a few cells showed a weak, most probably unspecific labelling (< 5 gold particles per cell).

The K1 polysaccharide specificity of the mouse monoclonal antibody MAb735D4 (Frosch *et al.*, 1985) was also examined with the whole-mount technique using cells of the strain LE392 (a derivative of *E. coli* K12) carrying either the plasmid pKT274 or the plasmid pGB33. Cells carrying the plasmid pKT274, which specifies the synthesis and the transport of K1 polysaccharide to the cell surface, revealed a strong gold labelling, whereas cells carrying the plasmid pGB33, a gamma–delta insertion mutant of pKT274 which specifies the synthesis (Fig. 7) but not the transport of K1 polysaccharide to the cell surface, were almost devoid of gold label (not shown). The specificity of this antibody was also confirmed with this procedure by using cells of *E. coli* D1737 K1 (positive control) and cells of the strain LE392 carrying no plasmid (negative control).

The mono- and polyclonal antibodies tested for their high specificity were then used for investigations with the whole-mount procedure, pre-embedding immunoelectron microscopy, as well as for on-section labelling with Lowicryl K4M and frozen sections.

B.   Combined application of the whole-mount procedure and pre-embedding immunoelectron microscopy to the study of cell surface antigens of *Yersinia enterocolitica* and *Escherichia coli*

1. Localization of ECA in Yersinia enterocolitica *strain 75, smooth and rough forms (Ye 75 S and Ye 75 R)*

The localization of ECA in these strains was studied mainly by the whole-mount technique (Acker *et al.*, 1981). For a better understanding of the results

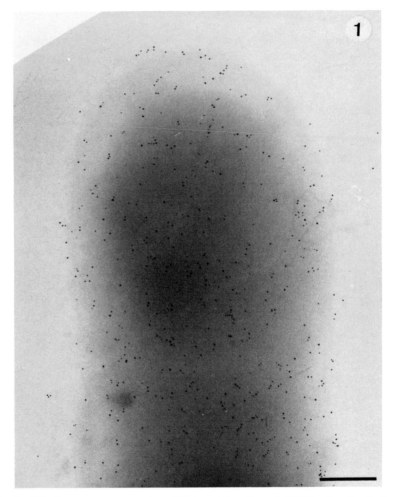

Fig. 1. Whole-mount preparation: ECA-positive *E. coli* F470 cell after indirect immunogold labelling using the MAb 898 (see text). Bar, 0.2 μm. From Acker *et al.* (1986).

obtained, a few aspects of the LPS structure and of the ultrastructure of the cell wall should be mentioned. The LPS of Ye 75 S (Wartenberg *et al.*, 1975; Acker *et al.*, 1980), like LPS of other Gram-negative bacteria, consists of lipid A, a basal polysaccharide core, and O-specific chains. Cells of this strain studied using conventionally prepared Epon thin sections showed a fine structure of the cell wall characteristic of Gram-negative bacteria, namely a murein layer and an outer membrane. Having been incubated with Ye 75 S antiserum before preparation for the thin-section technique, cells of this strain show an additional electron-translucent layer of about 20 nm thickness on the cell surface, bounded to the outside by a serum-induced, more electron-dense deposition (Fig. 2). Most probably, this distinct surface layer is formed by the O-specific chains of the LPS (Acker *et al.*, 1980), and was prevented from

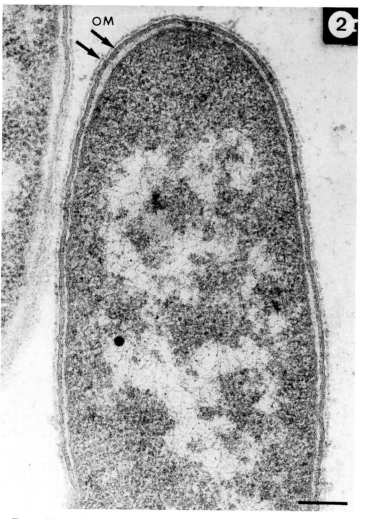

Fig. 2.   Epon thin section of a cell of the smooth strain Ye 75 S incubated with anti-Ye 75 S serum prior to embedding. Between the outer leaflet of the outer membrane (OM) and the serum depositions (arrow) a thin layer (light halo) can be seen, formed by O-specific chains of the LPS. Bar, 0.2 μm.

collapsing during the conventional dehydration process by antibody deposition before embedding. As expected, the rough mutant Ye 75 R, whose LPS lacks O-specific chains, did not show this thin electron-translucent layer; the serum depositions were in close contact with the outer leaflet of the outer membrane.

These two strains, Ye 75 S and Ye 75 R, were used to study the localization of ECA by the whole-mount procedure (Acker et al., 1981). Figure 3 shows a cell of Ye 75 S after incubation with rabbit anti-Ye 75 S serum and subsequently with ferritin-conjugated secondary antibody. This figure illustrates a representative labelling pattern of the examined cells, indicating the distribution of the O-chains of the LPS on the cell surface. In parallel experiments, the anti-Ye 75 S serum was replaced by ECA-specific serum. Almost no ferritin particles could be observed on Ye 75 S cells grown at 22°C. If the cells were grown at 40°C, however, most of the cells showed weak ferritin labelling. It is assumed that at 40°C at least part of ECA was exposed on the cell surface, whereas at 22°C it was covered by the thin surface layer shown in Fig. 2. This interpretation is also supported by the results of chemical analyses of LPS from Ye 75 S grown at 10, 22, 37 and 40°C (Acker et al., 1980; Wartenberg et al., 1983). At low temperatures (10°C) LPS of this strain contained about 40% (wt/wt) 6-deoxy-L-altrose, which forms the O-specific chain of this strain. By increasing the cultivation temperature, a significant decrease in this constituent was observed. LPS from cells grown at 40°C contained only about 12% 6-deoxy-L-altrose. The rough mutant Ye 75 R, which lacks O-specific chains completely, showed denser labelling with ferritin. These results indicate that ECA on the cell surface of Ye 75 S is shielded by O-specific chains of the LPS if grown at 22°C or at lower temperature. It becomes accessible when O-chains are lacking (R mutants) or when they are reduced in amount at higher growth temperature (Wartenberg et al., 1983).

2. Accessibility of ECA to antibodies in derivatives from a strain of E. coli *O8:K27 with different lesions in the O-chain and/or capsular polysaccharides synthesis*

In order to study how the O-chain and/or the capsular polysaccharides interfere with the accessibility of ECA to antibodies, the following derivatives of the wild-type strain *E. coli* E 56b (O8 : K27) were examined: D 280 (encapsulated S form), F 492 (non-capsulated S form), F 470 (non-capsulated R form), F 782 (encapsulated R form), F 1283 (non-capsulated R form, ECA-negative mutant) (see also Section II and Acker et al., 1982).

In a first series of experiments the accessibility of ECA to antibodies was examined on conventionally prepared Epon thin sections; cells of the above-mentioned derivatives were incubated with rabbit ECA antiserum and

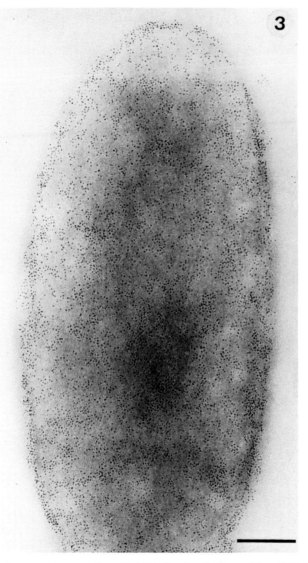

Fig. 3.   Whole-mount preparation: Ye 75 S after incubation with anti-Ye 75 S serum and ferritin-conjugated antibody. Bar, 0.2 µm. From Acker *et al.* (1980).

subsequently with ferritin-conjugated anti-rabbit antibodies to identify the
ECA antibodies on the cells. After washing, the encapsulated strains were
further incubated with anti-capsule serum to protect their capsules from
structural collapse and to visualize the relationship of the ECA to the outer
membrane and to the capsular layer. In all cells examined, the label was
localized deep within the capsular layer, associated with the outer membrane

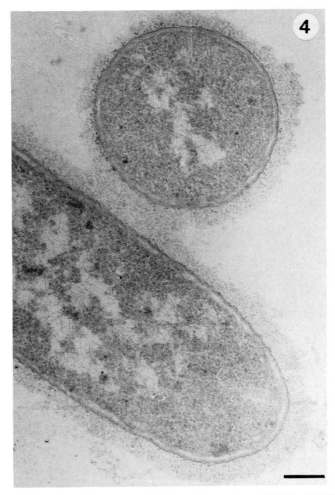

Fig. 4. Epon thin section: encapsulated R form *E. coli* F 782 (O8:K27) after
incubation with ECA-specific serum and labelling with ferritin-conjugated secondary
antibody; the labelled cells were further treated with anti-capsule serum to preserve the
capsule. The outer side of the outer membrane shows many ferritin particles indicating
ECA. Bar, 0.2 μm. From Acker *et al.* (1982).

surface (Fig. 4). All cells of the strains examined, except that of the ECA-negative mutant F 1283, revealed ferritin particles, but the degree of labelling differed from strain to strain. A relative quantification of the different labelling degree observed was made on whole-mount preparations (ferritin particles counted per μm^2 taken from the micrographs). The average number of particles determined per μm^2 was used as a quantitative measure of the accessibility of ECA. The number of ferritin particles per μm^2 decreased in the

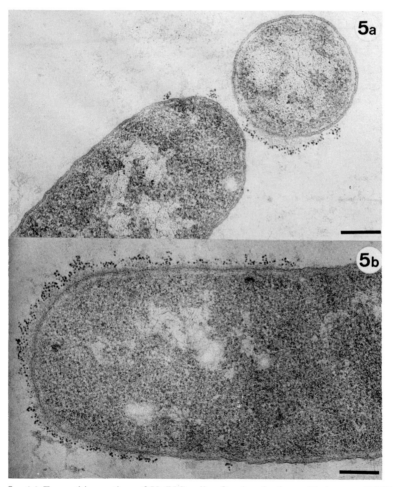

Fig. 5. (a) Epon thin section of Ye 75 S cells after incubation with human EGTA-serum (20 min at 37°C), then with rabbit anti-human C3, and finally with ferritin-conjugated anti-rabbit IgG for demonstration of deposited C3b (dark dots). (b) As in (a), but after 40 min incubation time. Bars, 0.2 μm. From Acker and Brade (1980).

order non-capsulated R (943) > encapsulated R (148) > non-capsulated S (66) > encapsulated S form (23 particles μm^{-2}). It is notable that neither the capsular nor the O polysaccharides apparently form such a dense net that ferritin conjugate cannot penetrate, in contrast to the O-specific chains of Ye 75 S which clearly prevent ferritin labelling (see above).

3. *Kinetic studies on the deposition of activated human complement (C) on the cell surface of the smooth strain* Yersinia enterocolitica *75 (Ye 75 S)*

It has been shown that Ye 75 S activated the complement system by both the alternative pathway and classical pathway (Acker and Brade, 1980). There have also been reports on the C activation, bacterial assays, C-induced ultrastructural alterations and deposition of C3 fragments (C3b) on the cell surface of Ye 75 S after incubation with serum or lysozyme-free serum. In this section only a few aspects of the combined applications of the whole-mount procedure and the conventional thin-section technique in kinetic studies on the C3b deposition on the cell surface of serum-incubated cells of Ye 75 S can be mentioned. The kinetic of C3b deposition was demonstrated with the indirect immunoferritin technique by examining cells after different periods of incubation with EGTA-serum (alternative pathway). Since the deposition of C3b started at single, discrete sites, the whole-mount procedure, which permitted the examination of a large number of cells in a short time, was the method of choice for the identification of these faint ferritin depositions. Embedding experiments were carried out in parallel, but thin sections were prepared from representative samples only in which ferritin label was already identified by the less laborious and far more effective whole-mount procedure. Examinations with the thin-section technique, however, were necessary for a more precise localization of C3b deposition in the cell envelope. From small sites or patches (Fig. 5a) spreading of deposited C occurred, until large areas of the cell surface or even the whole surface of the cells examined was covered after incubation periods of up to 60 min (Fig. 5b; and Acker and Brade, 1980).

C. **Post-embedding (post-sectioning) labelling using Lowicryl K4M and frozen sections**

1. *Single and double immunolabelling of ECA- and K1 antigen-immunoreactive sites on Lowicryl K4M resin sections*

As already mentioned, neither the whole-mount procedure nor the pre-embedding immunoelectron microscopy were suitable for labelling ECA on the cell surface of morphologically intact smooth cells of Ye 75 S (grown at 22°C), where this antigen was most probably covered by a distinct surface layer formed by the O-specific chains of the LPS (Fig. 2). Therefore, for the localization of ECA, deep in this layer, as well as for the intracellular localization of ECA- and K1 antigen-immunoreactive sites in *E. coli* strains, two alternative techniques were applied, namely the Lowicryl K4M and the cryosection techniques (see below).

(a) Single labelling on Lowicryl K4M thin sections of agar-enrobed cells dehydrated in ethanol

In Lowicryl K4M thin sections, both the cell envelope of Ye 75 S and the envelope of the other strains examined showed a fine structure similar to that described by Kellenberger and associates (Hobot *et al.*, 1984, 1985) for *E. coli* B cells also aldehyde-fixed and embedded with the progressive lowering of temperature technique. Post-sectioning labelling of such K4M thin sections through Ye 75 S cells using mono- or polyclonal antibodies revealed a similar weak labelling as shown in Fig. 6 for *E. coli* D 280, showing that ECA, inaccessible with the whole-mount procedure (Section III.B.1), was exposed in thin sections with its antigenicity preserved. In all thin sections of ECA-positive strains examined, good labelling of membrane-associated areas in the cytoplasm was observed (Fig. 6, black arrow). Significant labelling, however, was also identified in ribosome-containing areas of the cytoplasm (Fig. 6, white arrow), whereas ribosome-free areas revealed practically no labelling (Fig. 6, less electron-dense regions of the cell). Possible interpretations of this labelling pattern in connection with the biosynthesis of ECA, its assembly on antigen carrier lipid (undecaprenyl pyrophosphate; Rick *et al.*, 1985; Barr and Rick, 1987) and its transport into the outer membrane were recently discussed (Acker *et al.*, 1986).

Although the K1 capsule is the most extensively investigated of *E. coli* capsules (Ørskov *et al.*, 1977; Timmis *et al.*, 1985), detailed knowledge of the sites its biosynthesis are not available. It was known, however, that undecaprenol is also involved in K1 polysaccharide biosynthesis. Therefore we localized K1-immunoreactive sites on Lowicryl K4M thin sections through different *E. coli* strains and mutants using the monoclonal anti-K1 polysaccharide antibody 735 D4 (Acker and Bitter-Suermann, in preparation). As expected, thin sections of the encapsulated strain *E. coli* D 1737 K1 revealed, after indirect immunolabelling, gold particles outside the outer membrane. However, the gold particles showed a rather patchy or irregular distribution and were often in closer contact with the outer membrane, indicating that the capsule most probably collapsed during the preparation procedure. Gold particles were also identified regularly in membrane-associated areas in the cytoplasm, although to distinguish the intra- from extracellularly localized particles was not always easy.

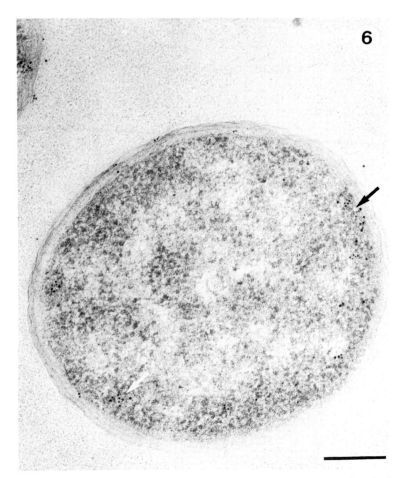

Fig. 6. Lowicryl K4M thin section: *E. coli* D 280 (wild-type) after indirect immunogold labelling using ECA-specific serum (see text). Bar, 0.2 μm. From Acker *et al.* (1986).

In order to overcome this difficulty, cells of the derivative LE 392 carrying the plasmid pKT 274, which specifies the synthesis and transport of K1 polysaccharide to the cell surface, or carrying the plasmid pGB 33, a mutant of pKT 274 which specifies the synthesis but not the transport of K1 polysaccharide, were also examined. Cells carrying the plasmid pKT 274 showed a labelling pattern similar to that of *E. coli* D 1737 K1 cells (see above). However, cells carrying the pGB 33 plasmid remained clearly unlabelled outside the outer membrane, but labelled intracellularly. As shown in Fig. 7, the vast majority of the gold particles was identified in membrane-associated

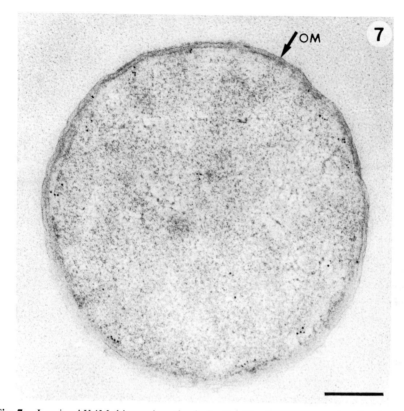

Fig. 7.   Lowicryl K4M thin section: the derivative *E. coli* LE 392 carrying the plasmid pGB 33 (transport defect for K 1 polysaccharide). After indirect labelling using the anti-K 1 polysaccharide MAb 735 D4, membrane-associated areas showed gold labelling whereas the outer membrane (OM) was devoid of gold particles. Contrasted with uranyl acetate and lead citrate. Bar, 0.2 µm. From unpublished work of Acker and Bitter-Suermann.

areas of the cytoplasm or even in more or less close contact with the cytoplasmic membrane. With the whole-mount procedure practically no label could be detected on whole cells, indicating the absence of K 1 polysaccharide on the cell surface.

The identification of label in membrane-associated areas in the cytoplasm using ECA- and K1 antigen-specific monoclonal antibodies is noteworthy. Further investigations using freeze substitution in combination with low temperature embedding in Lowicryl resins (Humbel and Müller, 1985; Acetarin *et al.*, 1986) are under way to verify the antigenic distribution patterns observed.

(b) Double labelling on Lowicryl K4M thin sections of gelatin-enrobed cells dehydrated in dimethylformamide (DMF)

Thin sections of glutaraldehyde-fixed cells of *E. coli* D 1737 K1, enrobed in gelatin and dehydrated in DMF, after indirect immunolabelling with anti-K1 polysaccharide MAb 735 D4, showed a relatively uniform distribution of gold particles outside the outer membrane when compared with the patchy distribution of gold particles on thin sections obtained from cells of the same strain but enrobed in agar and dehydrated in ethanol (previous paragraph). The uniform labelling observed indicated that the capsule was probably uncollapsed, similar to the situation shown for the first time with this novel procedure for the capsule of *E. coli* O9:K29 (Bayer *et al.*, 1985).

Applying this procedure to the simultaneous localization of ECA and K1 antigen on the same section of *E. coli* D 1737 K1 cells using 5- and 10-nm gold particles, respectively, a more precise topological relationship between the two antigens and the outer membrane could be observed when the capsule was opsonized with anti-K1 polysaccharide MAb 735 D4 on living cells in their growth medium. Figures 8a and 8b show thin sections from the same sample of such an opsonized cell. The thin section in Fig. 8a was stained with uranyl acetate and lead citrate for better visualization of the fine structure, including the opsonized capsule. When thin sections from such opsonized cells were incubated directly with gold-labelled secondary anti-species antibodies, almost no gold particles could be observed, indicating that the antigenicity of the primary antibody had not survived the preparation procedure. However, when thin sections of opsonized cells were labelled in sequence with the primary and secondary antibodies, a good and uniform labelling was observed. Figure 8b shows a representative micrograph of such cells labelled for ECA and K1 antigen according to the principle of Bendayan (1982), but with modifications (Section II.F.2) and stained with uranyl acetate to visualize more easily the smaller 5-nm gold particles. Note that the ECA, indicated by these small particles, is located deep within the capsule, close to the outer membrane surface. This observation is in good agreement with the results obtained by the pre-embedding method with Epon thin sections of ferritin-labelled cells of the encapsulated R strain *E. coli* F 782 K27 (compare Fig. 4 and Fig. 8b). Furthermore, small gold particles were also observed in membrane-associated areas of the cytoplasm, similar to the finding obtained with Lowicryl K4M sections of cells of *E. coli* D 280 enrobed in agar and dehydrated with ethanol (compare Fig. 6 and Fig. 8b). The identification of K1 polysaccharide-immunoreactive sites in membrane-associated areas in Fig. 8b, indicated by the larger gold particles, is in agreement with the labelling pattern obtained also for K1 polysaccharide-immunoreactive sites with the transport-defective derivative LE 392 carrying the plasmid pGB 33 (compare

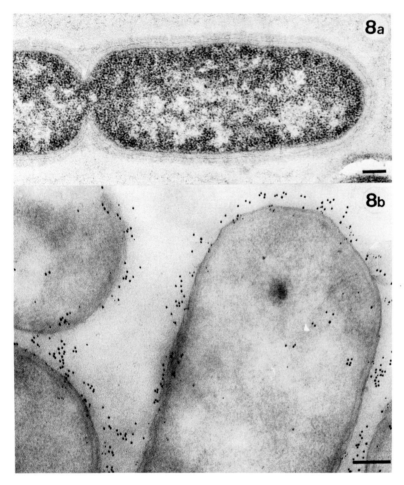

Fig. 8. (a) Lowicryl K4M thin section: gelatin-enclosed cells of *E. coli* D 1737 K1 dehydrated in DMF. The K1 capsule was stabilized with anti-K1 polysaccharide MAb 735 D4 before embedding. Uranyl acetate and lead citrate contrasting. (b) Sections obtained as for (a) were labelled on "face A" with MAb 898 for ECA (small gold particles) and on "face B" with MAb 735 D4 for K1 polysaccharide (bigger gold particles). Explanations given in the text. Contrasted with uranyl acetate only. Bars, 0.2 μm. From unpublished work of Acker and Bitter-Suermann.

Fig. 7 and Fig. 8b); the gold particles in the cytoplasm of two of the cells in Fig. 8b most probably represent unspecific background labelling.

2. *Localization of ECA- and K1 antigen-immunoreactive sites on thawed cryosections*

ECA- and K1 antigen-immunoreactive sites investigated on Lowicryl K4M resin sections (Acker *et al.*, 1986; and previous paragraph) were also localized on cryosections using mono- and polyclonal antibodies from the same batches.

In contrast to the observation obtained with the Lowicryl K4M thin sectioning technique, where no division of the cell envelope into distinct layers could be observed, thin cryosections of *E. coli* and *Yersinia enterocolitica* cells examined revealed, after contrasting with uranyl acetate–oxalate and treatment with uranyl acetate–methyl cellulose, the layered structure of the cell envelope shown in Fig. 9a. A similar division of the cell envelope in the outer and cytoplasmic membrane was previously observed by Dubochet *et al.* (1983) on cryosections of *Klebsiella pneumoniae* cells, fixed with osmium tetroxide prior to cryofixation. In our labelling experiments with mono- or polyclonal antibodies no notable difference of the labelling patterns could be observed. Figure 9b shows a cryosection of the ECA-positive strain *E. coli* F 470 after immunolabelling with ECA-specific monoclonal antibody MAb 989. The cell envelope and the membrane-associated areas in the cytoplasm revealed labelling patterns comparable to those observed on Lowicryl K4M sections, and they confirmed the occurrence of ECA-immunoreactive sites in the outer membrane and membrane-associated areas. However, it must be mentioned that a considerably higher degree of labelling was expected on cryosections than on Lowicryl K4M sections, since it was shown that the immunolabelling of resin sections with colloidal gold was restricted to the section surface (Roth, 1982), whereas with cryosections penetration of immunoreagents occurs to a different extent (Geuze *et al.*, 1979; Griffiths and Hoppeler, 1986). A similar unexpected low labelling on cryosections was observed with the K1 polysaccharide. Figure 10 shows a frozen section of a cell of the K1 encapsulated strain *E. coli* D 1737, enrobed in gelatin and contrasted with uranyl acetate–oxalate before treatment with methyl cellulose. The cell is surrounded by a light halo with a rather uniform thickness and a sharp delineation from the more electron-dense gelatin matrix. After labelling and contrasting, a low degree of labelling was observed. However, by omitting the above-mentioned contrasting procedure, the fine structure collapsed as expected, but strong labelling was observed on the cell periphery, indicating that the antigenicity of the K1 polysaccharide in the halo was preserved. This observation, together with the above-mentioned result

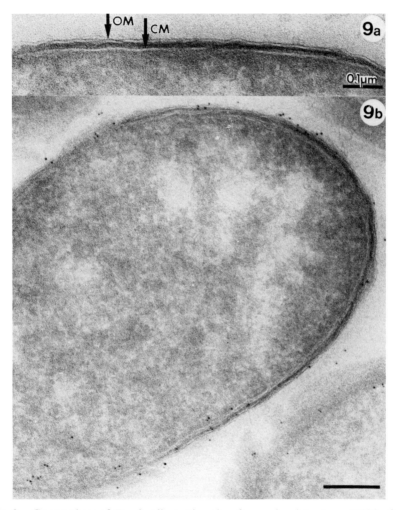

Fig. 9.   Cryosections of *E. coli* cells, sectioned as frozen droplets at *c.* −110°C after infusion with 1.1 M sucrose. Contrasted with uranyl acetate–oxalate and treated with methyl cellulose–uranyl acetate. (a) ECA-negative mutant F 1283; control section devoid of gold particles after indirect immunolabelling using ECA-specific MAb 898. Outer membrane (OM), cytoplasmic membrane (CM). (b) ECA-positive strain F 470 after labelling as in (a). (b) from Acker *et al.* (1986). Bar in (b), 0.2 μm.

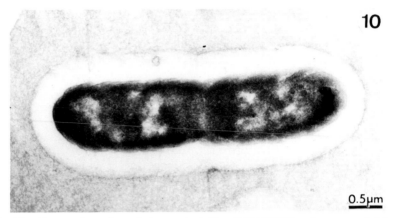

Fig. 10.   Cryosection of a gelatin-enrobed *E. coli* D 1737 K1 cell, surrounded by a halo outside the outer membrane (see text). Positively contrasted after Tokuyasu (1980). From unpublished work of Acker and Bitter-Suermann.

obtained with cryosections labelled for ECA, indicated that the fine structure obtained with the contrasting procedure used to generate images such as that shown in Fig. 9a was not compatible with the high degree of labelling observed with uncontrasted samples. In many labelling experiments a weak background was observed on contrasted cryosections, whereas uncontrasted sections, labelled in parallel, showed negligible background, especially when monoclonal antibodies were used. The weak background resulted most probably from relocation of label and/or labelled antigen during the contrasting procedure. This interpretation is supported by the observed dislocation of label on Lowicryl K4M sections when the contrasting procedure was performed immediately after the final water rinsing step. The loss and/or relocation of label on Lowicryl K4M sections was considerably reduced, if not suppressed, when well-dried sections were shortly contrasted with uranyl acetate only (Fig. 8b). A compromise between fine structure preservation of cryosections and relocation of label was obtained by application of the negative staining procedure (Tokuyasu, 1973) or, alternatively, the positive staining method of Tokuyasu. Additional investigations are needed to overcome the difficulties in combining the good fine structural preservation (Fig. 9a) and the high degree of labelling observed in uncontrasted cryosections.

IV.   Conclusions

From the literature cited and the examples presented in this chapter, it can be concluded that the ideal method for the localization of extra- and intracellular antigens is not yet available. Furthermore, it has been shown that the subcellular localization of the surface antigens presents considerable preparative difficulties because most of them are components of highly hydrated cell structures which collapse during conventional dehydration procedures. Therefore it is advisable to localize such antigens by combining the results of different methods before firm conclusions are drawn on the bacterial surface architecture of living cells. The neglected whole-mount procedure, a very simple and rapid technique, seems likely to continue to be an important tool for examining immunologically accessible surface antigens, especially when used in combination with the conventional thin-section technique. The main advantage that both procedures offer is that the antigen–antibody reaction can be allowed to take place directly in the fresh growth medium under *in vivo* conditions where the highest labelling efficiency can be expected. Samples screened with the whole-mount procedure can be fixed in glutaraldehyde and OsO_4, and stained *en bloc* before being embedded conventionally, thus combining strong labelling and optimal preservation of ultrastructure.

In combination with the Lowicryl K4M thin-section technique described, the whole-mount procedure carried out with cells fixed in different concentrations of glutaraldehyde solutions can be useful in determining which fixation conditions lead to an optimal compromise between labelling signal and ultrastructural preservation. This information can be applied to on-section localization of the same antigen in subsequent experiments. The localization of ECA on the outer membrane surface in both frozen-thawed and Lowicryl K4M resin sections is in good agreement with our previous observations on different smooth and/or encapsulated strains and mutants. This demonstrates that, with both on-section labelling methods applied, promising results may be obtained. However, additional information seems to be needed concerning the influence of the heavy-metal stains on the deposited immunoreagents, as could be observed in particular with frozen sections. Furthermore, the air-drying artefacts in frozen sections of the highy hydrated polysaccharide-containing surface layers probably limited the resolution much more than in well-fixed protein-containing cell structures.

The uniform label distribution outside the outer membrane of encapsulated cells, obtained by application of the recently modified Lowicryl K4M embedding procedure (Bayer *et al.*, 1985), indicated that this novel procedure may be useful in the study of the antigen architecture even of highly hydrated layers in an uncollapsed state. Thin sections obtained with this procedure were also successfully used for localization of ECA and K1 antigen on the same thin section. Experiments using freeze substitution in combination with low temperature embedding in Lowicryl resins (Humbel *et al.*, 1983; Humbel and Müller, 1985; Sitte *et al.*, 1985; Acetarin *et al.*, 1986) are under way to verify our results obtained by on-section labelling as described here. With the latter methods, as well as with other cryotechniques (see reviews by Pinto da Silva, 1984; Pinto da Silva and Kan, 1984; De Mey and Moeremans, 1986), only limited experience is available. However, labelling with these methods may be an important addition to the techniques applied so far and could also be valuable tools in studying the poorly understood mechanisms of biosynthesis and translocation of surface antigens, as well as their roles in adhesion, colonization of different surfaces and pathogenicity (Costerton et al., 1981; Brubaker, 1985; Joiner, 1985; Kasper, 1986).

Acknowledgement

The author wishes to thank Mrs Rita Bergfeld for her excellent technical assistance throughout the course of this work.

References

Acetarin, J.-D., Carlemalm, E. and Villiger, W. (1986). *J. Microsc.* **143**, 81–88.
Acker, G. and Brade, V. (1980). *Zbl. Bakt. Hyg., I. Abt. Orig. A* **248**, 210–229.
Acker, G., Wartenberg, K. and Knapp, W. (1980). *Zbl. Bakt. Hyg., I. Abt. Orig. A* **247**, 229–240.
Acker, G., Knapp, W., Wartenberg, K. and Mayer, H. (1981). *J. Bacteriol.* **147**, 602–611.
Acker, G., Schmidt, G. and Mayer, H. (1982). *J. Gen. Microbiol.* **128**, 1577–1583.
Acker, G., Bitter-Suermann, D., Meier-Dieter, U., Peters, H. and Mayer, H. (1986). *J. Bacteriol.* **168**, 348–356.
Armbruster, B. L. and Kellenberger, E. (1986). In *Ultrastructure Techniques for Microorganisms* (H. C. Aldrich and W. J. Todd, eds), pp. 267–295. Plenum Press, New York.
Ashford, A. E., Allaway, W. G., Gubler, F., Lennon, A. and Sleegers, J. (1986). *J. Microsc.* **144**, 107–126.
Barr, K. and Rick, P. D. (1987). *J. Biol. Chem.* **262**, 7142–7150.
Bayer, M. E. and Thurow, H. (1977). *J. Bacteriol.* **130**, 911–936.
Bayer, M. E., Carlemalm, E. and Kellenberger, E. (1985). *J. Bacteriol.* **162**, 985–991.
Bayer, M. E., Weed, D., Haberer, S. and Bayer, M. H. (1986). *FEMS Microbiol. Lett.* **35**, 167–170.
Beesley, J. E. (1984). In *Immunolabeling for Electron Microscopy* (J. M. Polak and S. van Norden, eds), pp. 289–300. Elsevier, Amsterdam.
Bendayan, M. (1982). *J. Histochem. Cytochem.* **30**, 81–85.
Beveridge, T. J. (1981). *Int. Rev. Cytol.* **72**, 229–317.
Brubaker, R. P. (1985). *Ann. Rev. Microbiol.* **39**, 21–50.
Costerton, J. W., Irwin, R. T. and Cheng, K.-J. (1981). *Ann. Rev. Microbiol.* **35**, 229–324.
De Mey, J. and Moeremans, M. (1986). In *Advanced Techniques in Biological Electron Microscopy III* (Koehler, ed.), pp. 229–271. Springer-Verlag, Berlin.
Dubochet, J., McDowall, A. W., Menge, B., Schmid, E. N. and Lickfeld, K. G. (1983). *J. Bacteriol.* **155**, 381–390.

Echarti, C., Hirschel, B., Boulnois, G. J., Varley, J. M., Waldvogel, F. and Timmis, K. N. (1983). *Infect. Immun.* **41**, 54–60.
European Molecular Biology Organization (1984). *Immuno Electron Microscopy: A Practical Course Manual.* University of Basel, Basel.
Freer, J. H. (1985). In *Immunology of the Bacterial Cell Envelope* (D. E. Stewart-Tull and M. Davies, eds), pp. 355–383. John Wiley, Chichester, New York, Brisbane, Toronto and Singapore.
Frosch, M., Görgen, I., Boulnois, G. J., Timmis, K. N. and Bitter-Suermann, D. (1985). *Proc. Natl Acad. Sci. USA* **82**, 1194–1198.
Galanos, C., Lüderitz, O., Rietschel, E. T. and Westphal, O. (1977). In *International Review of Biochemistry* (T. W. Goodwin, ed.), Vol. 14, pp. 239–335. University Park Press, Baltimore.
Geuze, J. J., Slot, J. W., Tokuyasu, K. T., Goedemans, W. E. M. and Griffith, J. M. (1979). *J. Cell. Biol.* **82**, 697–707.
Griffiths, G. (1984). In *The Science of Biological Specimen Preparation* (J.-P. Revel, T. Bernard and G. H. Haggis, eds), pp. 153–157. SEM Inc., AMF O'Hare, IL 60666, USA.
Griffiths, G. and Hoppeler, H. (1986). *J. Histochem. Cytochem.* **34**, 1389–1398.
Griffiths, G., Simons, K., Warren, G. and Tokuyasu, K. T. (1983). In *Methods in Enzymology*, (S. Fleischer and B. Fleischer, eds), Vol. 96, pp. 466–485. Academic Press, New York.
Griffiths, G., McDowall, A., Back, R. and Dubochet, J. (1984). *J. Ultrastruct. Res.* **89**, 65–78.
Hobot, J. A., Carlemalm, E., Villiger, W. and Kellenberger, E. (1984). *J. Bacteriol.* **160**, 143–152.
Hobot, J. A., Villiger, W., Escaig, J., Maeder, M., Ryter, A. and Kellenberger, E. (1985). *J. Bacteriol.* **162**, 960–971.
Humbel, B. and Müller, M. (1985). In *The Science of Biological Specimen Preparation* (J.-P. Revel, T. Bernard and G. H. Haggis, eds), pp. 175–183. SEM Inc., AMF O'Hare, IL 60666, USA.
Humbel, B., Marti, Th. and Müller, M. (1983). In *BEDO* (G. Pfefferkorn, ed.), Vol. 16, pp. 585–594. Antwerpen.
Jann, K. and Jann, B. (1983). *Prog. Allergy* **33**, 53–79. Karger, Basel.
Joiner, K. A. (1985). *Curr. Top. Microbiol. Immunol.* **121**, 99–133.
Kasper, D. L. (1986). *J. Infect. Dis.* **153**, 407–415.
Kuhn, H. M., Meier, U. and Mayer, H. (1984). *Forum Mikrobiol.* **7**, 274–285.
Lugtenberg, B. and van Alphen, L. (1983). *Biochem. Biophys. Acta* **737**, 51–115.
Mayer, H. and Schmidt, G. (1979). *Curr. Top. Microbiol. Immunol.* **85**, 99–151.
Ørskov, I., Ørskov, F., Jann, B. and Jann, K. (1977). *Bacteriol. Rev.* **41**, 667–710.
Ørskov, F., Ørskov, I., Sutton, A., Schneerson, R., Lin, W., Egan, W., Hoff, G. E. and Robbins, J. B. (1979). *J. Exp. Med.* **149**, 669–685.
Peters, H., Jürs, M., Jann, B., Jann, K., Timmis, K. N. and Bitter-Suermann, D. (1985). *Infect. Immun.* **50**, 459–466.
Pinto da Silva, P. (1984). In *Immunolabeling for Electron Microscopy* (J. M. Polak and S. van Norden, eds), p. 179. Elsevier, Amsterdam.
Pinto da Silva, P. and Kan, F. W. K. (1984). *J. Cell. Biol.* **99**, 1156–1161.
Plattner, H. and Zingsheim, H. P. (1987). In *Elektronenmikroskopische Methodik in der Zell- und Molekularbiologie*, pp. 125–142. Gustav Fischer, Stuttgart.
Rick, P. D., Mayer, H., Neymeyer, B. A., Wolski, S. and Bitter-Suermann, D. (1985). *J. Bacteriol.* **162**, 494–503.

Roth, J. (1982). In *Techniques in Immunochemistry* (G. R. Bullock and P. Petrusz, eds), Vol. 1, pp. 108–133. Academic Press, London.

Singer, S. J. (1959). *Nature (Lond.)* **183**, 1523–1524.

Sitte, M., Neumann, K. and Edelmann, L. (1985). In *The Science of Biological Specimen Preparation* (M. Müller, R. P. Becker, A. Boyde and J. J. Wolosewick, eds), pp. 103–118. SEM Inc., AMF O'Hare, IL 60666, USA.

Smit, J. and Todd, W. J. (1986). In *Ultrastructure Technique for Microorganisms* (H. C. Aldrich and W. J. Todd, eds), pp. 469–510. Plenum Press, New York.

Timmis, K. N., Boulnois, G. J., Bitter-Suermann, D. and Cabello, F. C. (1985). *Curr. Top. Microbiol. Immunol.* **118**, 197–210.

Tokuyasu, K. T. (1973). *J. Cell. Biol.* **57**, 551–565.

Tokuyasu, K. T. (1978). *J. Ultrastruct. Res.* **63**, 287–307.

Tokuyasu, K. T. (1980). *Histochem. J.* **12**, 381–403.

Troy, F. A. (1979). *Ann. Rev. Microbiol.* **33**, 519–560.

Vaara, M. and Nikaido, H. (1984). In *Handbook of Endotoxin* (E. T. Rietschel, ed.), Vol. 1, pp. 1–45. Elsevier, Amsterdam.

Valentine, R. C., Shapiro, B. M. and Stadtman, E. (1968). *Biochemistry* **7**, 2143–2152.

Wartenberg, K., Lysy, J. and Knapp, W. (1975). *Zbl. Bakt. Hyg., I. Abt. Org. A* **230**, 361–366.

Wartenberg, K., Knapp, W., Ahamed, N. M., Widemann, C. and Mayer, H. (1983). *Zbl. Bakt. Hyg., Abt. Org. A* **253**, 523–530.

8

Immunoelectron Microscopic Localization of Bacterial Enzymes: Pre- and Post-embedding Labelling Techniques on Resin-embedded Samples

MANFRED ROHDE*, HOLGER GERBERDING†,
THOMAS MUND† and GERT-WIELAND KOHRING†

* Gesellschaft für Biotechnologische Forschung mbH, (GBF),
Mascheroder Weg 1, D-3300 Braunschweig, Federal Republic of Germany
† Institut für Mikrobiologie der Georg-August-Universität Göttingen,
D-3400 Göttingen, Federal Republic of Germany

I. Introduction

Several methods are available for the localization of antigens by electron microscopy, all of which take advantage of the affinity properties of

METHODS IN MICROBIOLOGY
VOLUME 20   ISBN 0 12 521520 6

macromolecules such as antibody, avidin, enzymes, lectins or protein A. These macromolecules are coupled to appropriate markers in order to visualize their binding to specific antigenic or reaction sites. In earlier investigations, ferritin was used as an electron-dense marker (Singer, 1959; Oppenheim and Sacton, 1973; MacAlister et al., 1977; Wagner et al., 1980). However, ferritin showed some major disadvantages, in particular the tendency to bind non-specifically to cellular components (Morgan, 1972; Parr, 1979; Wientjes et al., 1980). Faulk and Taylor (1971) were the first to adsorb an antibody to colloidal gold instead of ferritin, and use this complex as a direct labelling probe to identify and localize surface antigens. An indirect labelling technique was introduced by Romano et al. (1974, 1975); they applied secondary antiserum directed against the IgG molecules of the primary antiserum; the IgG molecules of this secondary antiserum were complexed with colloidal gold. These complexes could be used to detect the primary antibody already bound to the antigen. Roth et al. (1978) showed that colloidal gold covered with protein A (which reacts specifically with the Fc-portion of an antibody) can be applied as a marker complex with high sensitivity and resolution. Since then new applications and further developments of the colloidal gold marker system have proliferated (Bendayan et al., 1980; Roth et al., 1981; Bendayan, 1982; Bendayan and Zollinger, 1983).

At present two main labelling procedures have been applied successfully—the pre-embedding and the post-embedding labelling methods. In the pre-embedding method, the immunoreagent is allowed to react with the antigen prior to the embedding of the sample. This approach produces several problems; the impermeability of the cytoplasmic membrane is a major drawback. In addition, the free diffusion of the immunoreagent within the cytoplasm may be hindered. This method is therefore used mainly for the localization of antigens located at the cell surface. An extension of the colloidal gold marker system was developed following the observations of Roth et al. (1978) that such complexes exhibit a low non-specific binding to structural components exposed at the surface of ultra-thin sections of resin-embedded samples. This was the introduction of the post-embedding labelling procedure. In this method, the sample is embedded in resin or other matrix material, and ultra-thin sections are prepared which are allowed to react with the immunoreagent and a suitable marker system.

Over the past few years we have applied the antibody–ferritin, the protein A–gold, and the antibody–gold complex techniques for the localization of bacterial enzymes using the pre-embedding and post-embedding labelling procedures. The aim of this chapter is to provide information about the preparation of colloidal gold, formation of protein–gold complexes, and embedding and further processing of bacteria with sufficiently well-preserved antigenicity of the protein to be labelled, as well as good ultrastructure of the cell. The application of such marker systems is demonstrated for the labelling of several types of bacterial enzymes in different compartments of the cell.

II. Methodology

A. Preparation of colloidal gold

1. Glassware

The properties of the glassware surface have an important effect on the initiation of the reduction process in the preparation of colloidal gold. Even small amounts of contaminants on the walls of the glassware can influence the formation of colloidal gold or can cause variations in the particle size. Therefore all glassware has to be extremely carefully cleaned, rinsed repeatedly with distilled water and siliconized.

2. Reagents

Every solution must be made from double-distilled and filtered (pore size 0.2 μm) water. Tetrachloroauric acid is commercially available as a crystalline substance (Merck, Darmstadt, FRG). It is recommended that a 1% aqueous stock solution is prepared from the entire ampoule in a well-sealed brown vial. The solution can be stored for several months at 4°C. The saturated solution of white phosphorus in ether is prepared with diethyl ether. White phosphorus sticks are cut (under water) into small pieces, dried by blotting them for a few seconds on filter paper, and then rapidly transferred into the diethyl ether. Saturation takes 1–2 days at room temperature. It is most important to store and handle this flammable and explosive solution with great care. For all the other solutions it is recommended that freshly prepared and filtered solutions are used.

3. Recipes for the preparation of colloidal gold solutions

(a) Reduction with phosphorus (Zsigmondy and Thiessen, 1925; Faulk and Taylor, 1971; Frens, 1973)

This procedure is useful for the preparation of colloidal gold particles of approximately 4–5 nm with relatively good reproducibility.

1.4 ml of 1% tetrachloroauric acid are added to 120 ml distilled water adjusted with 0.2 M K_2CO_3 to pH 7.2. 1 ml ether-saturated phosphorus solution is made from one part ether-saturated phosphorus and four parts of diethyl ether. The mixture is shaken for 10–15 min at room temperature. The appearance of a brownish red colour signals the beginning of the reduction process. Finally, the mixture is heated until it turns the wine-red colour of the colloidal gold.

(b) Reduction with ascorbic acid (Slot and Geuze, 1981)

Applying this procedure, the gold particle size varies between 8 and 10 nm.
1 ml of 1% tetrachloroauric acid and 1.5 ml of 0.2 M K_2CO_3 are mixed with 25 ml distilled water on ice. While stirring, 1 ml of 0.7% aqueous sodium ascorbate is added quickly. After addition of the ascorbate the colour changes to brownish purple–red. Finally, the volume is adjusted with distilled water to 100 ml and heated until the wine-red colour appears.

(c) Reduction with trisodium citrate (Frens, 1973; Slot and Geuze, 1981)

This method gives monodispersed colloidal gold solutions varying in particle size between 15 and 150 nm, depending on the amount of trisodium citrate added to a constant volume of tetrachloroauric acid. Following the protocol gold particles with 15 nm in size can be prepared.
1 ml of 1% tetrachloroauric acid is added to 100 ml distilled water. After boiling of the mixture 4 ml 1% aqueous trisodium citrate are rapidly added. To complete the reduction process the mixture is boiled for another 5 min until the wine-red colour appears.

(d) Reduction with tannic acid (Slot and Geuze, 1981)

Applying this procedure, monodispersed gold solutions with particle sizes of 4, 6, 8 or 12 nm can be made, depending on the amount of tannic acid added.
Solution 1 contains 1 ml of 1% tetrachloroauric acid in 79 ml distilled water. The reduction mixture, Solution 2, consists of 4 ml 1% trisodium citrate and 0 to 2 ml of 1% tannic acid. The more tannic acid is added, the smaller the gold particle size. It should be mentioned that reproducible results could only be obtained using tannic acid prepared from Aleppo nutgalls (this tannic acid is available from Mallinckrodt, St Louis, USA, code no. 8835).
The same volume 25 mM K_2CO_3 and 1% tannic acid are combined to adjust the pH. The total volume is adjusted to 20 ml with distilled water. Both solutions are heated to 60°C. Then the reducing mixture is added quickly to Solution 1 while stirring. The reduction process is completed by boiling until the solution turns the final wine-red colour.

B. Properties of colloidal gold and protein A

In this chapter only those properties of colloidal gold and protein A affecting the preparation of the IgG antibody–gold or protein A–gold complexes will be discussed.

Colloidal gold solutions are a wine-red colour in transmitting light. Such solutions have a single peak of absorption between 520 and 540 nm. The mean size of the colloidal gold particles is responsible for the characteristic shape and position of the peak; the peak moves to longer wavelengths with increasing mean particle diameter. Agglomeration of gold particles is easily detectable by a colour change to blue. An important surface characteristic of colloidal gold particles is their negative surface charge; colloidal gold can therefore interact with positively charged proteins. The stability of colloidal gold solutions is maintained by electrical repulsions. The addition of electrolytes to unprotected gold solutions causes an immediate colour change to blue due to agglomeration of gold particles (Pauli, 1949). This electrolyte-induced agglomeration can be prevented by the addition of proteins which bind to the colloidal gold particles by non-covalent electrostatic van der Waals' forces.

Protein A is produced by *Staphylococcus aureus* and consists of a single polypeptide chain with extended shape with a molecular weight of 42 000 (Forsgreen and Sjöquist, 1966; Björk *et al.*, 1972; Hjelm *et al.*, 1975). The characteristic property of protein A is the interaction with the Fc-portion of IgG antibody from different species (Hjelm *et al.*, 1972). Thus, no interference occurs with the antigen–antibody reaction used in immunocytochemical methods.

C. Complex formation of colloidal gold with IgG antibody and protein A

Under appropriate conditions of concentration and pH, proteins bind to colloidal gold by non-covalent electrostatic forces. The resulting complex is stable and does not affect the biological activity of the protein. It was shown that maximal binding occurs at pH values on the basic side of the pI of the macromolecule used (Geoghegan and Ackerman, 1977). For protein A this value is between pH 6.0 and 6.9 and for the IgG antibody around pH 8.9. Such pH adjustment is performed by adding $0.2 \text{ M } K_2CO_3$ for raising and 0.1 N acetic acid for lowering the pH value. For the determination of the minimal amount of IgG antibody or protein A for stabilizing a certain amount of colloidal gold, the method developed by Horisberger *et al.* (1975) and Horisberger and Rosset (1977) can be used.

0.1 ml of serial dilutions of the proteins are added to constant volumes of colloidal gold (1 ml), followed by an incubation for 15 min at room temperature. Addition of 0.1 ml of 10% sodium chloride reveals if the amount of protein added can stabilize the colloidal gold. Stabilized solutions retain the wine-red colour, whereas insufficiently stabilized solutions change to a blue colour due to agglomeration of the gold particles. For the preparation of the complexes the protein is used in 10% excess of the minimal stabilizing amount. Usually the colloidal gold solution is added to the protein, which should be dissolved in distilled water. After 5–10 min a 1% aqueous polyethylene glycol (PEG 20 000) solution is added. It is thought that PEG acts as a further stabilizing agent, thus lowering the rate of aggregate formation.

Further removal of free, non-complexed protein and of insufficiently stabilized gold particles is achieved by ultra-centrifugation. Depending on the size of the gold particles different centrifugation conditions are applied: for 2–4 nm gold particles 110 000g, 4°C, 1.5 h; 6–8 nm gold particles 80 000g, 4°C, 45 min; 12–15 nm gold particles 30 000g, 4°C, 20 min. During the centrifugation a small dark pellet of densely packed gold particles is formed which has to be discarded. The colourless supernatant, containing unbound protein, is also discarded. In addition a loose, intensely red-coloured band is formed at the bottom of the centrifuge tube. This material is resuspended in an appropriate buffer (for example, phosphate-buffered saline (PBS)) containing 0.2 mg PEG ml^{-1}. The centrifugation is repeated twice to remove unbound protein.

1. IgG antibody–gold complex

Affinity-purified polyclonal antibody samples contain mixtures of antibody having different ranges of isoelectric points. Therefore it can sometimes be difficult to determine the pH value giving optimal adsorption to the colloidal gold. Several protocols have been described (Romano *et al.*, 1974; Horisberger *et al.*, 1975; Geoghegan and Ackerman, 1977; Goodman *et al.*, 1981); the method of choice seems to be the procedure of de Mey *et al.* (1981). The IgG antibody solution is dialysed against 2 mM borax–HCl buffer, pH 9.0. Protein aggregates developing during the dialysis are removed by centrifugation at 100 000g, 4°C, 1 h. The colloidal gold solution, adjusted to pH 9.0, is added to a 10% excess of the optimal stabilizing amount of the IgG antibody. After 5–10 min bovine serum albumin (BSA) in distilled water, its pH adjusted with NaOH to 9.0, is added to give a final concentration of 1%. After centrifugation at conditions specified, depending on the desired particle size (see above), the clear supernatant is discarded and the IgG antibody–gold complexes are suspended in 20 mM Tris-buffered saline, pH 8.2, containing 1% BSA. This centrifugation is repeated twice. Finally, the complexes are suspended in Tris-buffered saline containing 1% BSA and sodium azide (0.01%).

2. Protein A–gold complex

For the preparation of protein A–gold complexes the method of Roth *et al.* (1978) is well established. 10% excess of the optimal stabilizing amount of protein A is dissolved in 0.2 ml distilled water. The colloidal gold, pH 6.9, is added to this solution, followed by an incubation for 10–15 min at room temperature. 1% PEG is added for further stabilization of the solution. The mixture is then centrifuged under appropriate conditions depending on the gold particle size required. The centrifugation is repeated twice. Finally, the protein A–gold complexes are suspended in phosphate-buffered saline, pH 6.9, containing 0.2 mg ml^{-1} PEG and sodium azid (0.01%). This stock solution can be stored in a refrigerator for at least six months without any significant loss of activity.

D. Processing of bacterial cells

When applying immunocytochemistry for localization studies of enzymes in bacterial cells, one is confronted with the problem of retaining the antigenicity of the enzyme and preserving reasonably well the bacterial ultrastructure. Thus, the fixatives used should stabilize all the cellular bacterial components in such a way that artificial diffusion and replacement of antigenic material or its extraction is prevented. Also fixatives should preserve the tertiary structure of the antigen to retain sufficient reactivity with the specific antibody.

1. Fixation

It has been reported by several authors that usual fixatives like formaldehyde and glutaraldehyde can be used for the fixation process (Roth *et al.*, 1978; Bullock and Petrusz, 1982; Roth, 1982; Bendayan, 1984; Kohring *et al.*, 1985; Rohde *et al.*, 1986; Ossmer *et al.*, 1986). In general, concentrations of the fixatives should be kept as low as possible. It is recommended that bacterial cells are fixed with a fixation solution containing 0.3% formaldehyde and 0.2% glutaraldehyde for 1 h on ice. In some cases it may be advantageous to fix the cells directly in the growth medium at the given temperature. This fixation solution sufficiently preserves the antigenicity and the cellular ultrastructure for subsequent localization using IgG antibody–gold or protein A–gold complexes. Nevertheless, strong fixation procedures with 2% glutaraldehyde and 1% osmium tetroxide have also been used successfully for the localization of enzymes (Bendayan and Zollinger, 1983; Rohde *et al.*, 1984). Sections of cells fixed with osmium tetroxide have to be treated with strong oxidizing agents, such as sodium metaperiodate or H_2O_2 (10%, 10–20 min), before immunolabelling.

2. Embedding in conventional resins

Immunocytochemistry is mostly performed on resin-embedded cells. Therefore bacterial cells have to be dehydrated in graded series of ethanol, acetone or methanol if epoxy resins like Epon, Spurr or araldite are used. Treatment of bacterial cells with such reagents can cause a significant loss of the antigenicity of the protein to be labelled. To avoid this we have taken advantage of the water-soluble glycol methacrylate (GMA) resin. When applying this resin the cells do not need to be dehydrated. Embedding is performed following the procedure of Leduc *et al.* (1963) and Leduc and Bernhard (1967). After fixation, samples are rinsed in ice-cold buffer and processed through different aqueous solutions of glycol methacrylate on ice: 20% GMA, 30 min; 40% GMA, 30 min; 60% GMA, 30 min; 80% GMA, 30 min; 97% GMA, 1 h; 1 part GMA and 1 part embedding mixture, 1 h; embedding mixture alone, 1 h, and overnight. The embedding mixture consists of seven parts of GMA and three parts of *N*-butyl methacrylate containing 2% benzoylperoxide and one part of divinyl benzene in 50% styrene. Polymerization is achieved in 3–5 days at 4°C using UV light (360 nm) or in 16–24 h at 60°C.

3. Embedding in low temperature resin Lowicryl K4M

Embedding in Lowicryl K4M resin is carried out according to Carlemalm *et al.* (1980), Roth *et al.* (1981) and Carlemalm *et al.* (1982). After fixation, the bacterial cells are rinsed with ice-cold buffer and processed as follows: 15% ethanol, 10 min, on ice; 30% ethanol, 30 min, on ice; 50% ethanol, 60 min, at −20°C; 70% ethanol, 60 min, at −35°C; 95% ethanol, 60 min, at −35°C; 100% ethanol, 2 h, at −35°C with several changes. Ethanol can be replaced by methanol. Further infiltration of samples is done with Lowicryl K4M resin (1 part) and 1 part 100% ethanol for 60 min and with 2 parts resin and 1 part 100% ethanol for another 60 min at −35°C. Subsequently samples are infiltrated with the resin alone overnight at −35°C. Polymerization is performed with UV light (360 nm) for 2–3 days at −35°C and for another 2 days at room temperature.

E. Treatment of the specific antibody

The main requirement for performing the pre-embedding and post-embedding labelling methods is the high specificity of the antiserum used. It is recommended that the antiserum is purified by affinity chromatography using, for example, protein A–sepharose CL-4B. The use of whole antisera should be avoided because of high background labelling. Purification of the antisera by affinity chromatography using protein A has the advantage that only intact antibodies or Fc-fragments are obtained in the resulting antibody solution. Thus, no Fab-fragments are present and, therefore, no masking of antigenic sites on the ultra-thin sections can occur without further binding of protein A–gold complexes because protein A will bind only to the Fc-portion of an antibody. On the other hand, IgG antibody–gold complexes will usually bind to the already bound Fab-fragments (see Fig. 11).

F. Pre-embedding labelling procedure

The main characteristic of this procedure is that the reaction between antibody and antigen is performed, in most cases, prior to fixation, dehydration and resin embedding of bacterial cells, protoplasts, vesicles or other cellular components. Unfortunately this method is restricted to the localization of such antigens, especially external antigens, to which the antibody has access. Antibody and electron-dense marker are not able, for example, to penetrate membranes. For preventing antigen migration it is recommended that a slight fixation is performed prior to the antigen–antibody reaction, otherwise false results might be obtained because of the movement of antigenic material in unfixed cells. Internal antigens can only be localized by first permeabilizing the cellular membrane with detergent to allow antibody and marker to penetrate this barrier. However, this treatment alters the ultrastructure of the cell to a great extent and, therefore, may lead to false labelling results.

As mentioned in the introduction, ferritin is useful as an electron-dense marker for immunocytochemical studies. An antibody is coupled to ferritin with highly reactive reagents like xylene-diisocyanate. During this reaction the reactivity of the antibody can be altered greatly. Additional loss of reactivity may result from the coupling of the ferritin molecule which is large with respect to the small antibody, thus masking the antigenic binding sites on the specific antibody (Morgan, 1972; Parr, 1979; Wientjes et al., 1980). To overcome these restrictions several authors (van Driel and Wicken, 1973; Wagner et al., 1980; Acker et al., 1981; Rohde et al., 1984) used an anti-antibody, in our case a goat-anti-rabbit antibody, which is coupled to the ferritin molecule. We use a commercially available antibody–ferritin complex (Miles-Yeda, Rehovot, Israel). These complexes were tested for reactivity with the specific polyclonal antibody using negative staining with uranyl acetate (4%, pH 4.5).

A labelling protocol is as follows: bacterial cells or cellular components are incubated with the specific antibody, given in excess, for 2 h at 30°C. After washing twice with a suitable buffer samples are incubated with the electron-dense marker, for example protein A–gold complex or ferritin coupled to a species-directed anti-antibody against the specific antibody, for 2 h at 30°C.

Again, unbound antibody is removed by several washing steps. Subsequently samples are embedded in 1.5% agar, fixed with glutaraldehyde and osmium tetroxide, dehydrated and embedded in resin. After polymerization of the resin ultra-thin sections are prepared and post-stained with uranyl acetate and lead citrate (Venable and Coggeshall, 1965) before being examined in the electron microscope.

Another version of a pre-embedding labelling procedure was developed to study external antigens on bacterial pili (Beesley *et al.*, 1984; Beesley and Betts, 1984). Small aliquots (0.5 μl) of the particle suspension to be examined are air-dried onto grids covered with a plastic film (Formvar, collodion). Then the grids are labelled by sequential incubation with the specific antibody and the marker probe. Finally, grids are negatively stained with a suitable negative stain like neutralized phosphotungstic acid or uranyl acetate. This method has several advantages. Firstly, it can be used on-grid, thus consuming only a minimal amount of the sample. Secondly, the label is easily detectable, and even low labelling levels can be identified. Thirdly, the method can be used for multiple labelling steps and for a quantitation of the labelling obtained.

G.   Post-embedding labelling procedure

The advantage of the post-embedding labelling method is that labelling of external and internal antigens is possible. This is achieved by sectioning of the resin-embedded cells. This procedure exposes the antigenic material at the surface of the section to the antibody present in an incubation mixture; the problem of impermeability of membranes, or diffusion problems like those encountered in the pre-embedding labelling methods, do not arise. Furthermore, this method allows the detection of more than one antigen on the same section by using, as well as more than one specific antibody, gold particles of different sizes.

Fixation of cells is done according to the pre-embedding labelling method and embedding is performed as described above. Ultra-thin sections, cut with glass knives, are mounted on uncovered or Formvar-coated nickel grids. It is recommended that the sections are incubated for 5 min at room temperature first on a drop of PBS containing 1% ovalbumin or 0.5% milk powder for masking non-specific binding sites, thus lowering the background labelling. Then the grids are transferred without further rinsing (only blotting for a few seconds on filter paper) to drops of dilution series of specific antibody in PBS. Alternatively, the soaking step in ovalbumin can be omitted. This should be tested in each case. The sections are incubated for various periods of time and at different temperatures. For example, incubation with the antibody can be performed overnight at 4°C or for 2 h at room temperature. However, incubation time should be as short as possible to avoid non-specific background labelling. Then the sections are rinsed in PBS to remove unbound antibody. Sometimes it can be useful to add 0.1% Tween 20 to the washing buffer. The washing step is done by a mild spray from a plastic bottle for 10–20 s and by floating the sections for 5 min each on drops of PBS. The sections can now be transferred to a drop of PBS containing, for example, 1% ovalbumin, for further masking of non-specific binding sites, or this step can be omitted. Finally, sections are incubated on drops of dilution series of the electron-dense marker at room temperature for 1–2 h. At the end of the incubation procedure sections are rinsed in PBS and distilled water before post-staining with uranyl acetate (4%, 3–5 min) and lead citrate (15–30 s) or with uranyl acetate alone (5–10 min) before examination in the electron microscope.

H. Immunocytochemical control experiments

The specificity of the immunocytochemical labelling procedures described has to be established by performing some of the following controls:

(a) Incubation of the sections with the electron-dense marker alone, omitting the incubation with the specific antibody. This control identifies the non-specific binding of the marker to the resin.

(b) Allowing the specific antibody to react with the corresponding antigen prior to the incubation of the sections, followed by the incubation with the marker verifying the specificity of the antigen–antibody reaction.

(c) Incubation of the sections with the specific antibody, followed by an incubation with IgG antibody or protein A alone, and then by the IgG antibody–gold and protein A–gold complexes; this control verifies the interaction between the specific antibody and the IgG antibody or the protein A.

(d) Incubation of the sections with the pre-immuno serum, purified the same way as the specific antibody, followed by the marker system; this reveals the non-specific binding of IgG antibody to the resin.

(e) Performing controls taking advantage of physiological conditions in which, for example, the enzyme to be localized is not expressed.

III. Examples of applications

A. Pre-embedding labelling

1. Membrane-attached CO dehydrogenase using antibody–ferritin complex

Carbon monoxide dehydrogenase (CO dehydrogenase) is the key enzyme in the CO metabolism of aerobic carboxydobacteria (Meyer and Schlegel, 1983;

Meyer and Rohde, 1984). After ultra-centrifugation of crude extracts of carboxydobacteria more than 50% of the total CO-oxidizing activity was found in the cytoplasmic fraction, suggesting that CO dehydrogenase is a soluble enzyme (Cypionka et al., 1980). The most important function of this enzyme is to feed electrons derived from the CO oxidation into the CO-insensitive branch of the respiratory chain for generation of ATP by electron transport phosphorylation (Cypionka and Meyer, 1983). Due to this fact, and the enzyme's inability to reduce soluble physiological electron acceptors, its location in the cytoplasm was difficult to understand. Therefore it was speculated that CO dehydrogenase might be attached to the cytoplasmic membrane.

For this purpose cells of *Pseudomonas carboxydovorans* were prefixed with 0·1% glutaraldehyde before passing the cells through a French pressure cell. Under these disruptive conditions fragments of the cytoplasmic membrane were predominantly formed; occasionally right side- and inside-out vesicles as well as protoplasts could be observed. In Fig. 1 a representative labelling pattern is depicted. Fragments of the cytoplasmic membrane and right-side-out vesicles are exclusively labelled at the cytoplasmic side of the membrane. Replacing the specific antibody by a non-specific one resulted in no detectable label at the cytoplasmic membrane (Fig. 1f). From the observation that ferritin label is located on the inner side of right-side-out vesicles (Fig. 1c), and on the outside of inside-out vesicles (Fig. 1d), and from the absence of label on protoplasts (Fig. 1e), it was concluded that CO dehydrogenase is attached to the inner aspect of the cytoplasmic membrane and that the enzyme is neither a periplasmic nor transmembraneous protein.

2. Immunonegative staining of membrane-attached methyl-CoM reductase and membrane-bound ATPase using antibody–gold complex

Methanogenic bacteria are able to utilize H_2 and CO_2, formiate, methanol, methylamines or acetate under anaerobic conditions (Balch et al., 1979). Conversion of these compounds leads to a common intermediate, methyl-coenzyme M (2-methylthio)ethanesulphonic acid, which is converted by methyl-CoM reductase to methane and coenzyme M (Wolfe, 1985). Recently it was demonstrated that this reaction generates a proton-motive force across the cytoplasmic membrane. This gradient is subsequently used for ATP synthesis (Blaut and Gottschalk, 1985). The presence of ATPases has been shown in a wide variety of methanogenic bacteria (Doddema et al., 1978; Daniels et al., 1984; Inatomi, 1986). Therefore it was postulated that methyl-CoM reductase might be attached to the cytoplasmic membrane, despite the fact that the enzyme was found in the soluble fraction after ultra-centrifugation of disrupted cells (Ellefson and Wolfe, 1981).

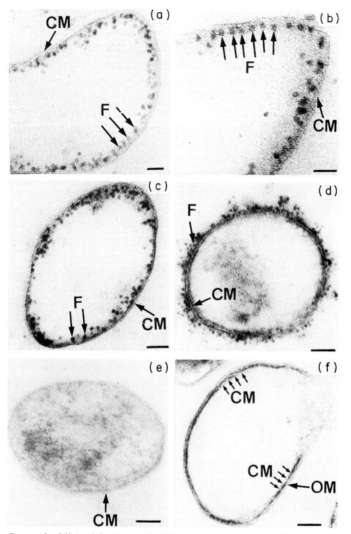

Fig. 1. Pre-embedding labelling of CO dehydrogenase. Cells of *Pseudomonas carboxydovorans* were prefixed with glutaraldehyde, passed through a French pressure cell, incubated with polyclonal antibody against CO dehydrogenase and incubated with antibody–ferritin complexes before embedding in resin. Bars represent 100 nm. (a) and (b) Cytoplasmic membrane fragments show ferritin (F) molecules, indicating the presence of CO dehydrogenase exclusively on the inner side of the cytoplasmic membrane (CM). (c) and (d) Dense labelling of the cytoplasmic side of the membrane of right-side-out (C) and inside-out (D) vesicles. (e) No labelling is detectable at the outer side of a protoplast. (f) Replacing the specific antiserum by a non-specific antiserum (normal goat IgG serum) resulted in a lack of label at the cytoplasmic membrane. OM, outer membrane.

We have investigated the localization of methyl-CoM reductase and ATPase (Mayer et al., 1987) by applying immunonegative staining, a version of the pre-embedding labelling method. We have taken advantage of a newly isolated methylotrophic bacterium, strain Gö 1, from which protoplasts and inside-out vesicles can be obtained easily. Studies have shown that using the same procedure as described for CO dehydrogenase resulted in an almost complete absence of labelling. This might have been caused by the many washing steps which are involved in this procedure, or by insufficient fixation. Therefore vesicles, mostly inside-out vesicles, were air-dried onto Formvar grids, incubated with specific polyclonal antiserum against the component C of the methyl-CoM reductase system from Methanosarcina barkeri, and subsequently incubated with antibody–gold complexes (gold particle size 15 nm). Figures 2a and 2b depict the labelling of methyl-CoM reductase on inside-out vesicles, indicative of a localization of this enzyme at the inner aspect of the cytoplasmic membrane. Furthermore, we demonstrated that methyl-CoM reductase might be very loosely attached to the membrane. Figure 2b depicts the result of a labelling experiment in which washed inside-out vesicles were used, incubated with the same concentrations of specific antiserum and antibody–gold complexes as for Fig. 2a. It is obvious that a much lower label intensity is detectable, indicating only a weak interaction of the enzyme with the cytoplasmic side of the membrane.

For labelling of the ATPase we used the same organism and the same procedure. As the enzyme has not been purified until now we incubated inside-out vesicles with polyclonal antiserum directed against the β-subunit of the F1-ATPase particle of Escherichia coli. The cross-reactivity with vesicles of Gö 1 was demonstrated by Western blotting. Figure 2c reveals that a dense population of gold particles is attached to the surface of inside-out vesicles. Figure 2d depicts a control experiment in which pre-immuno serum was taken instead of the specific antiserum, followed by the antibody–gold complexes. Only rarely are gold particles seen, indicating the specificity of the labelling obtained. A quantitative evaluation of the localization results of methyl-CoM reductase and ATPase revealed that methyl-CoM reductase covers approximately 7% of the inner cytoplasmic membrane surface, whereas ATPase molecules cover approximately 15% of the inner surface of the membrane. For this evaluation the number of bound gold particles per unit area of the vesicles was compared to the total number of particles per unit area seen in freeze-fracture replicas of the vesicles.

Our immunocytochemical studies provided evidence that the energy metabolism system in methanogenic bacteria might operate as follows: electrons flow from an electron donor to the membrane-associated methyl-CoM reductase system, at which the reduction of methyl-CoM takes place. This results in a proton translocation and subsequently in a proton gradient

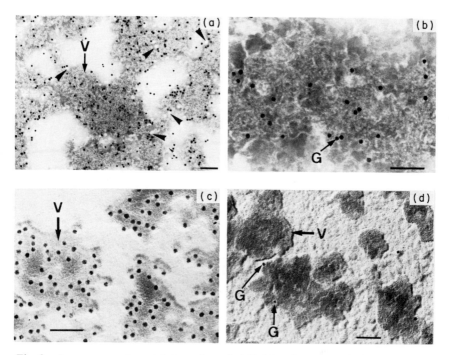

Fig. 2. Immunonegative staining of methyl-CoM reductase and ATPase of the methanogenic bacterium strain Gö 1. Protoplasts of the bacterium strain Gö 1 were produced by pronase treatment; they were passed through a French pressure cell giving rise to (mainly) inside-out vesicles. Vesicles were air-dried onto Formvar-covered grids, incubated with polyclonal antiserum specific against the component C of the methyl-CoM reductase system of *Methanosarcina barkeri*, followed by incubation with antibody–gold complexes (gold particle size 15 nm). Labelling of ATPase was done by incubation of the air-dried vesicles with polyclonal antiserum directed against the β-subunit of the F1-part of *E. coli* ATPase, followed by incubation with antibody gold complexes. Bars represent 100 nm.

(a) and (b) Inside-out vesicles (V) of strain Gö 1 are densely labelled with gold particles (G), indicating the presence of methyl-CoM reductase at the inner side of the cytoplasmic membrane. (b) represents the labelling pattern after washing the vesicles twice with buffer. A much lower degree of labelling is detectable, giving some evidence for a weak interaction of the enzyme with the membrane. Arrow heads mark methyl-CoM reductase complexes with associated label. (c) and (d) Labelling of ATPase on inside-out vesicles of strain Gö 1. Compared to the label in (a) a more intense label is detectable. Samples were unidirectionally metal-shadowed with Pt/Ir. (d) depicts a control experiment; incubation with the specific antiserum was omitted.

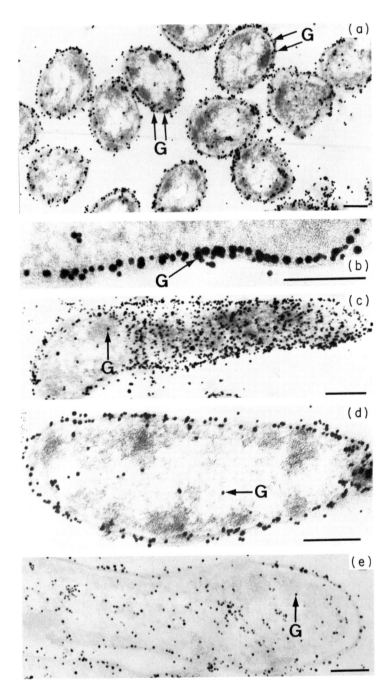

across the cytoplasmic membrane; this gradient is used for the generation of ATP.

B. Post-embedding labelling using protein A–gold complex and conventional embedding resin

1. Membrane-attached CO dehydrogenase

As described in an earlier section, we have demonstrated, using pre-embedding labelling, that CO dehydrogenase is located at the inner aspect of the cytoplasmic membrane. This method is restricted to antigens exposed at surfaces. It is not suitable for the localization of an antigen in the cytoplasm of a cell; the quantitation of the distribution of an enzyme in the cellular compartments is not possible. However, we were able to overcome this restriction by using the post-embedding method in which the antigen–antibody reaction is carried out on ultra-thin sections of cells (Rohde et al., 1984). For this purpose preservation of the protein antigenicity and ultrastructure of the cell are prerequisites for a sensitive and accurate label.

In our hands fixation with 0.3% formaldehyde and 0.2% glutaraldehyde or with 0.5% glutaraldehyde alone followed by osmium tetroxide (1%), and embedding in glycol methacrylate was a suitable procedure. After using osmium tetroxide as a fixative the sections have to be incubated with strong oxidizing reagents like sodium metaperiodate or H_2O_2 (10%, 10–20 min) to make the antigenic sites accessible to the antibody (Bendayan and Zollinger, 1983).

After treatment of sections with the polyclonal antiserum and the protein

---

Fig. 3. Localization of CO dehydrogenase using a conventional embedding resin. CO-grown cells of *Pseudomonas carboxydovorans* harvested from the exponential or the stationary growth phases were fixed with formaldehyde (0.3%) and glutaraldehyde (0.2%) and osmium tetroxide (1%) and embedded in glycol methacrylate. Ultra-thin sections were pretreated with sodium metaperiodate (10%, 10–20 min), incubated with polyclonal antiserum and protein A–gold complexes (gold particle size 3–5 nm). Post-staining of the sections was done with uranyl acetate for 5 min. Bars represent 100 nm.

(a)–(d) Cells from the exponential growth phase. Arrows marked G refer to bound gold particles, indicating the localization of CO dehydrogenase mainly in the area of the cytoplasmic membrane of the cells. (c) depicts a section through a cell made in such a way that a major part of the inner aspect of the cytoplasmic membrane is exposed. The bound gold particles in (b) represent the best resolution and the highest label intensity which could be obtained using 4 nm protein A–gold particles. (e) This cell from the stationary growth phase exhibits a lower degree of labelling in the cytoplasmic membrane area as compared to the extent seen in cells from the exponential growth phase; however, a much higher degree of labelling is observable in the cytoplasm.

A–gold complexes of CO-grown cells of *Pseudomonas carboxydovorans* a labelling pattern as shown in Fig. 3 was obtained. Again, most of the label is located near the cytoplasmic membrane, but label is also detectable in the cytoplasm. It should be mentioned that using the post-embedding labelling method applied as the only labelling technique did not allow the conclusion that CO dehydrogenase is located at the inner aspect of the membrane to be drawn. This conclusion could only be reached by comparison with the pre-embedding labelling results described earlier. Figures 3a–d depict the labelling observed with cells of the exponential growth phase, whereas Fig. 3e shows the distribution of gold particles, indicative of the presence of CO dehydrogenase, using cells of the stationary growth phase. A quantitative comparison of these two labelling patterns reveals that cells taken from the exponential growth phase show 87% of CO dehydrogenase to be associated with the cytoplasmic membrane, whereas 13% of the enzyme is located in the cytoplasm. The percentage of CO dehydrogenase attached to the membrane decreased from 87% to 50% in cells of the stationary growth phase, whereas the amount of enzyme in the cytoplasm increased from 13% to 50%. This example may demonstrate the usefulness of this method for a quantitative evaluation of the intracellular distribution of enzymes, in this case CO dehydrogenase, in different growth phases, thus giving evidence for a dynamic process of attachment/detachment of CO dehydrogenase to the cytoplasmic membrane. By comparing biochemical data, in particular the respiratory rate, our findings show that the electron flow from CO to O_2 in cells of *P. carboxydovorans* is controlled by the amount of CO dehydrogenase attached to the cytoplasmic membrane (Rohde et al., 1985). However, one should keep in mind that this labelling procedure does not discriminate between active and inactive states of an enzyme as long as the structures of the antigenic sites are preserved. Therefore we could not prove that the enzyme is in its active configuration at the cytoplasmic membrane and in its inactive state in the cytoplasm.

C. Post-embedding labelling using protein A–gold complex and low temperature embedding resin

1. Soluble and membrane-bound hydrogenase

This section aims to demonstrate some major methodical aspects of the protein A–gold method.

Alcaligenes eutrophus is an obligate aerobic, facultative chemolithoauto-trophic bacterium which expresses a soluble and a membrane-bound hydrogenase during autotrophic (H_2 and CO_2) and heterotrophic (fructose, glycerol) growth (Schneider and Schlegel, 1976; Schink and Schneider, 1979). A substrate shift from fructose to glycerol causes new synthesis of soluble and membrane-bound hydrogenase. This indicates that a derepression mechanism for the regulation of both hydrogenases is involved (Friedrich et al., 1981; Friedrich, 1982). We have applied the post-embedding labelling method using the low temperature embedding resin Lowicryl K4M and protein A–gold complexes in order to detect the localization of the soluble hydrogenase (Rohde et al., 1986) and membrane-bound hydrogenase in cells of A. eutrophus after a substrate shift, and to evaluate the amount of the enzyme under different growth conditions.

For a determination of the relative amounts of the hydrogenases in cells of A. eutrophus, the labelling procedure has to be standardized. First of all, a titration series of the antiserum and the protein A–gold complexes has to be carried out. It should be stated here that this is the method of choice for the determination of the dilution step which results in a specific labelling without a significant background labelling.

Figure 4 illustrates the results obtained for the soluble hydrogenase. It is obvious that at a 1:50 dilution of the antibody stock solution (Fig. 4c) more label is detectable compared to the 1:100 (Fig. 4a) and 1:75 (Fig. 4b) dilutions. After incubation of the sections with a higher amount of antibody (Fig. 4d and e) approximately the same density of gold particles occurred as in the case for the 1:50 dilution step. This proves that nearly all antigenic sites of the soluble hydrogenase had been labelled by the amount of antibody present in the 1:50 dilution. The determination of the optimal degree of dilution, i.e. that dilution where additional antibody does not result in a higher degree of labelling, is a prerequisite for the quantitative evaluation of an enzyme. Application of this determined amount of antibody and protein A–gold or antibody–gold complexes to sections of cells grown under different conditions will allow a comparison of the amounts of enzyme present under those different growth conditions to be made.

Figure 4 also illustrates that soluble hydrogenase is located in the cytoplasm of the cell. The DNA region, as well as the polyhydroxybutyric acid granules (PHB), are almost devoid of label. A comparison of Fig. 3 and Fig. 4 reveals that cells of A. eutrophus embedded in Lowicryl K4M show an excellent preservation of the cellular ultrastructure, especially the DNA region. On the other hand, cells of P. carboxydovorans embedded in glycol methacrylate (Fig. 3) exhibit only a poor preservation of the ultrastructure, demonstrating that low temperature embedding with Lowicryl K4M is in most cases superior to other resins with respect to the preservation of the cellular ultrastructure. However, the antigenicity of the two enzymes, soluble hydrogenase and CO dehydrogenase, is retained with both embedding procedures. Figure 5 depicts

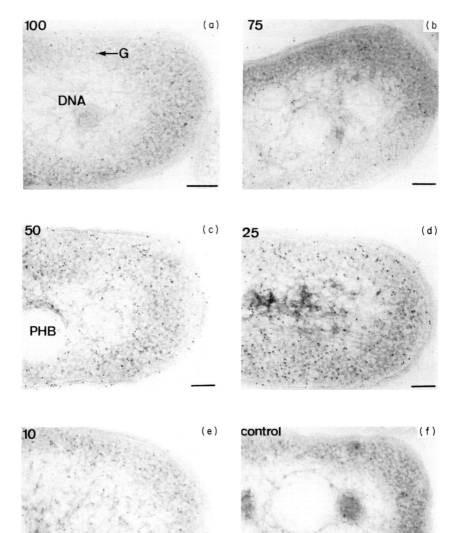

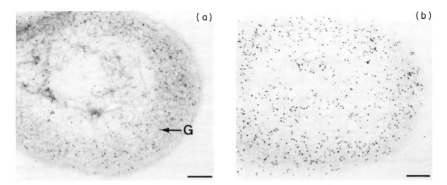

Fig. 5. Localization of soluble hydrogenase in heterotrophically and autotrophically grown cells of *Alcaligenes eutrophus*. Cells were treated as described in Fig. 4. The sections were incubated with the same amount of antibody (300 μg ml^{-1}) and the same concentration of the protein A–gold complexes (gold particle size 5 nm). Post-staining was done with uranyl acetate. Bars represent 100 nm.

(a) A heterotrophically grown cell of the late exponential growth phase shows gold particles (G), indicative for the localization of soluble hydrogenase, mainly in the cytoplasm. (b) An autotrophically grown cell is depicted in this panel. A much higher degree of labelling is evident here than in Fig. 5a.

cells of *A. eutrophus* which were grown autotrophically (Fig. 5b) or heterotrophically (Fig. 5a). Both samples were taken at the end of the exponential growth phase. A comparison of these two cells reveals that autotrophically grown cells contain a higher amount of soluble hydrogenase than heterotrophically grown cells. These immunocytochemical data support earlier biochemical investigations by Friedrich *et al.* (1981). Furthermore, we were able to demonstrate that in heterotrophically grown cells (fructose) the amount of soluble hydrogenase is very similar in all growth phases, whereas in

Fig. 4. Titration series of polyclonal antiserum directed against the soluble hydrogenase of *Alcaligenes eutrophus*. Heterotrophically grown cells (fructose) of *A. eutrophus* were fixed with 0.3% formaldehyde and 0.2% glutaraldehyde and embedded in the low temperature embedding resin Lowicryl K4M. Ultra-thin sections were treated with different dilutions of the polyclonal antiserum (stock solution contains 2.7 mg IgG ml^{-1}). Then the sections were incubated with the same number of protein A–gold complexes (gold particle size 5 nm) and post-stained with uranyl acetate (4%, 5 min). The dilution steps of the polyclonal antiserum are indicated in the panels. Bars represent 100 nm.

(a)–(c) A comparison of these three panels reveals that an increase in the degree of labelling is achieved after using higher amounts of the polyclonal antiserum. (d) and (e) In the range indicated, higher amounts of antiserum do not result in higher labelling intensity. (f) Control experiment: the incubation step with the polyclonal antibody was omitted. Only a few gold particles are detectable, demonstrating the specificity of the labelling obtained in the labelling experiments.

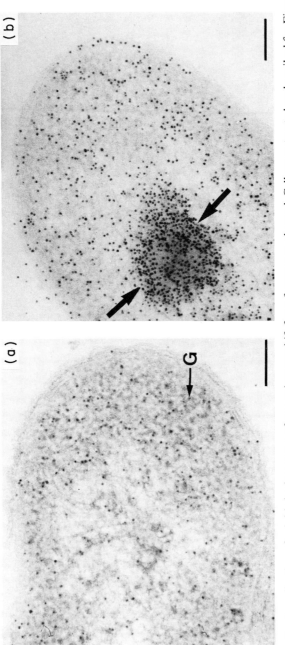

Fig. 6. Localization of soluble hydrogenase after a substrate shift from fructose to glycerol. Cells were treated as described for Figs 4 and 5. Bars represent 100 nm.

(a) A heterotrophically grown cell of the late exponential growth phase. (b) In this micrograph a cell is shown 5 h after a substrate shift from fructose to glycerol was performed. Compared to (a) an enormous increase in labelling is detectable. A dense aggregate containing soluble hydrogenase in high local concentration is marked with arrows.

autotrophically grown cells a decrease in the degree of labelling occurs in the stationary growth phase.

After a substrate shift from fructose to glycerol, the high amount of newly synthesized soluble hydrogenase in the cytoplasm is obvious (Fig. 6b compared to Fig. 6a), demonstrating the derepression mechanism of the soluble hydrogenase. Figure 6b illustrates that the amount of label after the substrate shift experiment nearly equals that found in autotrophically grown cells (Fig. 5b). Furthermore, newly synthesized hydrogenase is also located to a great extent in inclusion bodies (marked with arrows in Fig. 6b). This incorporation of overproduced enzymes in inclusion bodies is also seen in genetically manipulated bacteria which express an enzyme in high amounts (Williams et al., 1982).

Figure 7 illustrates the rate of formation of membrane-bound hydrogenase during heterotrophic growth on fructose and glycerol. Our immuno-cytochemical labelling data indicate that the enzyme level increases slowly during growth on fructose between the early and the late exponential growth phase (compare Figs 7b and d). The bound gold particles, indicative of the localization of the membrane-bound hydrogenase, reveal that an increase is detectable mainly at the cell periphery. On the other hand, glycerol-shifted cells show also an increase of the enzyme in the cytoplasm (Fig. 7g). This might be due to the fact that the possible binding sites for the enzyme in/at the membrane are covered by enzyme molecules that are already bound; thus, the newly synthesized protein accumulates in the cytoplasm. The few gold particles in the DNA region of the cell (Fig. 7g) could be interpreted as newly synthesized protein molecules. These observations give evidence for a derepression mechanism, in this case of the regulation of the membrane-bound hydrogenase, supporting earlier biochemical investigations (Friedrich et al., 1981). This coincides with the findings for the soluble hydrogenase as mentioned before (Friedrich et al., 1986; Rohde et al., 1986). Figure 7e and f illustrates one of the main problems involved in post-embedding labelling studies. By applying only this method it is difficult to identify whether an enzyme is membrane-attached, partially membrane-integrated, or a trans-membraneous protein. As Figs 11 and 12 illustrate, the antibody–marker complex may be found a certain distance away from the antigen. During the labelling procedure the antibody–marker complex is bound to the antigen in a three-dimensional manner because the sections are floating on drops of the corresponding antibody and marker solutions. This changes during air drying of the sections. A two-dimensional complex is formed. The antibody–marker complexes will then tip over, resulting in their attachment to the surface of the section. Accordingly, the antigen–antibody binding site is in a certain area around the gold particle; the diameter of this area depends on the size of the applied marker probe.

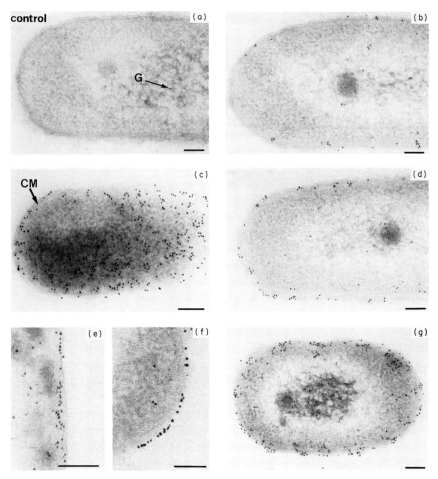

Fig. 7. Localization of membrane-bound hydrogenase in *Alcaligenes eutrophus.*
Cells were grown heterotrophically on fructose or glycerol and treated as described in
Fig. 4, except that the sections were incubated with non-diluted antiserum (2.6 mg IgG
ml^{-1}). Bars represent 100 nm.
  (a) Control experiment: the incubation step with the polyclonal antiserum was
omitted. G, gold particles. (b) Cell grown heterotrophically on fructose; harvested from
the early exponential growth phase. (c) and (d) Heterotrophically grown cells of the late
exponential growth phase. A higher labelling intensity in the near vicinity of the
cytoplasmic membrane is detectable. Sectioning carried out close to the cell surface (c);
in this section the cytoplasmic membrane area of the cell is exposed at the right side,
and parts of the cytoplasm at the left side. The labelling pattern obtained indicates that
the enzyme is located at the cytoplasmic membrane. CM, cytoplasmic membrane.
(e)–(g) Glycerol-shifted cells. (e) and (f) demonstrate the width of the area at the cell
periphery in which antibody–marker complexes can be located, assuming that their
antigen-binding sites are in the cytoplasmic membrane proper.

2. EcoRI restriction endonuclease and methylase

Restriction enzyme systems are assumed to be mechanisms for protecting the bacterial cell against invading foreign DNA. A cell containing such an enzyme has to protect its own DNA against self-destruction by modification of the newly synthesized DNA with methylases. EcoRI restriction endonuclease and methylase have been characterized with respect to both structure and function (Rubin and Modrich, 1977; Modrich and Roberts, 1982; Johannssen et al., 1984; Pingoud et al., 1984). As a type II restriction enzyme system EcoRI has two different protein entities, one with endonuclease and one with methylase activity (Collins and Mayer, 1980). To investigate the in situ localization of both proteins in Escherichia coli Bs 5 the protein A–gold method was applied. It was carried out as post-embedding labelling, or immunonegative staining with protoplasts of E. coli was used (Kohring et al., 1985; Kohring and Mayer, 1987). Figure 8a clearly indicates the localization of most endonuclease molecules at the cell periphery. The enzyme could be found on the surface of

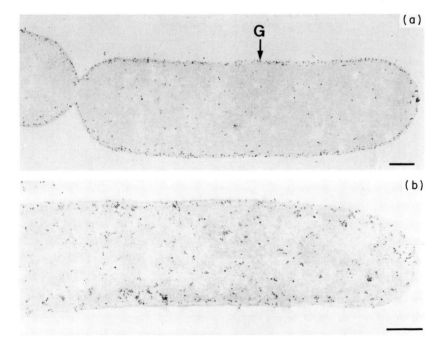

Fig. 8. Localization of EcoRI restriction endonuclease and methlyase in *Escherichia coli* Bs 5. Cells were fixed with 0.3% formaldehyde and 0.2% glutaraldehyde and embedded in Lowicryl K4M. Ultra-thin sections were treated with polyclonal antisera and protein A–gold complexes. Bars represent 100 nm. (a) EcoRI restriction endonuclease is mainly located at the cell periphery, indicated by the bound gold particles (G). (b) EcoRI methylase activity is located to a high degree in the cytoplasm.

protoplasts, but not on the surface of untreated whole *E. coli* cells (Kohring *et al.*, 1985). When the post-embedding labelling was done on ultra-thin sections of slightly plasmolysed cells, a large amount of label was detected in the periplasmic space. These results coincide with the finding that the major part of the enzyme activity was measured in the soluble fraction after generation of protoplasts by lysozyme treatment and mild osmotic shock. A quantitative evaluation of the post-embedding labelling obtained revealed 70–90% of endonuclease protein at the cell periphery and 10–30% in the cytoplasm.

The labelling experiments for the localization of EcoRI methylase (Fig. 8b) showed 60–70% of the enzyme in the cytoplasm and 30–40% at the cell periphery. Because most of the gold particles near the cell envelope were found inside the cytoplasmic membrane, the amount determined in the cytoplasm represents a minimum value.

3. Methyl-CoM reductase

As mentioned in an earlier section, methane synthesis as the final reaction of methanogenesis is catalysed by methyl-CoM reductase. For *in vitro* enzyme activity a complex system of four protein components (A1, A2, A3, C) and various cofactors, including component B, FAD and ATP, are required (Wolfe, 1985). Component C has been demonstrated to be the methyl-CoM reductase proper. It consists of three different types of subunits in an $\alpha_2\beta_2\gamma_2$ composition (Ellefson and Wolfe, 1980, 1981). Two different models have been suggested for the mechanism of energy conservation in methanogenic bacteria: (a) It was demonstrated for *Methanosarcina barkeri* that ATP synthesis occurs according to the chemiosmotic principles (Blaut and Gottschalk, 1984, 1985). Therefore it was postulated that component C might be associated with the cytoplasmic membrane. (b) On the other hand, uncoupler experiments provided evidence that ATP synthesis is independent of an electrochemical proton gradient in *Methanobacterium thermoautotrophicum* (Schönheit and Beimborn, 1985) and *Methanococcus voltae* (Grider *et al.*, 1985). These results are consistent with a model which proposes that the methanogenic enzymes are soluble, and that an energy-rich intermediate is involved during ATP synthesis (Lancaster, 1986). We have applied the post-embedding labelling method to determine whether component C in different methanogenic bacteria is a cytoplasmic enzyme, or whether it might be associated with the cytoplasmic membrane (Ossmer *et al.*, 1986).

A comparison of Fig. 9a and b reveals that in *M. voltae* the bound gold particles show the location of methyl-CoM reductase at the cell periphery, whereas the label in cells of *M. thermoautotrophicum* is randomly distributed all over the cell. A quantitative evaluation reveals that 75% of the bound gold particles are located in the cytoplasm and 25% at the cell periphery. In rare cases (Fig. 9c) the gold particles were found to be predominantly located at the cell envelope. Furthermore, the labelling pattern in Fig. 9b was independent of the growth phase. In addition, the distribution of methyl-CoM reductase in cells of *M. thermoautotrophicum* did not change by application of different fixation conditions (anaerobic at room temperature, or in the growth medium at 65°C).

These immunocytochemical data fit in with the concept that methyl-CoM reductase may be involved in the generation of an electrochemical proton gradient across the cytoplasmic membrane in *M. voltae*. In another immunocytochemical study (Aldrich *et al.*, 1987) the randomly distributed label in cells of *M. thermoautotrophicum* is interpreted as being due to an overproduction of component C in growth medium with an excessive supply of nickel. Thus, it should be noted that the growth conditions can have an influence on both the amount and the location of an enzyme in the cell.

D. Post-embedding labelling using antibody–gold complex and low temperature embedding resin

1. Hydroxyglutaryl-CoA dehydratase and glutaconyl-CoA decarboxylase

In this section some additional important methodical aspects of the post-embedding labelling procedure will be described. For this purpose we have chosen a soluble enzyme, hydroxyglutaryl-CoA dehydratase, and a membrane-bound enzyme, glutaconyl-CoA decarboxylase. These two enzymes were localized in cells of *Acidaminococcus fermentans* using antibody–gold complexes with different gold sizes. Ultra-thin sections were incubated with the same amount of antibody and nearly the same number of gold particles as measured spectrophotometrically at 520 nm. The bound gold particles prove that hydroxyglutaryl-CoA dehydratase is located in the cytoplasm, being a soluble enzyme, whereas glutaconyl-CoA decarboxylase is located mainly in the near vicinity of the cytoplasmic membrane or at the cytoplasmic membrane (Fig. 10d). The degree of labelling obtained (Fig. 10) clearly indicates that a higher label intensity can be achieved with smaller antibody–gold complexes (5 nm, Fig. 10b) than with larger (15 nm) gold particles (Figs 10a and c), and that better resolution can be obtained using smaller gold particles (compare Figs 10c and d; in Fig. 10d protein A–gold complexes were used).

A quantitative evaluation of the amount of these two enzymes reveals that a much higher number of enzyme molecules is detected in the cell using smaller gold particles compared to larger gold particles. Figure 11 might give an explanation for these different results. Applying small (3 nm) protein A–gold or antibody–gold complexes nearly every bound polyclonal antibody is

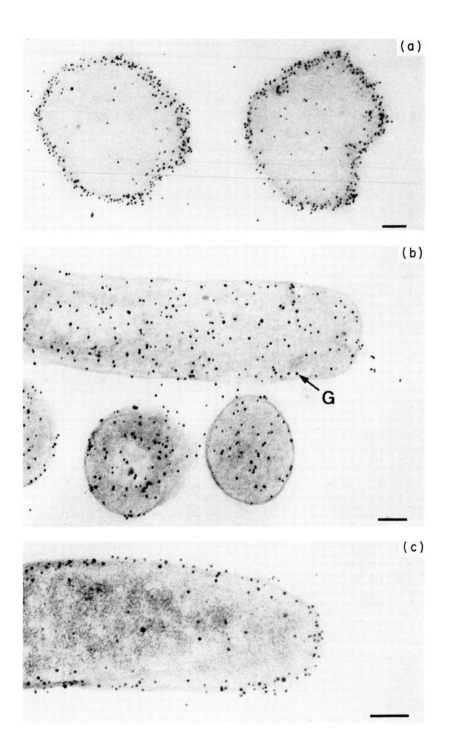

(a)

(b)

G

(c)

complexed with a marker; thus, in most cases, a one-to-one ratio of marker to antigen will be the result because these small gold particles mostly bind one marker molecule only (Romano and Romano, 1977). This situation changes drastically if larger gold particles are applied because these gold particles are able to bind more than one marker molecule (Geoghegan and Ackerman, 1977; Roth *et al.*, 1978; Roth, 1982). Figure 11 shows, for example, that 5, 3 or 2 bound polyclonal antibodies are visualized by only one marker using 15 nm gold particles. In addition, other bound polyclonal antibodies are not detected by the marker probe due to steric hindrance.

This leads to the general conclusion that a quantitation of the amount of an enzyme in a cell only reflects the number of enzyme molecules which are visualized, depending on the size of the marker probe applied. This may not be the actual number of enzyme molecules in the cell. As mentioned above, the application of 2–3 nm marker probes might result in a one-to-one ratio of marker to antigen, but even then the effect of steric hindrance between markers in close apposition is uncertain.

An interpretation of Fig. 10c (indicative of the localization of glutaconyl-CoA decarboxylase) might lead to the conclusion that this enzyme is not randomly distributed in the cytoplasmic membrane, but could be concentrated in certain areas. Incubation of the sections with smaller gold particles proves that glutaconyl-CoA decarboxylase is in fact located in the near vicinity of the cytoplasmic membrane and all around the entire cell.

As mentioned in an earlier section, it is difficult to conclude, merely from post-embedding labelling studies, whether an enzyme is located at the cytoplasmic membrane, in the periplasmic space or at the outer membrane. Theoretically, the best resolution which can be achieved is about 15 nm. This is based on the dimension of the IgG antibody (around 10 nm), the protein A molecule (2 nm) and the applied gold particle size (Fig. 12). The maximum span

---

Fig. 9. Localization of methyl-CoM reductase in *Methanococcus voltae* and *Methanobacterium thermoautotrophicum*. The bacteria were grown in complex medium, fixed with 0.3% formaldehyde and 0.2% glutaraldehyde at 37°C (a), or at room temperature (b and c), and embedded in Lowicryl K4M. Ultra-thin sections were incubated either with a mixture consisting of polyclonal antisera raised against the α-subunit, β-subunit and the complete component C of the methyl-CoM reductase system (a), or with polyclonal antiserum against purified complete component C (b and c), followed by incubation with protein A–gold complexes (gold particle size between 4 and 6 nm). Post-staining was done with lead citrate (10%, 4 min) and uranyl acetate (4%, 5 min). Bars represent 100 nm.

(a) Cells of *M. voltae*. The position of the bound gold particles indicates that the enzyme is located in the near vicinity of the cytoplasmic membrane. (b) and (c) Cells of *M. thermoautotrophicum* exhibit a random distribution of the gold markers (G) all over the cell. Only in rare cases was a labelling pattern achieved as for the cells of *M. voltae* (c).

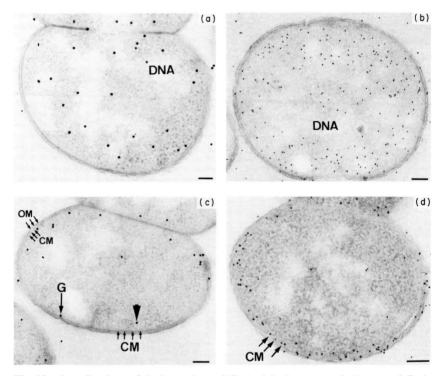

Fig. 10. Localization of hydroxyglutaryl-CoA dehydratase and glutaconyl-CoA decarboxylase using antibody–gold complexes with different sizes. Cells of *Acidaminococcus fermentans* were fixed and embedded as described in Fig. 4. Ultra-thin sections were treated with the same amount of antibody (300 µg IgG ml^{-1} for hydroxyglutaryl-CoA dehydratase and 250 µg IgG ml^{-1} for glutaconyl-CoA decarboxylase) and nearly the same number of antibody–gold complexes per ml. Bars represent 100 nm.

(a) Localization of hydroxyglutaryl-CoA dehydratase using 15-nm antibody–gold complexes. The labelling reveals that this enzyme is located exclusively in the cytoplasm. (b) After incubation of the sections with the same amount of antiserum as in (a) and 5-nm antibody–gold complexes, a higher degree of label is obvious. The DNA region of the cell is devoid of label, indicating the specificity of the labelling obtained. (c) Localization of glutaconyl-CoA decarboxylase using 15-nm antibody–gold complexes. Most of the gold particles (G) can be detected in the near vicinity of the cytoplasmic membrane (CM) or bound to the membrane, indicating that this sodium transport enzyme is located in or at the membrane. (d) Higher resolution is achieved by replacing the 15-nm antibody–gold complexes (c) by 5-nm protein A–gold complexes. Again, denser labelling is detectable using this smaller marker probe. In contrast to the 15-nm gold, the cell is labelled at or near the membrane all around the cell periphery.

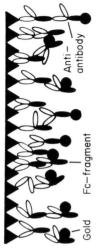

Fig. 11. Scheme illustrating the principles of the protein A–gold and antibody–gold method. (The situation before air drying of the sections is depicted.)

Protein A–gold complex, 3 nm

| 15 nm |

Antibody–gold complex, 15 nm

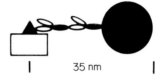

| 35 nm |

Fig. 12. Comparison of the achievable resolutions using protein A–gold complex (gold particle size 3 nm) and antibody–gold complex (gold particle size 15 nm).

between the antigen and the marker which uses a specific antibody and an anti-antibody coupled to 15-nm gold particles is about 35 nm (Fig. 12). Therefore the gold particle marked with an arrow head in Fig. 10c might indicate a glutaconyl-CoA decarboxylase molecule which is located at the cytoplasmic membrane. Higher resolution could be obtained by replacing the antibody–gold complex (15 nm) by the protein A–gold complex (3 nm); thus, the maximum span between antigen and gold particle is approximately 15 nm. Using this system, gold particles can be detected which appear to be directly bound to the cytoplasmic membrane (areas marked with arrows in Fig. 10d), proving the better resolution of this marker probe.

From these observations a general rule can be deduced: larger gold particles allow for an overall evaluation of the labelling, especially when antigens with a high copy number are to be localized. For proteins present in small amounts, especially membrane proteins, small gold particles should be used. For localization studies which require a high sensitivity and spatial resolution, the marker probe of choice is the protein A–gold complex applying 2–4 nm gold particles.

IV.   Concluding remarks and technical problems

The pre-embedding labelling procedure and the post-embedding labelling method are useful means for the electron microscopic localization of bacterial antigens. The post-embedding procedure is a particularly reliable method for the quantitation of the amount of antigen in a cell. Another advantage is that the antibody has direct access to the antigen exposed by sectioning of the cells; therefore this method does not require permeabilization of membranes and does not encounter problems of restricted diffusion in the cytoplasm of a cell, as does the pre-embedding labelling procedure. As with all other immuno-cytochemical techniques, one is confronted with the problem of the preservation of the antigenicity together with a good preservation of the cellular ultrastructure. However, we have demonstrated in this chapter that combinations of different fixation conditions and embedding in different resins allows reliable results to be obtained. The use of low temperature embedding resins preserves excellently the ultrastructure of the cell and the antigenicity of a protein. Furthermore, these resins, together with meth-acrylate resins, exhibit only a very low non-specific binding tendency of the IgG antibody or the different gold complexes to the resin.

The protein A–gold method provides a highly specific and sensitive labelling. Also antiserum raised in a wide variety of mammals can be used. In addition, the preparation of the colloidal gold and the protein A–gold complexes is a simple procedure. The non-covalent nature of the binding of proteins to colloidal gold, without diminishing the activity of the bound proteins, allows for the use of different protein–gold complexes (antibody–gold, avidin–gold, enzyme–gold, protein A–gold) for detecting an antigen or protein in bacterial cells. Depending on what question has to be answered, the resolution, sensitivity and intensity of the labelling can be altered by using different sizes of gold particles. Furthermore, the gold particles are easily detectable on ultra-thin sections and therefore allow a quantitative estimation of the immunolabel. Thus the amount of an enzyme in the cell can be estimated and compared to the amount of this enzyme present, for example, under different growth conditions. Due to the particulate character of the colloidal gold a labelling protocol can be established for a simultaneous localization of two antigens on the same section using different sizes of marker probes.

Nevertheless, the described procedures suffer from some limitations and technical problems. First of all, the procedures are carried out as post-embedding labelling; the label is restricted only to the surface of sections, and thus only the antigens exposed by the sectioning process can be detected. This can cause problems, especially when antigens present in low amounts, like most of the membrane-bound enzymes, are to be labelled. Sometimes it can be an advantage to etch the sections, for example with sodium methylate, prior to the labelling procedure. Thus, more structural components are exposed than were exposed by cutting the cells without further treatment. As a precondition, the high specificity of the antibody to the antigen has to be established by immunochemical methods because the method relies completely on the ability of the specific antibody to react with the resin-embedded antigen. Incubation of sections with insufficiently purified antiserum or with the whole antiserum will result, in most cases, in high background labelling. The antiserum should be purified using a protein A–sepharose column to make sure that no Fab-fragments will mask antigenic sites without further binding of the protein A–gold complexes (Fig. 11). A significant problem is the level of non-specific binding to the resin. Once a high background labelling is obtained, either the resin can be changed or sections can be soaked (for example in ovalbumin) before performing the immunolabel protocol. Some other problems can arise by following the described labelling protocol. Drying of the sections should be avoided because this will cause non-specific adsorption and clustering of the colloidal gold. Incubation steps should be done by floating the grids on drops of the different reagents. Such a treatment will often result in an increase of the non-specific label.

References

Acker, A., Knapp, W. and Mayer, H. (1981). *J. Bacteriol.* **147**, 602–611.
Aldrich, H. C., Beimborn, D. B., Bokranz, M. and Schönheit, P. (1987). *Arch. Microbiol.* **147**, 190–194.
Balch, W. E., Fox, G. E., Magrum, L. J., Woese, C. R. and Wolfe, R. S. (1979). *Microbiol. Rev.* **93**, 260–296.
Beesley, J. E. and Betts, M. P. (1984). In *Proceedings of the 8th European Congress on Electron Microscopy*, Vol. 3, pp. 1595–1596.
Beesley, J. E., Day, S. E. J., Betts, M. P. and Thorley, C. M. (1984). *J. Gen. Microbiol.* **130**, 1481–1487.
Bendayan, M. (1982). *Biol. Cell* **43**, 153–156.
Bendayan, M. (1984). *J. Electron Microsc. Tech.* **1**, 243–270.
Bendayan, M. and Zollinger, M. (1983). *J. Histochem. Cytochem.* **31**, 101–109.
Bendayan, M., Roth, J., Perrelet, A. and Orci, L. (1980). *J. Histochem. Cytochem.* **28**, 149–160.
Björk, I., Petersson, B.-A. and Sjöquist, J. (1972). *Eur. J. Biochem.* **29**, 579–584.
Blaut, M. and Gottschalk, G. (1984). *Eur. J. Biochem.* **141**, 217–222.
Blaut, M. and Gottschalk, G. (1985). *Trends Biochem. Sci.* **10**, 486–489.
Bullock, G. R. and Petrusz, P. (eds) (1982). *Techniques in Immunocytochemistry I.* Academic Press, London and New York.
Carlemalm, E., Garavito, M. and Villiger, W. (1980). In *Proceedings of the 7th European Congress on Electron Microscopy*, Vol. 2, pp. 656–657.
Carlemalm, E., Garavito, M. and Villiger, W. (1982). *J. Microscopy* **126**, 123–143.
Collins, J. and Mayer, H. (1980). *Arzneimittel-Forschung* **30**, 541–547.
Cypionka, H. and Meyer, O. (1983). *J. Bacteriol.* **156**, 1178–1187.
Cypionka, H., Meyer, O. and Schlegel, H. G. (1980). *Arch. Microbiol.* **127**, 301–307.
Daniels, L., Sparling, R. and Sprott, G. D. (1984). *Biochim. Biophys. Acta* **768**, 113–163.
de Mey, J., Moeremans, M., Gemeus, G., Moydens, R. and de Brabander, M. (1981). *Cell Biol. Int. Rep.* **5**, 889–899.
Doddema, H. J., Hutten, T. J., van der Dift, C. and Vogels, G. D. (1978). *J. Bacteriol.* **136**, 19–23.
Ellefson, W. I. and Wolfe, R. S. (1980). *J. Biol. Chem.* **255**, 8388–8389.
Ellefson, W. L. and Wolfe, R. S. (1981). *J. Biol. Chem.* **256**, 4259–4262.

Faulk, W. P. and Taylor, G. M. (1971). *Immunocytochemistry*, **8**, 1081–1083.
Frens, G. (1973). *Nature (Phys. Sci.)* **241**, 20–22.
Forsgreen, A. and Sjöquist, J. (1966). *J. Immunol.* **97**, 822–827.
Friedrich, C. G. (1982). *J. Bacteriol.* **149**, 203–210.
Friedrich, C. G., Friedrich, B. and Bowien, B. (1981). *J. Gen. Microbiol.* **122**, 69–78.
Friedrich, B., Kortlücke, C., Hogrefe, C., Ebertz, G., Silber, B. and Warrelmann, J. (1986). *Biochimie* **68**, 133–145.
Geoghegan, W. D. and Ackerman, G. A. (1977). *J. Histochem. Cytochem.* **25**, 1187–1200.
Goodman, S. L., Hodges, G. M., Trejdosiewicz, L. K. and Livingston, D. C. (1981). *J. Microscopy* **123**, 201–213.
Grider, B. P., Carper, S. W. and Lancaster, J. R. (1985). *Proc. Natl Acad. Sci. USA* **82**, 6793–6796.
Hjelm, H., Hjelm, K. and Sjöquist, J. (1972). *FEBS Lett.* **28**, 73–76.
Hjelm, H., Sjödahl, J. and Sjöquist, J. (1975). *Eur. J. Biochem.* **57**, 395–403.
Horisberger, M. and Rosset, J. (1977). *J. Histochem. Cytochem.* **25**, 295–305.
Horisberger, M., Rosset, J. and Bauer, H. (1975). *Experientia* **31**, 1147–1148.
Inatomi, Y. (1986). *J. Bacteriol.* **167**, 837–841.
Johannssen, W., Schütte, H., Mayer, H. and Mayer, F. (1984). *Arch. Microbiol.* **140**, 265–270.
Kohring, G.-W. and Mayer, F. (1987). *FEBS Lett.* **216**, 207–210.
Kohring, G.-W., Mayer, F. and Mayer, H. (1985). *Eur. J. Cell Biol.* **37**, 1–6.
Lancaster, J. R. (1986). *FEBS Lett.* **199**, 12–18.
Leduc, E. H. and Bernhard, W. (1967). *J. Ultrastruct. Res.* **19**, 196–199.
Leduc, E. H., Marinozzi, V. and Bernhard, W. (1963). *J. R. Microsc. Soc.* **81**, 119–130.
MacAlister, T. J., Irvin, R. T. and Costerton, J. W. (1977). *J. Bacteriol.* **130**, 318–328.
Mayer, F., Jussofie, A., Salzmann, M., Lübben, M., Rohde, M. and Gottschalk, G. (1987). *J. Bacteriol.* **169**, 2307–2309.
Meyer, O. and Rohde, M. (1984). In "Microbiol Growth on C_1 Compounds", *Proceedings of the 4th Symposium, American Society for Microbiology, Washington, DC*, pp. 26–33.
Meyer, O. and Schlegel, H. G. (1983) *Ann. Rev. Microbiol.* **37**, 277–310.
Modrich, P. and Roberts, R. J. (1982). In *Nucleases*, pp. 109–154. Cold Spring Harbor, New York.
Morgan, C. (1972). *Int. Rev. Cytol.* **32**, 291–317.
Oppenheim, J. D. and Sacton, M. R. J. (1973). *Biochim. Biophys. Acta* **298**, 297–322.
Ossmer, R., Mund, T., Hartzell, P. L., Konheiser, U., Kohring, G.-W., Klein, A., Wolfe, R. S., Gottschalk, G. and Mayer, F. (1986). *Proc. Natl Acad. Sci. USA* **83**, 5789–5792.
Parr, E. L. (1979). *J. Histochem. Cytochem.* **27**, 1095–1102.
Pauli, W. (1949). *Helvet. Chim. Acta* **32**, 795–810.
Pingoud, A., Urbanke, C., Alves, J., Ehlbrecht, H.-J., Zabeau, M. and Gualerzi, C. (1984). *Biochemistry* **23**, 5697–5702.
Rohde, M., Mayer, F. and Meyer, O. (1984). *J. Biol. Chem.* **259**, 14 788–14 792.
Rohde, M., Mayer, F., Jacobitz, S. and Meyer, O. (1985). *FEMS Microbiol. Lett.* **28**, 141–144.
Rohde, M., Johannssen, W. and Mayer, F. (1986). *FEMS Microbiol. Lett.* **36**, 83–86.
Romano, E. L. and Romano, M. (1977). *Immunochem.* **14**, 711–715.
Romano, E. L., Stolinski, C. and Hughes-Jones, N. C. (1974). *Immunocytochemistry* **11**, 521–522.
Romano, E. L., Stolinski, C. and Hughes-Jones, N. C. (1975). *Br. J. Haematol.* **30**, 507–516.

Roth, J. (1982). *Histochem. J.* **14**, 791–801.
Roth, J., Bendayan, M. and Orci, L. (1978). *J. Histochem. Cytochem.* **26**, 1074–1081.
Roth, J., Bendayan, M., Carlemalm, E., Villiger, W. and Garavito, M. (1981). *J. Histochem. Cytochem.* **29**, 663–669.
Rubin, R. A. and Modrich, P. (1977). *J. Biol. Chem.* **252**, 7265–7272.
Schink, B. and Schneider, K. (1979). *Biochim. Biophys. Acta* **567**, 315–324.
Schneider, K. and Schlegel, H. G. (1976). *Biochim. Biophys. Acta* **452**, 66–80.
Schönheit, P. and Beimborn, D. B. (1985). *Eur. J. Biochem.* **148**, 545–550.
Singer, S. J. (1959). *Nature* **183**, 1523–1525.
Slot, J. W. and Geuze, H. J. (1981). *J. Cell Biol.* **90**, 533–536.
van Driel, D. and Wicken, A. J. (1973). *J. Ultrastruct. Res.* **43**, 483–497.
Venable, J. H. and Coggeshall, R. (1965). *J. Cell Biol.* **25**, 407–408.
Wagner, B., Wagner, M., Kubin, V. and Ryc, M. (1980). *J. Gen. Microbiol.* **118**, 95–105.
Wientjes, F. B., van't Riet, J. and Nanninga, N. (1980). *Arch. Microbiol.* **127**, 39–46.
Williams, D. C., van Frank, R. M., Muth, W. C. and Burlett, J. P. (1982). *Science* **215**, 687–689.
Wolfe, R. S. (1985). *Trends Biochem. Sci.* **10**, 396–399.
Zsigmondy, R. and Thiessen, P. A. (1925). In *Das Kolloidale Gold.* Akademische Verlagsgesellschaft, Leipzig.

9

Localization of Macromolecular Components by Application of the Immunogold Technique on Cryosectioned Bacteria

JAN W. SLOT and HANS J. GEUZE

Department of Cell Biology, Medical School, University of Utrecht,
Utrecht, The Netherlands

ANTON J. WEERKAMP

Department of Oral Biology, Dental School, University of Groningen,
Groningen, The Netherlands

I. Introduction

Immunocytochemical approaches are used to add detail at the molecular level to morphological studies. Specific immunoglobulins, usually of the IgG type,

METHODS IN MICROBIOLOGY
VOLUME 20 ISBN 0-12-521520-6

are bound *in situ* to antigenic macromolecules, to which they are directed (further indicated as antigens). An immunomarker attached to the specific IgG can be observed by EM, and indicates where the antigen is located. There are many different ways to carry out immunocytochemical localizations, all of which operate within technical limitations. Restrictions are set on the one hand by the need for preservation of the structural details and for immediate and complete immobilization of the antigen molecules. These requirements can be met by chemical fixation of cells or tissues using reagents like glutaraldehyde and formaldehyde. On the other hand, the denaturizing effect of these fixatives on antigens can reduce their binding with the specific IgG, resulting in a conspicuous decrease of the immunolabelling. In addition, fixation induces a decrease in penetrability of the cell material for immunoreagents, which again may detract from the immunolabelling. In other words, good morphology and faultless localization require a tissue treatment that can destroy a significant part of the immunoreaction.

Compromises in this conflicting field are achieved in different ways. Ultra-cryotomy is one approach that has been successfully applied to a series of EM localization studies in bacteria (Beesley, 1984; Schilstra *et al.*, 1984; Tommassen *et al.*, 1985; Bosch *et al.*, 1986; Voorhout *et al.*, 1986; Weerkamp *et al.*, 1986, 1987). With this method accessibility of antigens is optimal, since the reaction is done on ultra-thin sections of non-embedded material. Damage to the antigens during tissue processing is minimal. The technique is compatible with very mild fixation, and other denaturizing steps like dehydration in organic solvents and embedding in resins are not involved. The ultrastructure is well preserved and can be visualized very well. This combination of conditions means that cryosections in general offer probably the best chances for high quality of the immunoreaction and the ultrastructural image. On the other hand, it is technically more demanding on workers and equipment than other approaches, such as post-embedding methods, that have been dealt with in the previous chapter (see Rohde *et al.*, this volume).

With cryosections, as with other on-section labelling methods, the need for penetration of the immunomarkers is minimal. Therefore particulate immunomarkers, like ferritin (Singer, 1959) and colloidal gold (Faulk and Taylor, 1971), can be used instead of enzyme immunomarkers, like peroxidase, which diffuse more easily into cellular material and on which pre-embedding methods largely depend. Particulate markers are preferable because they mark immunoreactions more precisely than enzymes and provide a better quantitative image.

In this chapter we will first deal with the preparation of immunogold markers. These do very well on cryosections (Geuze *et al.*, 1981) and have several advantages. Gold particles are highly electron dense. They can be prepared in different sizes, which opens up the possibility of distinct labelling of two antigens in one section (double labelling). Gold markers can be prepared very easily. Gold labelling patterns can be enhanced by silver treatment following a simple procedure (Holgate *et al.*, 1983). This enabled us to observe immunoreactions in cryosections using the light microscope with high resolution and sensitivity (Slot *et al.*, 1986; Geuze *et al.*, 1987).

In Section III we will describe how ultra-thin cryosections are prepared and immunolabelled. Section IV of the chapter will deal with some observations of labelled bacteria and the differences in labelling patterns obtained using immunogold markers after different procedures.

II. Immunogold markers

Spherical gold particles can be prepared in colloidal sols. The lyophylic nature of the sols makes them unstable in the presence of electrolytes. However, when hydrophylic colloids like proteins are added under proper conditions (Geoghegan and Ackerman, 1977), these bind spontaneously to the gold spheres. After binding of sufficient protein a gold sol becomes hydrophylic and stable in the presence of electrolytes. The binding of proteins to gold is fairly strong and, when optimal binding conditions are chosen (De Mey *et al.*, 1981; Horisberger and Clerc, 1985; see also Sections II.C and II.D), practically irreversible. The proteins usually maintain their biological activities. These properties make stabilized gold sols convenient tools for affinity cyto-chemistry with the electron microscope (EM). Thus, immunoglobulins, lectins and specific ligands for several receptor systems, bound to colloidal gold particles, are increasingly popular markers used to delineate intracellular locations and pathways of a variety of endogenous and exogenous substances (De Mey, 1983; Horisberger, 1983; Roth, 1983). Gold markers were first used for immunocytochemistry by Faulk and Taylor (1971). They attached gold particles directly to the specific IgG molecules, before they were allowed to react with the antigens in the tissue. This direct method was later largely replaced by two indirect methods. In indirect methods the gold is put on the specific IgG molecules after binding of the latter to the antigen in the tissue. The marker is conjugated with a carrier molecule which binds specifically to the specific IgG in the tissue during a second incubation. In one indirect method, the Ig–gold method (Romano *et al.*, 1974), the carrier is a second IgG directed to the specific IgG. The other common indirect method is the pA–gold method, in which staphylococcal protein A is used as the carrier. This method was introduced by Romano and Romano (1977), and later in on-section techniques by Roth and colleagues (1978).

A. Preparation of gold sols

Colloidal gold sols are prepared by condensation of metallic gold from a supersaturated solution which is created by reduction of Au^{3+}, usually in the form of gold trichloride. The circumstances under which reduction and sol formation take place determine two important characteristics of the resulting gold particles: (a) their size, henceforth expressed as the average particle diameter (APD); and (b) their size variability, i.e. the coefficient of variation (CV) of the diameter of particles in a sol. When the CV exceeds 15% sols are considered to be heterodisperse, and when the CV is smaller than 15% the designation homodisperse is given to a sol.* The size of the gold particles is critical. Large particles can be observed at low magnification and are suitable for scanning EM observations, whereas small particles, which favour the reaction sensitivity, are appropriate for high resolution transmission electron microscopy. Homodispersity of the markers is not only convenient for microscopical observation, but is also important for double labelling techniques (Horisberger and Rosset, 1977; Geuze *et al.*, 1981).

Many different methods have been used for the preparation of gold sols. The nature and concentration of the agents used for Au^{3+} reduction are crucial for determining the APD and CV of the sols, but factors like reaction temperature and pH are also important. We will describe two methods; both are relatively simple and together they allow gold sols with APD ranges from nearly 3 nm up to 150 nm to be prepared reproducibly. Still smaller gold markers can be prepared by other methods, and recently even < 1 nm particles were used for immunolabelling (Hainfield, 1987). Markers less than 3 nm in size are not easy to find in cryosections stained with uranyl, and we therefore restrict ourselves to this 3–150 nm range.

Gold sols with ADPs ranging from approximately 14 nm to nearly 150 nm can be prepared by using citrate for the reduction of Au^{3+} (Frens, 1973).

1. Citrate gold procedure (Frens, 1973)

- Take 100 ml of a 0.01% $HAuCl_4$ solution in distilled water.
- Heat until boiling.
- Add a variable volume (4–0.32 ml) 1% trisodium citrate . $2H_2O$ and mix.

---

*There exists some confusion about the meaning of the terms mono-, iso- and homodisperse, which were originally introduced by colloid chemists to indicate that sols are homogeneous with respect to particle size. In cytochemistry, however, their meaning is not consistent and sometimes monodisperse is used to indicate that a preparation of gold markers contains no significant amount of aggregates. Such aggregates can be formed during the preparation procedure of protein–gold complexes and exist of two or more gold particles with adhering protein that stick together. We will use *homo-* or *heterodisperse* for sols that contain particles within a narrow size range (CV < 15%), or that are variable in size (CV > 15%), respectively. The term *monodisperse* will be reserved for protein–gold preparations without aggregates.

The mixture is boiled until the colour turns bright orange–red where a high amount of citrate is added, or violet when this concentration drops to low values. The reaction time is dependent on the citrate concentration, but is always less than 15 min. The APD of the resulting sols is approximately 14 nm when 4 ml of the citrate solution is used, and rises slowly when less citrate is added. A maximum APD of approximately 150 nm is described. Sols in the range of 15–25 nm are homodisperse, but when the APD increases further they become heterodisperse.

For high resolution studies, smaller markers are required in many cases. Most preparation methods for sols with smaller particles are much less satisfactory in terms of reproducibility, control of particle size and homodispersity. In addition, the most common method, based on the use of white phosphorus to reduce Au^{3+} (Zsigmondy and Thiessen, 1925; Slot and Geuze, 1981), is dangerous. We have studied in detail the reaction conditions for sol formation using a mixture of tannic acid (TA) and citrate (C) as reducing agents (Slot and Geuze, 1985). Early this century TA was already used for preparing gold sols (Garbowski, 1903), and some years ago Muhlpfordt (1982) introduced a tannic acid–citrate (TA–C) mixture as a substitute for white phosphorus in a procedure for making heterodisperse 5 nm sols. By selecting the right conditions we succeeded in making these TA–C sols homodisperse and with any optional APD in the range from nearly 3 to 15 nm.

2. Tannic acid–citrate gold procedure (Slot and Geuze, 1985)

To make 100 ml of a TA–C sol, two solutions are made.

(A) The gold chloride solution (250-ml beaker):
  • 1 ml 1% $HAuCl_4$.
  • 79 ml distilled water.
(B) The reducing mixture (50-ml beaker):
  • 4 ml 1% tri-Na-citrate . $2H_2O$.
  • A variable volume 1% TA (Aleppo tannin, Mallinckrodt, St Louis, Code 8835).
  • 25 mM K_2CO_3, to correct the pH of the reducing mixture. A volume is added equal to the volume of 1% TA used. When less than 1 ml of 1% TA is used no carbonate needs to be added.
  • Distilled water to make the volume of (B) up to 20 ml.

(A) and (B) are heated to 60°C. Then add (B) quickly to (A) while stirring. Red sols are formed within a second when high concentrations of TA are added. The reaction time increases gradually when lower concentrations of TA are used until, in the absence of TA, it lasts ±60 min before sol formation is complete. After the sol formation is finished, which is evident from the red colour, the sols are heated until boiling.

When in the procedure the volume of 1% TA added rises from 0.01 to 5 ml, the APD of the sols gradually falls from 15 nm to nearly 3 nm. For example, if 0.015, 0.1, 0.5 or 3 ml 1% TA is added to the reducing mixture, APDs of about 14, 9, 6 or 3.5 nm can be expected. Such sols are sufficiently homogeneous so that they can be used together as markers in multiple staining experiments.

The solutions used in both procedures are made up freshly except for the gold chloride solution. The latter is made in stock at 1% strength by dissolving the contents of a 1-g capsule, commercially available from several suppliers, in 100 ml distilled water. The solution is cleaned by centrifugation (10 min × 1500g) and can be stored in the refrigerator for at least a year.

The source of TA is critical. Only Aleppo tannin gives the reported results. This quality is warranted by Mallinckrodt under their code 8835.

Normally cleaned laboratory glassware can be used.

B.   Protein binding to gold

Colloidal gold sols are unstable in the presence of electrolytes. When electrolytes are added the colour turns blue and finally the gold precipitates. When a protein is added in sufficient concentration, it is adsorbed to the gold particles, rendering the sol stable in the presence of electrolytes. The minimum amount that can stabilize 10 ml of a gold sol was originally called the *gold number* of a protein (Zsigmondy and Thiessen, 1925). The gold number, or stabilization concentration, is different for each protein and can be established as follows (Zsigmondy and Thiessen, 1925; Horisberger and Rosset, 1977).

In small test tubes 250-μl samples of gold sol are mixed with samples of a protein solution. For example, volumes measuring from 5 to 50 μl of a 50 μg ml^{-1} protein solution can be added. After one minute 25 μl 10% NaCl is added. The lowest protein concentration that prevents the red to blue colour change occurring (which can be judged visually) is taken as the stabilization concentration, and from that value the gold number can be calculated.

In TA–C sols the colour change is slow and masked by the brownish colour contributed to the sol by the tannic acid. This is particularly noticeable with the finest sols, where a relatively high concentration of tannic acid has been used to prepare the sol. We found that this can be obviated by adding low concentrations of H_2O_2 (< 0.2%) to the test sol, or by doing the test overnight. Low electrolyte concentration in the protein solution used for binding to gold is usually recommended in order to prevent aggregation of gold particles upon addition of the protein. However, Lucocq and Baschong (1986) recently reported that protein preparations in various commonly used buffers could be complexed with gold without obvious aggregation.

The binding of protein to gold is pH-dependent. According to Geoghegan and Ackermann (1977), in general stable complexes can be achieved at a pH equal to, or slightly higher than, the isoelectric point of the protein involved (see also Sections II.C and II.D).

C. pA–gold preparation

Binding of staphylococcal protein A to gold particles has been studied recently by Horisberger and Clerc (1985). They found that a sol binds more protein A close to the pI (pH 5.1), but that the binding of protein A is more stable at higher pH. On the other hand, the tighter binding at higher pH (e.g. pH 7) may have a denaturizing effect on the protein and make the probes less reactive. Therefore it is advised to do the binding at pH 6. Furthermore, the results of Horisberger and Clerc showed that gold sols can bind much more protein A than indicated by the gold number. However, this "excess" protein A is bound less tightly and may dissociate during the first weeks after the preparation. The dissociated protein A will compete for reactive sites with the pA–gold and lower the immunolabelling intensity. Possibly this may explain the instability of pA–gold preparations reported earlier (Beesley et al., 1984). To some extent we had similar experiences, but since we have modified the binding procedure, according to the finding of Horisberger and Clerc, as described below, our probes are stable for long periods (see Section II.E). It is apparently important to do the binding at pH $\pm$ 6 and to add no more protein A to the sol than the stabilization concentration as determined by Zsigmondy's visual colour change test (see Section II.B). Recently a more sensitive test, based on EM observation of aggregate formation upon adding salt, was introduced by Lucocq and Baschong (1986). With that method they found much higher stabilization concentrations than with Zsigmondy's test. Apparently Zsigmondy's gold number indicates a rather arbitrary protein concentration, where the gold reaches some but not complete stability in the presence of salt. Fortunately this gold number value seems to coincide very well with the concentration beyond which, at least in the case of protein A, protein is bound less stably to the gold. Therefore Zsigmondy's test is appropriate to determining the concentration of protein A that should be added to a gold sol, whereas the proposed EM test is most probably not.

1. Protein A binding to gold

- Protein A (Pharmacia, Sigma or LKB) is dissolved in distilled water at $1 \, mg \, ml^{-1}$.
- A gold sol is prepared as described in Section II.A and used the same or the next day.

- The pH of the sol is adjusted to 6 with 0.1 N NaOH. Do not use the pH meter for unstabilized sols, because the sol will clog up the electrode. Therefore take a ± 5 ml sample of the sol, which is usually at a pH 5–5.5 when made after the above procedure, stabilize it with excess protein A (± 5 μg per ml sol), and determine the concentration of NaOH required to bring the pH to 6. Then add that concentration of hydroxide to the unstabilized sol.
- Determine the stabilization concentration (gold number) of protein A in small samples as described above (Section II.B).
- The pA–gold probe can now be prepared by adding this concentration of protein A to the sol at pH 6, while stirring.
- A few minutes after the addition of protein A the pH is brought to 7–7.5 by adding 0.1 N NaOH. Then add bovine serum albumin (BSA), at a final concentration of 0.1%, to be sure that the gold particles are stabilized maximally. For BSA we use Sigma fraction V. Make a 10% stock solution in distilled water, neutralize it with NaOH, and centrifuge at 100 000g for 1 h. Add 0.02% sodium azide and store at 4°C.

2. Purification of pA–gold complex

The most commonly used procedure to purify the pA–gold is centrifugation. We found the best result by gradient centrifugation as follows (Slot and Geuze, 1981):

- The preparation is concentrated by centrifugation: e.g. 5 nm gold, 45 min at 125 000g_{av}; 10 nm gold, 45 min at 50 000g_{av}. The resulting pellet is composed of a large loose part and a small tightly packed part. If the centrifuge speed is too high the tight pellet becomes large and an increasing amount of aggregates of gold particles is introduced into the preparation.
- Remove the supernatant without disturbing the pellet and resuspend the loose part of the pellet in PBS so that the volume is about 1/25 of the original amount of sol used.
- The concentrated probe is now purified by layering it over a 10–30% continuous glycerol (or sucrose) gradient in phosphate-buffered saline (PBS, pH 7.2 = 8 g NaCl, 0.2 g KCl, 1.44 g NaHPO$_4$.2H$_2$O and 0.23 g NaH$_2$PO$_4$.H$_2$O in 1 litre) + 0.1% BSA. The gradient is centrifuged for 45 min at 41 000 rpm (5 nm gold) or for 30 min at 20 000 rpm (10 nm gold) in a SW 41 rotor (Beckman Instruments) or at the appropriate speed in any other type of rotor.
- The dark red band is then collected. It contains essentially a monodisperse pA–gold preparation without free protein A.

Alternatively, purification can be carried out by column chromatography after concentration by osmosis, as suggested by Wang *et al.* (1985). We performed this procedure as follows:

- Take 20 ml of a crude pA–gold preparation prepared as above, except that now 0.01% BSA is added, and put it in a dialysis bag. The bag is immersed in 30% polyethylene glycol (PEG) 20 000 while agitating. The degree of concentration is followed by weighing, and after approximately 2 h the 10–20 × concentrated preparation is collected.

- For purification of this concentrate we use a 14 × 0.7 cm column (ACA 22 gel, LKB) which is equilibrated with PBS + 0.1% BSA. Samples of about 250 µl of the concentrate can be applied to the column. When run at a flow rate of ~3 ml h^{-1} the red gold complex comes off after about 1 h.

- Collect 250-µl fractions. Pool the three or four most concentrated ones.

Purification on a column is a convenient alternative when adequate centrifugation equipment is not available. Free protein A is completely removed, but we found the separation of single markers from aggregates not as good as after gradient centrifugation. This is of less importance for pA–gold preparations, in which usually only a minor part of the gold particles occurs as aggregates. However, gold complexes with other proteins do contain significant amounts of aggregates. We observed this, for instance, for certain IgG preparations. Then separation on a gradient may be preferable.

D. Ig–gold preparation

IgG preparations, except for monoclonals, are composed of many fractions with a variety of pI values. De Mey and colleagues (1981) found it advisable to bind IgG to gold at pH 9, which is on the alkaline side of the pI range. To prevent aggregation of IgG and the subsequent occurrence of clumps in the IgG–gold preparation, special care has to be taken during the preparation of the IgG fraction from antisera. We prepared goat anti-rabbit IgG as follows (Slot and Geuze, 1984): 4 ml of the goat antiserum is adsorbed to 2 ml CNBr-activated Sepharose gel to which rabbit immunoglobulin is bound, for 2 h at room temperature. Then the gel is washed with PBS and anti-rabbit immunoglobulin is eluted with 4 ml 3 M KCNS. The protein concentration (A^{280}) is adjusted to 1 mg ml^{-1} in 3 M KCNS and the solution is dialysed at room temperature against 2 mM borax, pH 9. If some precipitation occurs one has to dilute the IgG solution prior to dialysis. Essentially the same procedure was followed to purify swine anti-rabbit IgG (Section IV.B).

Complexing of immunoglobulins to gold and subsequent purification of Ig–gold fractions by gradient centrifugation or on columns are identical to the pA–gold procedure, except that coupling occurs at pH 9. Therefore both the IgG preparation and the sol have to be adjusted to pH 9.

E. Storage of gold markers

Preparations of gold markers, prepared as described above with 0.02% sodium azide added, can be stored for months in the refrigerator without noticeable loss of reactivity. Ultimately, however, the probes may lose activity due to the dissociation of the proteins from the gold particles (Horisberger and Clerc, 1985). For long-term storage they can be kept frozen in 15–20% glycerol in small samples (Slot and Geuze, 1985). The glycerol concentration is 15–20% in preparations when collected from gradients. We had also good experience with storage at $-20°C$ in 45% glycerol, which has the advantage that the solution remains fluid (Slot and Geuze, 1981).

III. Ultra-cryotomy and immunolabelling

The ultra-cryotomy approach was first tried in the 1950s (Fernandez-Moran, 1952). Initially, ice crystal damage was a major obstacle for achieving a well-preserved ultrastructure. In non-fixed material this difficulty still exists. Indeed, technical improvements in both cryosectioning (Christensen, 1971) and freezing (Livesey and Linner, 1987) make good preservation of structure in thin cell layers or cell suspensions feasible (McDowall et al., 1983), but in more bulky tissue freezing damage cannot be avoided. Moreover, these non-fixed sections are not suitable for immunolabelling procedures, unless the material is, subsequent to freezing, resin-embedded according to relatively complicated freeze-drying and vacuum-embedding techniques (Liversey and Linner, 1987).

Although it might induce some structural artefacts (Dubochet et al., 1983), chemical fixation is the generally accepted way to treat tissues (cells) for immunolocalization of macromolecules. With fixed material the use of cryoprotectants became possible, which facilitated ultra-cryotomy considerably. For this purpose, initially glycerol (Bernard and Leduc, 1967), but later, and more successfully, sucrose (Tokuyasu, 1973) are used. By simple freezing methods, frozen tissue samples in the vitrified state (i.e. without ice crystal formation) can now be prepared (Dubochet et al., 1985), from which, by use of modern ultra-cryotomes, thin sections can easily be prepared. This method of ultra-cryotomy for immunocytochemical purpose was largely designed by Tokuyasu (1973, 1978, 1980) and Tokuyasu and Singer (1976), and is known as the Tokuyasu method.

A. Fixation

Fixation is usually done with paraformaldehyde or glutaraldehyde, or mixtures of both aldehydes, in 0.1 M sodium phosphate buffer, pH 7.4 (PB). If the immunoreaction is glutaraldehyde-sensitive, one can use a straight formaldehyde fixative (2–4%). Reasonable results can be achieved with short (1 h) formaldehyde fixation, provided that the sections are given extra support by methyl cellulose (Section III.F). However, this has the result that small gold markers are not readily observed. Alternatively, the tissue may be left in formaldehyde overnight, or even for longer periods up to several weeks. This will improve the preservation of the ultrastructure in cryosections, but part of the immunoreaction may gradually be sacrificed.

For studying soluble compounds (e.g. cytosolic or secretory proteins) strong (2%) glutaraldehyde fixation for at least one hour is needed to prevent extraction of proteins from the sections. On the other hand, the immuno-staining of substances attached to membranes or cytoskeleton is often enhanced by the extraction of surrounding soluble proteins, thus allowing more exposure of the substances of interest to the immunoreagents. Membrane proteins may therefore be better studied in formaldehyde-fixed specimens.

To prevent redistribution of soluble proteins, it is important that fixation is rapid. Therefore perfusion fixation is preferable when working with tissues. Cell suspensions, like cultured bacteria, can be mixed at room temperature with double strength of the desired fixative, 1:1 (v/v). The cells are gently centrifuged down and resuspended in final strength fixative.

B. Specimen blocks

Good cryosections can be made more easily when the specimen has the right shape, for instance cubic or right-angled. To make such blocks from cell suspensions or tissues of loose texture, we usually embed in 10% gelatin. The gelatin is fixed afterwards. A convenient method is as follows:

- Spin fixed cells down and resuspend them in 10% gelatin in PBS at 37°C. Spin down again and remove the supernating gelatin completely, so that a concentrated pellet of cells in gelatin is left behind.
- Take the pellet, mix the cells gently with the remaining gelatin, and transfer it to the inner side of the cover of a pair of plastic Petri dishes (ϕ 3 cm) which has been covered by parafilm. Press the bottom side of the smaller dish onto it so that a thin slab is formed, the thickness of which is determined by the small (∼0.5 mm) rims that usually occur on the bottom of plastic dishes (Fig. 1). Until this point the procedure is carried out at room temperature and should be done quickly to keep the gelatin fluid.
- Put the system in the refrigerator. The gelatin solidifies within 15 min, whereupon the dishes can be separated. The gelatin remains attached to the smaller dish.
- Cover the gelatin with a drop of fixative (the same as used for the initial fixation of the cells) and loosen it carefully from the dish.

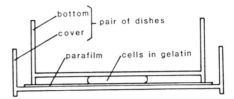

Fig. 1. Embedding of cells in 10% gelatin. See text for details.

Tissues are embedded in a similar way. Small fragments are incubated for 10 min in 10% gelatin at 37°C and further processed as cell pellets.

Cubic specimen blocks can easily be prepared from these gelatin slabs by cutting in two perpendicular directions with a razor blade. The size of the blocks is important; usually small blocks facilitate sectioning. In complicated tissues one might need to survey relatively large sections and use blocks that are as large as possible. In homogenous specimens, however, like cell suspensions, there is no need for large sections and it is suggested that squares are made approximately 0.2 × 0.2 mm.

C. Mounting and freezing

Tokuyasu introduced sucrose as a cryoprotectant (Tokuyasu, 1973; Tokuyasu and Singer, 1976). He achieved good results by using concentrations in the range 0.6–1 M (Tokuyasu, 1980). However, unless precautions are taken during freezing, ice crystals can still be formed. We used 2.3 M sucrose instead (Geuze and Slot, 1980). At that concentration only vitreous ice is formed, as was recently shown by Dubochet et al. (1985). These authors demonstrated that at sucrose concentrations > 1.6 M no ice crystals are formed during freezing.

Tokuyasu (1986) recently introduced polyvinylpyrrolidone (PVP), mixed with sucrose, as the infusion material for very low fixed or highly hydrated specimens. The addition of PVP seems to create in general more favourable cutting conditions. It lends plasticity to the blocks, which allows the cutting of very thin and well-stretched sections. We experienced very good sectioning conditions when using a mixture suggested by Tokuyasu (personal communication) that contains approximately 20% PVP + 1.7 M sucrose. To make 100 ml, a paste of 20 g of PVP, 4 ml of 1.1 M Na_2CO_3 and 80 ml of 2 M sucrose in PB is made in a container. The container is covered and left at room temperature overnight, during which time minute air bubbles in the paste escape into the air, leaving behind a clear solution. A solution made in this manner will be nearly neutral in pH, but if desired the pH may be further adjusted with 1 N NaOH. Specimen blocks infused with this mixture can be frozen in liquid nitrogen and cut at −110°C.

For thorough infusion the specimen blocks are left for a minimum of 1 h (sucrose) or 2 h (PVP-sucrose) in the cryoprotectants while rotating on a roundabout at room temperature. Then the blocks are mounted onto a specimen holder of the ultra-cryotome. This can be done under a dissection microscope, working quickly and using a fibre optic (cold) light source in order to minimize drying. Excess sucrose solution should be removed, leaving a thin film behind underneath the block, sufficient to glue it to the metal surface of the specimen holder, which should be degreased and slightly roughened by the use of fine sandpaper. The holder is then put in liquid nitrogen, in which it is transferred to the cryochamber of the microtome.

If the cutting surface of the block is not square or right-angled, necessary for getting optimal cutting conditions, the block can be trimmed inside the cryochamber with scalpel knives. Then the specimen holder should be fixed so that two sides of the block are perfectly parallel to the knife edge.

D. Sectioning

Thin cryosections are cut on dry knives. Glass knives can be used, but ultra-cryotomy requires sharper knives than the conventional ones used for sectioning of plastic-embedded material. The best way to prepare glass knives is by the symmetrical break method developed by Tokuyasu and Ukamara (1959) and more recently described in detail by Griffiths et al. (1983). Diamond knives, mounted on special holders for cryotomy, can also be used with good results. Recently Griffiths found that sectioning properties considerably improve after vaporizing tungsten onto the knives (Griffiths et al., 1984), a method introduced earlier by Roberts (1975). Tungsten treatment is particularly helpful when knives tend to build up electrostatic charge, a phenomenon that can make the handling of the sections in the cryochamber very troublesome. The tungsten layer should be very thin, or else the sections do not move smoothly over the glass surface.

Tools to handle the sections inside the cryochamber are an eyelash and a wire loop, both mounted on top of a 15-cm wooden stick. Good wire loops can be prepared from 0.2 mm thick stainless steel wire and have a diameter of 2 mm. With the eyelash, sections or ribbons can be straightened when they curl backwards during cutting, and they can be removed from the knife edge. Then they can be picked up quickly by the loop, filled with 2.3 M sucrose solution in PB, and transferred to glass slides or grids.

First cut sections ∼0.5 μm thick, which can be appreciated by light microscope after quick staining, for instance for 1 min in a solution containing 1% borax, 1% methylene blue and 1% azur II. After advancing to a suitable level in the tissue block, change to ultra-thin (50–100 nm).

The sucrose drops with ultra-thin sections are put on carbon-coated Formvar grids, with the sections downwards, so that these are pressed against the grids and covered by a layer of sucrose. The grids are placed with the sections downwards on plates (ϕ 3 cm) of 2% gelatin in PB on ice, so that the sucrose can diffuse away. On these plates the grids can be stored for a few hours or overnight until immunostaining starts. The gelatin plates are kept on moistened filter paper in a pair of dishes (ϕ 9 cm) at 4°C so that the gelatin stays solid. Before immunostaining proceeds this moisture chamber is transferred to room temperature. The gelatin will then become fluid and the grids can be picked up.

E. Immunostaining

During immunoincubation and washing procedures, grids pass through various solutions, always floating on drops of them. The back of the grids should stay dry. Fresh sheets of parafilm are used to put the drops on, as they provide a clean and flat surface. Proteins should be added to the incubation and washing solutions in order to lower non-specific binding of the immunoreagents. We usually add 0.1% BSA (Sigma, fraction V) + 0.4% gelatin (Merck) (BSA/gel).

It is convenient to transfer grids from drop to drop using small loops. However, relatively large quantities of adhering fluid are then co-transferred. This may introduce significant dilutions of the immunoreagents (steps (c) and (e) below), especially since these are often used in very small drops. Therefore at these critical stages it is better to use forceps for transfer and excess fluid can quickly be removed with adsorbent paper. Care should be taken, however, that the section-side of the grid never becomes dry during the procedure. This would cause severe damage to the fine structure; in addition, when a dry grid is placed on the surface of the next drop it may become contaminated with dirt from the surface of that drop and, when it contains immunoreagents, with non-specifically bound label.

1. Immunolabelling procedure

(a) 0.02 M glycine in PBS, 10 min. Glycine is added to quench free aldehyde groups. Alternatively, one can use 0.1–0.2% sodium borohydride in PBS ($\sim$ 10 min). This also inactivates aldehyde groups, but in addition borohydride is reported to restore part of the immunoreactivity which is lost after glutaraldehyde fixation (Eldred *et al.*, 1983; de Vries *et al.*, 1985). The borohydride solution is very unstable and loses reactivity within a few minutes. Therefore the solution should be made immediately before use.

(b) PBS + BSA/gel, a few minutes.

(c) Specific antibody at a concentration $< 25\,\mu g\,ml^{-1}$ in PBS + BSA/gel if affinity-pure preparations are used. Drops as small as 5 µl are sufficient. Incubate ~ 30 min at room temperature.

(d) PBS + BSA/gel, ~ 4 times, 1 min each wash.

(e) pA–gold or Ig–gold, approximately 30 min, room temperature. Dilute the gold markers immediately before use in PBS + BSA/gel. The concentration of the probes is usually expressed by their optical density at 520. Smaller particles can be diluted more. We use 4, 6 and 9 nm markers at OD_{250} 0.1, 0.15 and 0.2, respectively.

(f) Wash with PBS + BSA/gel, 4×5 min.

F. Staining and embedding

After immunolabelling, two important steps have to be done to prepare the sections for EM observation. First, heavy metal staining is needed to visualize ultrastructural features, and second, the sections have to be embedded in a substance that can support the structures during drying. Several procedures are described by Tokuyasu (1978) and his co-workers, including osmication and lead staining followed by plastic embedding (Keller et al., 1984). The latter procedure can lead to beautiful delineation of cell structures, yielding pictures reminiscent of osmium–lead staining patterns in conventional EM sections. We achieved the best results, particularly in respect of reproducibility, with Tokuyasu's hydrophylic embedding in methyl cellulose (MC), preceeded by uranyl acetate staining. In the original procedure membrane structures were first stabilized with uranyl acetate at neutral pH; then the sections were stained with acidic uranyl acetate and finally embedded in MC. This treatment resulted in a rather weak positive contrast. Most of the uranyl ions were extracted during washing in distilled water. This weak contrast was advantageous when working with low contrast immunomarkers like ferritin. Later, after gold markers were introduced for use on cryosections (Geuze et al., 1981), much higher contrast could be introduced into the sections without losing sight of the labelling patterns. Therefore Griffiths modified the procedure (Griffiths et al., 1982). He added uranyl acetate to the MC so that during drying the uranyl acetate is partly adsorbed to structures (e.g. ribosomes, filaments) and partly settles, along membranes, in a typical negative staining pattern. In addition, he changed to higher MC concentration, resulting in a thicker supporting film of MC and hence improved structural preservation.

1. Uranyl acetate staining and methyl cellulose embedding

(a) After the immunolabelling procedure, wash the grids, 4×1 min, on drops of distilled water.

(b) Uranyl–acetate–oxalate, pH 7–7.5, for 5 min. This is made by mixing equal parts of 0.3 M oxalic acid with 4% aqueous uranyl acetate and bringing the pH to about 7.5 with 5% NH_4OH. Use pH paper since the heavy metal will block the pH electrode. This procedure, which adds little contrast, is assumed to stabilize membrane structures against the subsequent low pH aqueous uranyl acetate (Tokuyasu, 1978).

(c) Wash three times on drops of distilled water, a few seconds each.

(d) Place the grids on drops containing 0.3% uranyl acetate + 1.8% MC on ice for 10 min, after first touching the surface of two other drops of the same solution for a few seconds. To prepare the uranyl acetate–MC solution, make up a 2% solution of 25 centipoise methyl cellulose (Methocel® from Fluka AG, Buchs, Switzerland). Add the powder to water preheated to 95°C. Mix with a magnetic stirrer at 95°C for a few minutes and then cool the solution on ice. Mix for at least 4–8 h at 0–4°C. Leave for a further 3–4 days at 4°C. Then mix the MC solution with 3% uranyl acetate, 9:1, and centrifuge at high speed (e.g. 60 000 rpm in a Beckman 60Ti or 70Ti) for 90 min at 4°C. The supernatant is decanted and can be used for approximately 4 weeks if stored in the dark at 4°C. We use routinely 0.3% uranyl acetate, but other concentrations are optional if stronger or weaker contrast is desired.

(e) Loop out the grid using a stainless steel loop, 3–3.5 mm in diameter, of 0.3–0.5 mm thick wire.

(f) Remove excess fluid by touching the surface of filter paper. Dry the grid, still on the loop. The final thickness of the methyl cellulose film is critical with respect to both contrast and fine structure preservation. Optimal films have gold-to-blue interference colours. This most critical step in the procedure may initially need repeated attempts, comparing each time the macroscopic appearance and the image of the ultrastructure observed in the EM.

IV. Observations and discussion

A. Appreciation of gold labelling patterns in cryosections

On-section labelling, as described in this chapter for cryosections, offers the possibility of labelling antigens everywhere both inside and outside cells. This is an improvement over what can be achieved with pre-embedding methods. Early studies demonstrated that the bacterial cell wall is an impregnable diffusion barrier. Antigens of Gram-positive species inside the cells and also in the cell membrane (Van Driel *et al.*, 1973) and inner layers of the cell wall (Linssen *et al.*, 1973) are not accessible to immunoreagents when cells are incubated in suspension. However, intracellular labelling of cell surface antigens in *Streptococcus salivarius* (Fig. 2) was readily achieved in cryosections (Weerkamp *et al.*, 1986). Similarly, in the Gram-negative *Escherichia coli*, detection of antigens at the surface of the outer membrane was problematic after immunolabelling of whole cells, whereas labelling of the same proteins in cryosections was regular (Voorhout *et al.*, 1986).

One advantage of the cryosection method over working with sections of embedded material (post-embedding methods) in immunocytochemistry is its rapidity. The entire procedure from fixation to microscopy takes about one day. Probably more important, though, is the fact that cryosections combine the most sensitive immunoreaction with a good representation of the ultrastructure. The accessibility of antigens is not affected by embedding material, and the possible denaturation of antigens during dehydration and embedding is avoided. On the other hand, careful interpretation of the density of labelling in cryosections requires consideration of the degree of penetration of label into the sections. Penetration merely depends on the density of the cell material surrounding the antigen molecules. For instance, at the very cell surface penetration may be complete, while labelling of intracellular antigens is often restricted to the section surface (Slot and Geuze, 1983; Freundl *et al.*, 1986; Voorhout *et al.*, 1986; Fig. 3). Therefore labelling densities are not necessarily proportional to antigen concentrations. We found that im-munogold labelling penetrates into cryosections of glutaraldehyde-fixed gelatin blocks at concentrations <10% gelatin, whereas at concentrations >10% only labelling at the section surface occurred (Posthuma *et al.*, 1987). Therefore, since we cut cells embedded in 10% gelatin, differences in antigen accessibility at the cell surface and intracellularly may be partly negated in our procedure.

Nevertheless, some variation in penetration remains. In *E. coli*, where we localized the elongation factor EF-Tu (Schilstra *et al.*, 1984), lower labelling efficiency was observed in dense intracellular granules than in the loose surrounding cytoplasm. These differences could be removed by introducing 30% polyacrylamide (Slot and Geuze, 1982) as a secondary matrix before preparing cryosections. In these cryosections of 30% polyacrylamide-embedded specimens, the labelling density appeared to reflect reliably the antigen concentrations present in cell structures (Posthuma *et al.*, 1987). Recently we used this system to develop an immunocytochemical method for determining intracellular protein concentrations (Chang *et al.*, in press Posthuma *et al.*, 1988; Slot *et al.*, in press).

In the stereo view of an immunolabelled cryosection (Fig. 3) it can be observed that the gold particles occur in a different plane of the section from the structure image of the specimen. It is probable that during drying of the MC–uranyl mixture (see Section III.F) most of the uranyl moves towards the

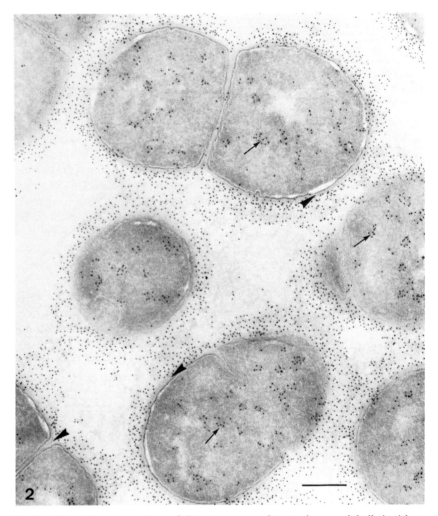

Fig. 2. Ultra-thin cryosection of *Streptococcus salivarius* immunolabelled with an antiserum against the cell surface protein, antigen C and pA–gold (6 nm) (Procedure 2 in Table I). Since this procedure tags one IgG with one gold, the clustered appearance of the intracellular labelling (arrows) indicates that antigen C occurs in rather large complexes at the site. At the arrowheads it appears that the greyish cell wall layer is not significantly labelled, as is the case between dividing cells. Bar, 0.2 μm.

deepest layers of the sections—those facing the supporting Formvar film. Consequently, these deeper layers are predominantly stained by uranyl, whereas, as already discussed above, the gold markers are often confined to the upper section surface. This may sometimes affect the accuracy of the labelling. For instance, in grazing or oblique sections a structure may not reach the upper section surface so that it is not accessible for immunolabelling, but it is visualized by uranyl. This happened apparently to the non-labelled part of the mesosome structure in Fig. 3. Another example of faulty labelling is

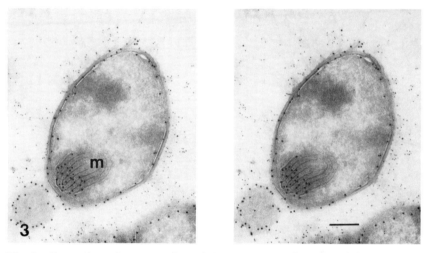

Fig. 3. Stereoview of a cryosection of *Streptococcus salivarius* which was double immunolabelled in four steps: (1) anti-antigen C; (2) pA–gold (6 nm), immediately followed by free protein A (50 μg ml^{-1}); (3) anti-lipoteichoic acid (LTA); (4) pA–gold (9 nm). m, mesosome. Bar, 0.5 μm.

shown in Fig. 4. LTA seems to occur inside the cells, at short distance from the cell membrane. This is probably not the case (Van Driel *et al.*, 1973). It is more likely that the intersections of the membrane with the lower and upper section surface are not exactly superposed. This may be due to sectioning obliquely to the cell surface, but it could be amplified by shrinkage during drying. The shrinkage affects mainly the upper layers of the section, the lower layers being fixed to the Formvar film. Deviations between structure image and labelling pattern can be minimized by cutting sections as thin as possible. In addition, embedding of the sections in thick MC films can help to avoid much of the shrinkage during drying. Furthermore, it is obvious that for reliable interpretation of labelling patterns, particularly over the various cell surface layers, one should study many different cell profiles.

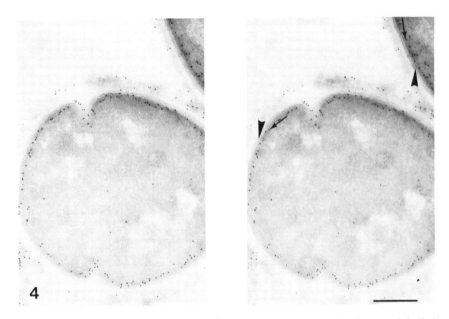

Fig. 4. Stereoview of a cryosection of *Streptococcus pyrogenius*, immunolabelled with anti-LTA, followed by pA–gold (6 nm). The narrow labelling zone just inside the cell membrane (arrows) probably marks the intersection line of the cell membrane and the upper section surface. The membrane structure is predominantly visualized near the under section surface (arrowheads). The intersections are not completely superposed, which explains why the labelling and structure images of the membrane do not coincide. In this species some LTA reactivity occurs at the cell surface. Bar, 0.2 μm.

B.   Choice of Ig–gold or pA–gold

The most commonly used immunogold markers are Ig–gold and pA–gold. IgG preparations against immunoglobulins from a large variety of species are commercially available, so that the indirect (two-step) Ig–gold method can be used to detect practically every specific antibody, no matter what species it is raised in. Protein A binds to IgG from several mammalian species (e.g. rabbit, swine, guinea pig), whereas it binds to none or only some of the IgG subclasses in others.

Therefore pA–gold has a much wider activity range than one particular Ig–gold probe, but it cannot be used in the regular indirect (two-step) procedure when a protein A-insensitive specific IgG is involved. In these cases one has to introduce an extra incubation step, preceeding the pA–gold incubation, with a protein A-sensitive IgG against the specific antibody. This three-step pA–gold procedure is needed, for example, for pA–gold detection of many mouse monoclonals.

We compared the labelling characteristics of these three labelling procedures: (1) the two-step Ig–gold; (2) the two-step pA–gold; and (3) the three-step pA–gold procedure. We used swine anti-rabbit IgG (SwaR) as a second antibody in procedures (1) and (3), and 5-nm gold particles either on protein A or on SwaR. When labelling LTA in bacterial cell membranes the two-step pA method (2) resulted in a narrow band of gold particles over the cell border (Fig. 5b). Inside the cells there were only a very few particles, which were mostly distributed singly. When the two other procedures, (1) and (3), were used for LTA labelling (Figs 5a and c), the cell border labelling was much more diffuse, intracellular labelling was much higher, and gold particles often occurred in clusters. Protein A binds in a relatively constant fashion to the Fc region of IgG molecules, whereas SwaR can bind everywhere and with more than one molecule to the specific IgG. In addition, protein A is smaller than IgG. These factors together may explain why labelling with the two-step pA–gold method is more precise and why it usually tags a specific IgG molecule with a single gold, whereas the other procedures, that include a secondary IgG, often label in clusters of a few gold particles.

We have quantitated for each procedure the immunolabelling in cryo-sections of 10% gelatin blocks in which the antigen amylase was dispersed (Table I). The labelling density for amylase is expressed as clusters (including singles) and as individual gold particles per μm^2. Background labelling was measured in parts of the sections that did not contain amylase. Two important facts emerged from these data (Table I, (3)). First, the three-step pA–gold procedure yielded the highest labelling both in number of clusters per μm^2 (i.e. number of antigen molecules recognized) and in number of gold particles per cluster (i.e. signal amplification). Both factors together resulted in a yield of gold particles per μm^2 which was ~ 6 times higher than in the two-step pA–gold procedure (Table I, (2)). The two-step Ig–gold procedure (Table I, (1)) gave results that were intermediate in both respects. The other important fact is that background labelling after the two-step pA–gold procedure is lower than after both other procedures, even when expressed as a percentage of the specific labelling (see also the intracellular labelling in Fig. 3). An explanation for this low background level of the two-step pA–gold method may be that protein A recognizes only the Fc part of the IgG molecules. The Fc part of specifically bound Ig molecules can be expected to be exposed for protein A (gold) binding. It is not clear how non-specifically bound Ig molecules are attached to the substrate, but it is conceivable that some of them do not expose the Fc part so that steric hindrance prevents binding of protein A (gold). The other procedures are based on secondary antibodies which probably recognize and bind to primary antibodies in any position.

pA–gold can be used in a simple double labelling procedure (Fig. 3). Sections are first labelled for one antigen according to the two-step method.

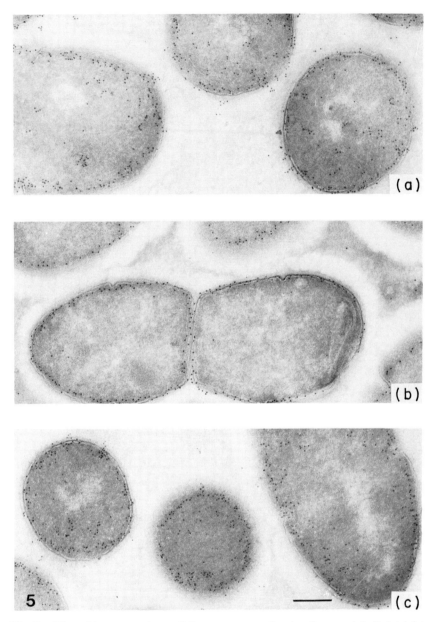

Fig. 5. Ultra-thin cryosections of *Streptococcus salivarius*. Immunolabelled (a) for rabbit anti-LTA and swine anti-rabbit IgG–gold (5 nm); (b) for rabbit anti-LTA and pA–gold (5 nm); and (c) for rabbit anti-LTA, swine anti-rabbit IgG and pA–gold (5 nm). Bar, 0.2 μm.

Then, after incubation for a few minutes with excess free protein A ($\sim 50\,\mu g$ ml^{-1}), they are subjected again to a two-step pA–gold labelling for a second antigen using a gold probe of different size (Geuze et al., 1981; Slot and Geuze, 1984). The procedure can be extended in a similar way for marking more than two antigens in one section (Slot and Geuze, 1985). Ig–gold cannot be used in this procedure unless a rather harsh treatment between the two labelling sequences is introduced to inactivate any free IgG binding sites of the first Ig–gold probe (Wang and Larsson, 1985). The two-step labelling with IgG

TABLE I
Immunogold labelling of amylase in 100 nm thick gelatin sections

Labelling procedure	Clusters per μm² (background[a])	Gold particles per μm² (background[a])	Gold particles per cluster
(1) 1 rabbit anti-amylase 2 swine anti-rabbit–gold	8.8 (6.8%)[a]	16.1 (5.1%)[a]	1.8
(2) 1 rabbit anti-amylase 2 protein A–gold	7.1 (3.1%)	8.2 (3.3%)	1.1
(3) 1 rabbit anti-amylase 2 swine anti-rabbit 3 protein A–gold	17.8 (7.1%)	51.1 (5.9%)	2.9

Cryosections were made from 10% gelatin blocks containing 1 mg ml^{-1} rat pancreas amylase and fixed in 2% glutaraldehyde.
[a] Non-specifically bound label, expressed as a percentage of the specific label, was measured in section parts that did not contain amylase.

instead of protein A as marker carrier can be used for double labelling, but then the antibodies against the two antigens involved have to be raised in different species, so that IgG–marker complexes with different specificities can be used together (Dutton et al., 1979; Tapia et al., 1983). The Ig–gold double labelling is more demanding in terms of the preparation of the immuno-reagents, but an advantage is that the two labelled IgG preparations can be applied simultaneously, which can particularly be important when the two antigens studied occur close together. In such a case, using sequential pA–gold double labelling, the first reaction may affect the second by steric hindrance. In the Ig–gold simultaneous procedure both reactions hinder each other to the same extent.

In conclusion, the two-step pA–gold labelling gave the best results in respect of resolution and low background. Also the distribution of one gold particle per specific IgG is preferable, particularly for a quantitative appreciation of the labelling, and the method offers a simple but very useful double-label

procedure. For these reasons we use pA–gold routinely in our immunocytochemical studies. If we want to boost the signal, or when we have to deal with
protein A-insensitive antibodies, the same pA–gold probe is used in the three-
step procedure.

References

Beesley, J. E. (1984). In *Immunolabeling for Electron Microscopy* (J. M. Polak and I. M. Varndell, eds), pp. 289–303. Elsevier, Amsterdam.

Beesley, J. E., Orpin, A. and Adlam, C. (1984). *Histochem. J.* **16**, 151–163.

Bernard, W. and Leduc, E. H. (1967). *J. Cell Biol.* **34**, 757–771.

Bosch, D., Leunissen, J., Verbakel, J., de Jong, M., van Erp, H. and Tommassen, J. (1986). *J. Mol. Biol.* **189**, 449–455.

Chang, L.-Y., Slot, J. W., Geuze, H. J. and Crapo, J. D. *Cell Biol.* (in press).

Christensen, A. K. (1967). *Anat. Rec.* **157**, 227.

Christensen, A. K. (1971). *J. Cell Biol.* **51**, 772.

de Mey, J. (1983). In *Immunocytochemistry* (J. M. Polak and S. Van Noorden, eds), pp. 82–112. Wright, Bristol, London and Boston.

de Mey, J., Moeremans, M., Geuens, G., Nuydens, R. and de Brabander, M. (1981). *Cell Biol. Int. Rep.* **5**, 889–899.

de Vries, J. W. A., Willemsen, R. and Geuze, H. J. (1985). *Eur. J. Cell Biol.* **37**, 81–88.

Dubochet, J., McDowall, A. W., Menge, B., Schmid, E. N. and Lickfeld, K. G. (1973). *J. Bacteriol.* **155**, 381–390.

Dubochet, J., Adrian, M., Lepault, J. and McDowell, A. W. (1985). *Trends Biochem. Sci.* **10**, 143–146.

Dutton, A., Tokuyasu, K. T. and Singer, S. J. (1979). *Proc. Natl Acad. Sci. USA* **76**, 3392–3396.

Eldred, W. D., Zucker, C., Karten, H. J. and Yazulla, S. (1983). *J. Histochem. Cytochem.* **31**, 285.

Faulk, W. P. and Taylor, G. M. (1971). *Immunocytochemistry* **8**, 1081–1083.

Fernandez-Moran, H. (1952). *Arkiv. f. Fysik.* **4**, 471–491.

Frens, G. (1973). *Nature (Phys. Sci.)* **241**, 20–22.

Freundl, R., Schwartz, H., Stierhof, Y.-D., Gamon, K., Hindennach, I. and Henning, U. (1986). *J. Biol. Chem.* **261**, 11355–11361.

Garbowski, L. (1903). Berichte der Deutschen Chemischen Gesellschaft, **36**, Band I, 1215–1220.

Geoghegan, W. D. and Ackerman, G. A. (1977). *J. Histochem. Cytochem.* **25**, 1187–1200.

Geuze, H. J. and Slot, J. W. (1980). *Eur. J. Cell Biol.* **21**, 93–100.

Geuze, H. J., Slot, J. W., Scheffer, R. C. T. and van der Ley, P. A. (1981). *J. Cell Biol.* **89**, 653–665.

Geuze, H. J., Slot, J. W., Yanagibashi, K., McCrackem, J. A., Schwartz, A. L. and Hall, P. F. (1987). *Histochemistry* **86**, 551–557.

Griffiths, G., Brands, R., Burke, B., Louvard, D. and Warren, G. (1982). *J. Cell Biol.* **95**, 781–792.

Griffiths, G., Simons, K., Warren, G. and Tokuyasu, K. T. (1983). *Methods Enzymol.* **96**, 466–483.

Griffiths, G., McDowall, A., Back, R. and Dubochet, J. (1984). *J. Ultrastruct. Res.* **89**, 65–78.
Hainfeld, J. F. (1987). *Science* **236**, 450–453.
Holgate, C. S., Jackson, P., Cowen, Ph. N. and Bird, C. C. (1983). *J. Histochem. Cytochem.* **31**, 938–942.
Horisberger, M. (1983). *Trends Biochem. Sci.* **8**, 395–397.
Horisberger, M. and Clerc, M.-F. (1985). *Histochemistry* **82**, 219–223.
Horisberger, M. and Rosset, J. (1977). *J. Histochem. Cytochem.* **25**, 295–305.
Keller, G.-A., Tokuyasu, K. T., Dutton, A. H. and Singer, S. J. (1984). *Proc. Natl Acad. Sci. USA* **81**, 5744–5747.
Linssen, W. H., Huis in 't Veld, J. H. J., Poort, C., Slot, J. W. and Geuze, H. J. (1973). In *Electron Microscopy and Cytochemistry* (E. Wisse, W. Th. Daems, I. Molenaar and P. van Duijn, eds), pp. 193–196. North-Holland, Amsterdam.
Livesey, S. A. and Linner, J. G. (1987). *Nature* **327**, 255–256.
Lucocq, J. M. and Baschong, W. (1986). *Eur. J. Cell Biol.* **42**, 332–337.
McDowall, A. W., Chang, J. J., Freeman, R., Lepault, R. Walter, C. A. and Dubochet, J. (1983). *J. Microsc. (Oxford)* **131**, 1–9.
Muhlpfordt, H. (1982). *Experientia* **38**, 1127–1128.
Posthuma, G., Slot, J. W. and Geuze, H. J. (1987). *J. Histochem. Cytochem.* **35**, 405–410.
Posthuma, G., Slot, J. W. and Geuze, H. J. (1988). *Euro. J. Cell. Biol.* **46**, 327–335.
Roth, J. (1983). In *Techniques in Immunocytochemistry* (G. R. Bullock and P. Petrusz, eds), Vol. II, pp. 217–284. Academic Press, London.
Roth, J., Bendayan, M. and Orci, L. (1978). *J. Histochem. Cytochem.* **26**, 1074–1081.
Roberts, I. M. (1975). *J. Microscopy* **103**, 113–119.
Romano, E. L. and Romano, M. (1977). *Immunochemistry* **14**, 711–715.
Romano, E. L., Stolinski, C. and Hughes-Jones, N. C. (1974). *Immunochemistry* **11**, 521–522.
Schilstra, M. J., Slot, J. W., van der Meide, P. H., Posthuma, G., Cremers, A. F. M. and Bosch, L. (1984). *FEBS Lett.* **165**, 175–179.
Singer, S. J. (1959). *Nature (London)* **183**, 1523–1524.
Slot, J. W. and Geuze, H. J. (1981). *J. Cell Biol.* **90**, 533–536.
Slot, J. W. and Geuze, H. J. (1982). *Biol. Cell* **44**, 325–328.
Slot, J. W. and Geuze, H. J. (1983). In *Immunohistochemistry* (A. C. Cuello, ed.), pp. 323–346. John Wiley and Sons, Chichester.
Slot, J. W. and Geuze, H. J. (1984). In *Immunolabeling for Electron Microscopy* (J. M. Polak and I. M. Varndell, eds), pp. 129–142. Elsevier, Amsterdam.
Slot, J. W. and Geuze, H. J. (1985). *Eur. J. Cell Biol.* **38**, 87–93.
Slot, J. W., Geuze, H. J., Freeman, B. A. and Crapo, J. D. (1986). *Lab. Invest.* **55**, 363–371.
Slot, J. W., Posthuma, G., Chang, L.-Y., Crapo, J. D. and Geuze, H. J. (in press).
Tapia, F. J., Varndell, I. M., Probert, K., de Mey, J. and Polak, J. M. (1983). *J. Histochem. Cytochem.* **31**, 977–981.
Tokuyasu, K. T. (1973). *J. Cell Biol.* **57**, 551–565.
Tokuyasu, K. T. (1978). *J. Ultrastruct. Res.* **63**, 287–307.
Tokuyasu, K. T. (1980). *Histochem. J.* **12**, 381–403.
Tokuyasu, K. T. (1986). *Proc. XIth Int. Congr. EM*, Kyoto, Japan, pp. 42–43.
Tokuyasu, K. T. and Singer, S. J. (1976). *J. Cell Biol.* **71**, 894–906.
Tokuyasu, K. T. and Ukamara, S. (1959). *J. Biophys. Biochem. Cytol.* **6**, 305–308.
Tommassen, J., Leunissen, J., van Damme-Jongsten, M. and Overduin, P. (1985). *EMBO J.* **4**, 1041–1047.

Van Driel, D., Wicken, A. J., Dickson, M. R. and Krox, K. W. (1973). *J. Ultrastruct. Res.* **43**, 483–497.

Voorhout, W. F., Leunissen-Bijvelt, J. J. M., Leunissen, J. L. M. and Verkleij, A. J. (1986). *J. Microscopy* **141**, 303–310.

Wang, B.-L. and Larsson, L.-I. (1985). *Histochemistry* **83**, 47–56.

Wang, B.-L., Scopsi, L., Hartvig Nielsen, M. and Larsson, L.-I. (1985). *Histochemistry* **83**, 109–115.

Weerkamp, A. H., Handley, P. S., Baars, A. and Slot, J. W. (1986). *J. Bacteriol.* **165**, 746–755.

Weerkamp, A. H., van der Mei, H. C. and Slot, J. W. (1987). *Infect. Immun.* **55**, 438–445.

Zsigmondy, R. and Thiessen, P. A. (1925). *Das Kolloidale Gold.* Akademische Verlagsgesellschaft, Leipzig.

10

Localization of Bacterial Enzymes by Electron Microscopic Cytochemistry as Demonstrated for the Polar Organelle

HORST-DIETMAR TAUSCHEL

Hinterdorfstrasse 29, D-7637 Ettenheim 3, Federal Republic of Germany

I. Introduction

The movement of bacteria, in particular the rotation of their flagella, is still not understood, and the question of what is the driving force has not yet been adequately answered. We can still only speculate as to where the energy needed for flagella rotation is generated, and we do not know exactly what kind of energy it is. Attempts to look for mitochondrial equivalents by cytochemical methods have sometimes shown some tellurite and tetranitro-blue tetrazolium (TNBT) reducing sources near the flagella base (Iterson and Leene, 1964a,b). In polar flagellated bacteria, a highly ordered complex structure is often seen (see Tauschel and Drews, 1969a). It is frequently associated with or is in the vicinity of the flagella base (Murray and Birch-Andersen, 1963; Ritchie *et al.*, 1966; Cohen-Bazire and London, 1967; Tauschel and Drews, 1969a,b; Vaituzis and Doetsch, 1969; Tauschel, 1985). There has been speculation about its function and its connection to the flagella apparatus. Due to the structural similarity of this complex organelle to the

METHODS IN MICROBIOLOGY
VOLUME 20 ISBN 0 12 521520 6

spherical basal structures of flagella observed in the phototroph bacterium *Rhodopseudomonas palustris*, it has been hypothesized that these spherical basal bodies have a morphological connection with and originate from the polar structure (Tauschel and Drews, 1969b). As this structure has mostly been observed in the polar cell region, it has been called polar membrane or polar organelle. This is presented schematically in association with or as a part of the flagella apparatus in Fig. 1. Until recently nothing was known about the function or the enzymatic activity of this polar organelle. The work of Vaituzis (1973) indicated that ATPase might be associated with this structure. More detailed information about this polar organelle comes from my own work. Therefore I would like to describe the cytochemical methods used in characterizing the ATPase and cytochrome oxidase of this organelle. My findings might help in future studies to isolate and characterize this structure further; as yet isolation and identification studies have been unsuccessful. In addition, the results might help to identify the mechanism of flagella rotation. A hypothesis resulting from the described data will be discussed.

II. Methodology

To characterize the polar organelle by electron microscopical cytochemistry, successful attempts for the localization of ATPase and cytochrome oxidase have been made. In addition, these methods seem to be specific for the demonstration of the polar organelle, at least in *Sphaerotilus natans* and *Rhodopseudomonas palustris*.

A. Cytochrome oxidase

1. Mechanism of cytochemical reaction

Seligman *et al.* (1968) introduced DAB (shortened name: 3,3'-diamino-benzidine; full name: 3,3',4,4'-tetraminobiphenyl: MW 214.27) as a cyto-chemical stain for cytochrome oxidase in mitochondria. This reagent is now much used for demonstration of this enzyme. It catalyses the following reaction (A) within the respiratory chain (Brunori and Wilson, 1982):

(A) $4\,H^+ + O_2 + 4\text{cyt}\,c^{2+} \rightarrow 2\,H_2O + 4\text{cyt}\,c^{3+}$

DAB is oxidized by this reaction due to the interaction with the cytochrome oxidase complex of the terminal end of the respiratory chain (B, C), and then it polymerizes further (C; Seligman *et al.*, 1968; Anderson *et al.*, 1975; Lewis, 1977a):

(B)

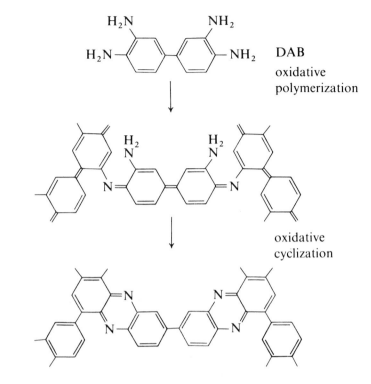

DAB
oxidative
polymerization

oxidative
cyclization

During this reaction the polymer (DAB)$_n$ accumulates at its place of generation due to its insolubility. However, during the oxidative polymerization of DAB a sufficient amount of primary amino groups still remains, which enables the reaction with OsO$_4$ during fixation (Lewis, 1977a). This interaction leads to the electron-scattering cytochemical stain "osmium black" (Hanker et al., 1964, 1967; Hayat, 1981), which is, in this case, mainly a coordination polymer of osmium with the DAB polymer (C; Anderson et al., 1975):

(C)

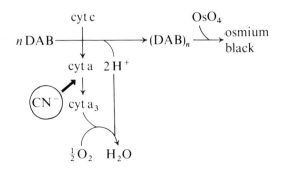

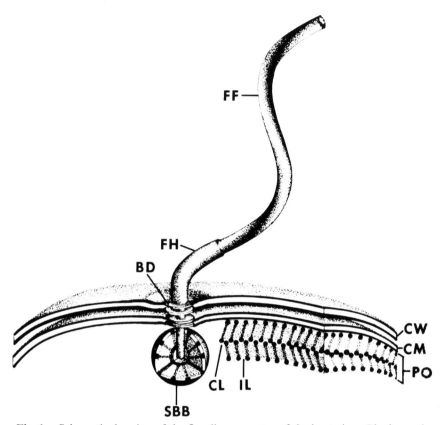

Fig. 1. Schematic drawing of the flagella apparatus of the bacterium *Rhodopseudomonas palustris*: it consists of the flagella filament (FF), the flagella hook (FH), the basal discs (BD) located in both the cell wall (CW) and the cytoplasmic membrane (CM), the rod (proximal part of the flagellum) which is inserted in the spherical basal bodies (SBB), and the polar organelle (PO). The PO is separated from the cytoplasm by the inner boundary layer (IL) and contains the central layer (CL).

"Osmium black" is a fine, amorphous polymer, insoluble in water, organic solvents used for dehydration, and resins. Therefore it is an excellent stain and indicator for enzyme activity.

The first point of interaction of DAB with the cytochrome oxidase complex is cytochrome c (C; Seligman *et al.*, 1968, 1973; Hirai, 1971). Therefore it can also be used to discover the localization of cytochrome c. The cytochemical reaction for cytochrome oxidase can be blocked specifically by cyanide and azide (Seligman *et al.*, 1968; Reith and Schüler, 1972). These substances inhibit the terminal reaction of the respiratory chain, and therefore the formation of osmium black (C).

The use of KCN not only allows the termination of the cytochrome oxidase activity, but also enables the differentiation of O_2-dependent oxidoreductase activity of the respiratory chain from that of photochemically driven reactions in chloroplasts and photosynthetically active microorganisms (Lauritis et al., 1975; Stevens et al., 1977; Gierczak et al., 1982).

2. Buffers and reagents

- DAB (3,3'-diaminobenzidine tetrahydrochloride) (E. Merck, Darmstadt, FRG).
- Catalase from beef liver, stabilized crystal suspension in water, $20\,mg\,ml^{-1}$ (Boehringer Mannheim GmbH, Mannheim, FRG).
- KCN (potassium cyanide) (E. Merck, Darmstadt, FRG).
- PBS (phosphate-buffered saline), pH 7.2 (Dulbecco and Vogt, 1954):

NaCl	8.00 g
KCl	0.20 g
Na_2HPO_4	1.15 g
KH_2PO_4	0.20 g
$CaCl_2$	0.10 g
$MgCl_2 \cdot 6\,H_2O$	0.10 g
Distilled water to make 1000 ml	

- Sodium phosphate buffer, pH 7.0, made from 20 mM Na_2HPO_4 with pH adjustment by means of the pH meter with 20 mM NaH_2PO_4.
- Formaldehyde 4% phosphate buffered: 4% formaldehyde solution buffered with 20 mM sodium phosphate buffer, pH 7.0, made up by mixing 1 part of acid-free 37% formaldehyde solution (E. Merck, Darmstadt, FRG) with 8.25 parts of the 20 mM sodium phosphate buffer.

3. Experimental procedure

For demonstration of the cytochrome oxidase the method of Seligman et al. (1968) has been modified by changing the buffer concentration and the pH of the phosphate buffer for the fixation mixture, as well as for the reaction mixture. Furthermore, cytochrome c and sucrose have been omitted from the reaction mixture, and the concentrations of DAB and of the catalase have been changed. The procedures for the (relatively) specific demonstration of the cytochrome oxidases of the polar organelle and of the cytoplasmic membrane in S. natans are represented by the flow diagrams in Table I.

TABLE I

Flow diagram presenting the steps for demonstration of the cytochrome oxidases of the polar organelle and of the cytoplasmic membrane in *Sphaerotilus natans*

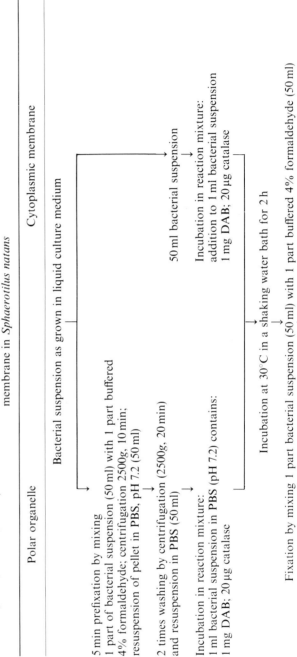

Polar organelle   Cytoplasmic membrane

Bacterial suspension as grown in liquid culture medium

5 min prefixation by mixing
1 part of bacterial suspension (50 ml) with 1 part buffered
4% formaldehyde; centrifugation 2500g, 10 min;
resuspension of pellet in **PBS**, pH 7.2 (50 ml)

2 times washing by centrifugation (2500g, 20 min)
and resuspension in **PBS** (50 ml)

Incubation in reaction mixture:
1 ml bacterial suspension in **PBS** (pH 7.2) contains:
1 mg **DAB**; 20 µg catalase

50 ml bacterial suspension

Incubation in reaction mixture:
addition to 1 ml bacterial suspension
1 mg **DAB**; 20 µg catalase

Incubation at 30°C in a shaking water bath for 2 h

Fixation by mixing 1 part bacterial suspension (50 ml) with 1 part buffered 4% formaldehyde (50 ml)

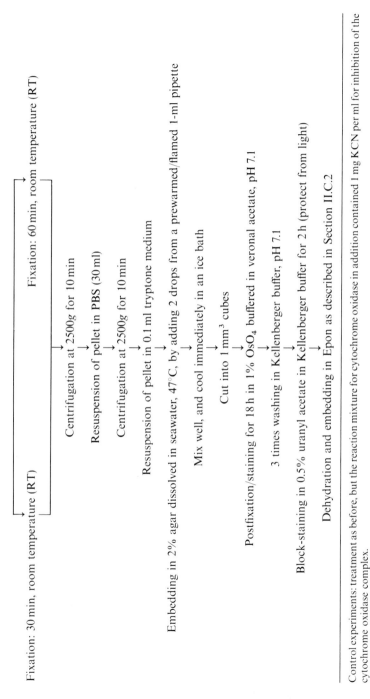

Fixation: 30 min, room temperature (RT)

Fixation: 60 min, room temperature (RT)

Centrifugation at 2500g for 10 min

Resuspension of pellet in PBS (30 ml)

Centrifugation at 2500g for 10 min

Resuspension of pellet in 0.1 ml tryptone medium

Embedding in 2% agar dissolved in seawater, 47°C, by adding 2 drops from a prewarmed/flamed 1-ml pipette

Mix well, and cool immediately in an ice bath

Cut into 1 mm^3 cubes

Postfixation/staining for 18 h in 1% OsO$_4$ buffered in veronal acetate, pH 7.1

3 times washing in Kellenberger buffer, pH 7.1

Block-staining in 0.5% uranyl acetate in Kellenberger buffer for 2 h (protect from light)

Dehydration and embedding in Epon as described in Section II.C.2

Control experiments: treatment as before, but the reaction mixture for cytochrome oxidase in addition contained 1 mg KCN per ml for inhibition of the cytochrome oxidase complex.

B. ATPase

1. Mechanism of cytochemical reaction

The basic mechanism of the ATPase reaction is the hydrolysis of ATP and the capturing of orthophosphate by lead ions according to the method of Wachstein and Meisel (1957). Insoluble lead phosphate is formed which precipitates at the place of enzyme reaction. The studies of Tandler and Solari (1968, 1969) demonstrated that the precipitate formed from orthophosphate and lead is lead hydroxyapatite ($Pb_5(PO_4)OH$). This precipitate can be used for the demonstration of ATPase activity due to its electron-scattering properties. The problems, difficulties and pitfalls associated with this cytochemical method are extensively reviewed by Firth (1978) and Lewis (1977b), and will not be discussed further in this chapter.

2. Buffers and reagents

- Tris-maleate buffer, 100 mM, pH 7.2, charcoal-filtered (modified after Hayat, 1972):
  Trizma maleate (mono[tris(hydroxymethyl)aminomethane]maleate) (Sigma Chemical Co., St Louis, MO, USA): dissolve 23.72 g in 800 ml distilled water; adjust to pH 7.2 with 2 N HCl; top up with distilled water to give 1000 ml; resuspend 4 g charcoal in buffer and let stay for 10 min; remove charcoal by filtration.

- Tris-maleate-buffered incubation medium: to make up the complete reaction mixture, mix the following solutions:

Solutions	Volume (ml)	Final concentration (mmol litre^{-1})
Tris-maleate buffer (100 mM, pH 7.2)	2.0	40
$MgSO_4$ (80 mM)	0.5	8
NaCl (500 mM)	0.1	10
KCl (400 mM)	0.1	8
$CaCl_2$ (100 mM)	0.1	2
ATP (40 mM)	1.0	8
Pb-(II)acetate (100 mM)	0.1	2
Bacterial suspension	1.1	

$$\Sigma = 5.0$$

- Tris-acetate buffer, 40 mM, pH 7.0:
  Trizma base (MW 121.14; Sigma Chemical Co., St Louis, MO, USA): dissolve 4.846 g in 900 ml distilled water; adjust to pH 7.0 with glacial acetic acid; top up with distilled water to give 1000 ml.

• Tris-acetate-buffered incubation medium *without* ATP, Pb^{2+}:

Compound	MW	Concentration (g litre^{-1})	Concentration (mmol litre^{-1})
Trizma base	121.14	4.846	40
$MgCl \cdot 6 H_2O$	203.33	0.813	4
NaCl	58.45	5.845	100
KCl	75.55	1.511	20

Dissolve compounds in 900 ml distilled water; adjust to pH 7.0 with glacial acetic acid; top up with distilled water to give 1000 ml.

• Tris-acetate-buffered incubation medium *with* ATP, Pb^{2+}:
Immediately before use add lead nitrate to the medium mentioned above and mix thoroughly to prevent precipitation (solution must be clear). Then add ATP:

Compound	MW	Concentration (g litre^{-1})	Concentration (mmol litre^{-1})
$Pb(NO_3)_2$	331.20	0.166	0.5
$ATP-Na_2H_2 \cdot 3 H_2O$	605.2	2.421	4

• Formaldehyde (4%), phosphate-buffered. For details see Section II.A.2.

3. Experimental procedure

ATPase reaction was performed using a modification of the method of Wachstein and Meisel (1957), as outlined in the flow diagrams of Table II.

C. Electron microscopical procedures

1. Buffers and reagents

• Tryptone medium (Ryter and Kellenberger, 1958)

Bacto-Tryptone (Difco Laboratories, Detroit, Michigan, USA)	1.0 g
NaCl	0.5 g
Distilled water to make	100 ml

Sterilize solution in autoclave and store at 4°C

• Veronal acetate buffer (Ryter and Kellenberger, 1958)

Sodium acetate	1.94 g
Sodium veronal/barbital	2.94 g
NaCl	3.40 g
Distilled water to make	100 ml

TABLE II

Flow diagram of preparation steps for cytochemical demonstration of ATPase in *Sphaerotilus natans* and *Rhodopseudomonas palustris*

Sphaerotilus natans	*Rhodopseudomonas palustris*
100 ml bacterial culture suspension	Bacterial culture suspension
Centrifugation 3000g, 10 min	Centrifugation 7700g, 10 min
Resuspension of pellet in 100 mM *Tris-maleate* buffer, pH 7.2 (5 ml)	Resuspension of pellet in 10 ml *Tris-acetate*-buffered incubation medium *without* ATP and Pb^{2+}, pH 7.1
Prefixation with equal volume (5 ml) phosphate-buffered 4% formaldehyde (pH 7.0) for 5 min (final conc. 2%)	
Centrifugation 3000g, 10 min	Centrifugation 7700g, 10 min
Resuspension of pellet in *Tris-maleate*, pH 7.2 (10 ml)	Resuspension of pellet again in *Tris-acetate*-buffered medium *without* ATP and Pb^{2+}
Centrifugation 3000g, 10 min	
Resuspension of pellet in *Tris-maleate*, pH 7.2	To give right turbidity for direct EM studies, mix some drops of bacterial suspension into complete *Tris-acetate*-buffered incubation mixture containing ATP + Pb^{2+}
Mixing of 1.1 ml of the bacterial suspension with 3.9 ml of the complete *Tris-maleate*-buffered incubation medium	

Incubation for 5 h at 30°C in a water bath

$\longrightarrow$

Drop of cell suspension placed on Formvar-coated, carbon-reinforced EM grid

$\rightarrow$

Direct evaluation in TEM *without* further staining

Incubation at 30°C in a water bath, 2 h

$\rightarrow$

Centrifugation 2500g, 10 min

$\rightarrow$

Resuspension of pellet in 0.1 ml tryptone medium

$\longrightarrow$

Postfixation for 18 h in 1 ml 1% OsO$_4$ buffered in Kellenberger buffer, pH 7.1

$\rightarrow$

Dilution with 8 ml Kellenberger buffer, pH 7.1

$\rightarrow$

Centrifugation 2500g, 10 min

$\rightarrow$

Discard supernatant and resuspend pellet in adherent liquid

$\rightarrow$

Embedding in 2% agar dissolved in seawater, 47°C, by adding 2 drops from a prewarmed/flamed 1-ml pipette

$\rightarrow$

Mix well and cool immediately in an ice bath

$\rightarrow$

Cut into 1 mm^3 cubes

$\rightarrow$

Block-staining in 0.5% uranyl acetate in Kellenberger buffer for 2 h (protect from light)

$\rightarrow$

Dehydration and embedding in Epon as described in Section II.C.2

- Kellenberger buffer (Ryter and Kellenberger, 1958)

Veronal acetate buffer (see above)	5.0 ml
0.1 N HCl	7.0 ml
1 M $CaCl_2$	0.25 ml
Distilled water	13.0 ml
Adjust buffer to pH 7.1	

- 1% OsO_4 solution (Ryter and Kellenberger, 1958)

OsO_4	1.0 g
Kellenberger buffer	100 ml
Adjust to pH 7.1 with NaOH	

- 0.5% uranyl acetate for block staining

Uranyl acetate	0.05 g
Kellenberger buffer	10 ml

- Seawater, 20%

Seawater (prepared from commercial sea salt)	20 ml
Distilled water	80 ml

- Agar (2%) in seawater

Agar	2.0 g
Seawater, 20% (see above)	100 ml

Adjust to pH 7.0–7.1 with NaOH; sterilize in autoclave; aliquots should be stored in test tubes at 4°C

- 2% uranyl acetate, aqueous, for staining of thin sections

Uranyl acetate	0.2 g
Distilled water	10 ml

Dissolve in the dark one day before use; filtrate solution and centrifuge immediately for use at c. 5500g for 20 min

- Lead citrate (Reynolds, 1963)

Lead nitrate, $Pb(NO_3)_2$	1.33 g
Sodium citrate, $Na_3(C_6H_5O_7) \cdot 2H_2O$	1.76 g
Distilled water	30 ml

Dissolve components in a 50-ml volumetric flask by shaking for about 20 min. Then add 8.0 ml 1 N NaOH and fill up to 50 ml with distilled water. Mix by inversion. The solution should be clear and have a pH of 12.0.

2. Specimen preparation for EM

After block staining with 0.5% aqueous uranyl acetate, the bacteria–agar cubes were further treated as outlined:

Dehydration in a graded series of ethanol
70% alcohol	10 min
80% alcohol	10 min
90% alcohol	15 min
95% alcohol	15 min
100% alcohol (dried over beads of molecular sieves)	2 × 30 min
Propylenoxide	2 × 30 min
Propylenoxide: Epon (1:1)	60 min
Epon	overnight

Embedding in Epon and polymerization at 60°C

After ultra-thin sectioning, stain sections with
2% aqueous uranyl acetate	20 min
Lead citrate	10 min

III. Cytochemical localization of ATPase and cytochrome oxidase

A. ATPase

Preparations of whole cells of *Rhodopseudomonas palustris* for ATPase reaction (Section II.B.3) show in almost all cells uniformly electron dense deposits in a subpolar location in otherwise unstained cells (Fig. 2). These mostly ovoid-shaped structures have been interpreted as being the polar organelles of these subpolarly flagellated bacteria (Tauschel, 1987a). They have been made visible by the ATPase reaction leading to lead phosphate, which presumably forms the electron dense precipitates at the place of hydrolysis by ATPase. No other places of ATPase reaction within these bacterial cells can be demonstrated by this method.

Similar results were obtained in the subpolarly flagellated swarm cells of *Sphaerotilus natans* with slight prefixation (Section II.B.3); the ATPase reaction can only be seen at the regions of the polar organelle in thin sections (Figs 3b and c). Here ATPase is associated with the central layer of the polar organelle, as well as with the inner boundary layer. The inner boundary layer shows the strongest enzymatic activity, which leads to heavy electron dense precipitates located underneath this structure and directed towards the cytoplasm (Fig. 3c). The reaction not only enables the localization of this enzyme system at special substructures of the polar organelle, but also enables the

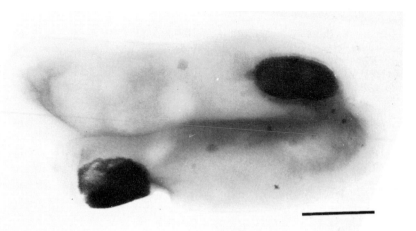

Fig. 2. *Rhodopseudomonas palustris*: demonstration of the near-surface, subpolar location of the polar organelle by ATPase reaction in whole, unstained cells. Bar, 0.5 μm.

visualization of the fine structure of this complex organelle, which under usual fixation and staining procedures is usually poorly preserved and difficult to recognize (Fig. 3a). The ATPase reaction may also help in identifying regions at the cytoplasmic membrane where the polar organelle is structurally not detectable in thin sections (Tauschel, 1985).

The cytochemical reaction product is primarily associated with the inner boundary layer of the polar organelle (Fig. 3b). The extension of the electron dense precipitate into the cytoplasmic matrix is probably due to the short-range diffusion of inorganic phosphate prior to capture by lead ions.

Trials to localize ATPase in whole negatively stained cells of *Bacillus licheniformis* treated for ATPase by the Wachstein–Meisel (1957) method showed areas of lead phosphate, some of which seemed to be associated with the flagella insertion area within the cells (Vaituzis, 1973). However, the electron dense precipitate had no defined structure. Also, in thin sections of bacteria treated for ATPase by the method mentioned above, some undefined electron dense deposits of reaction products were observed inside the cytoplasmic membrane. These were associated with the area of flagella insertion in the peritricheously flagellated *Bacillus licheniformis* and the polarly flagellated *Vibrio metchnikovii* (Vaituzis, 1973). However, the lead phosphate deposits did not seem to be directly associated with a specific subcellular structure, although it is known that *Vibrio* spp. possess a well-developed polar organelle which is usually located within this area (Ritchie *et al.*, 1966; Vaituzis and Doetsch, 1969). As the polar organelle is a very fragile

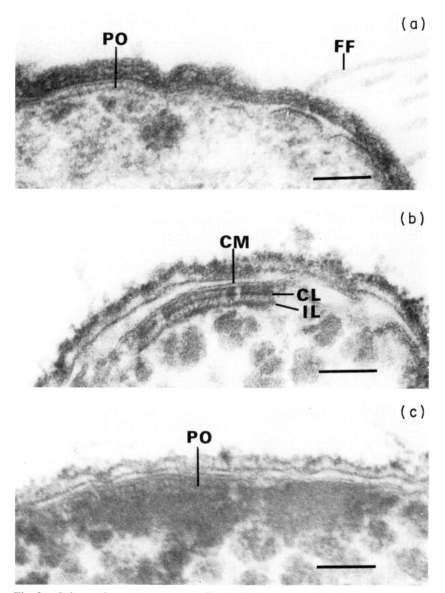

Fig. 3. *Sphaerotilus natans* swarm cells. (a) Thin section of polar organelle after Ryter–Kellenberger fixation. CL, central layer; CM, cytoplasmic membrane; FF, flagella. (b) Polar organelle treated for ATPase. Details of the polar organelle are clearly visible. CL, central layer; CM, cytoplasmic membrane; IL, inner layer. (c) Strong ATPase reaction in region of the polar organelle (PO) as demonstrated by electron-dense precipitates. Bar, 0.1 μm.

structure, it is possible that under the chosen fixation (6.5% glutaraldehyde for 30 min; Vaituzis, 1973) and preparation conditions the polar organelles might have lost their structural integrity; however, they still might possess some ATPase activity.

In general, mitochondrial and bacterial ATPases are very sensitive to the fixatives used and to the fixation time (Burnham and Hageage, 1973; Lewis, 1977b). Therefore no prefixation prior to cytochemical ATPase reaction, or only a very brief fixation with low concentrations of aldehyde fixatives, might be the recipe for successful demonstration of both structural and functionally well-preserved polar organelles, as has been shown in this chapter.

B. Cytochrome oxidase

Interesting results have been obtained in *S. natans* swarm cells by the cytochrome oxidase reaction. Dependent on whether a short prefixation was used or not, two types of cytochrome oxidases have been demonstrated.

Short prefixation with 2% formaldehyde (final concentration) prior to the DAB reaction yields a very specific cytochrome oxidase reaction occurring only within the polar organelle (Fig. 4a). The osmiophilic reaction products are accurately located within the spaces of the polar organelle, which are usually traversed by ladder-like spokes. The central layer appears unstained, separating the two reaction zones for cytochrome oxidase (Figs 4a and b). In contrast to the ATPase reaction, no diffusion of the reaction product DAB–osmium black is seen, which therefore delineates clearly the polar organelle.

The most impressive preservation and demonstration of the polar organelle's fine structure is seen under the control conditions for cytochrome oxidase in the presence of potassium cyanide (Fig. 4c), at least as far as the central layer and the inner boundary layer of the polar organelle are concerned. The cytochrome oxidase reaction is completely inhibited under these conditions, leaving electron-transparent spaces in the regions of the polar organelle spokes, whereas the boundary layers are stained.

A second type of cytochrome oxidase is associated with the inside of the cytoplasmic membrane. This enzyme can be identified in cells of this bacterial species without any kind of prefixation, i.e. in the native state (Fig. 5). A clearly defined zone of black osmiophilic reaction products frames the whole cytoplasm of swarm cells (Fig. 5a). This zone appears unstained, electron-transparent and structureless under control conditions in the presence of KCN (Fig. 5b). A membrane-associated cytochrome oxidase has also been demonstrated in the filamentous sheath-forming cells of *S. natans*, under exactly the same reaction conditions. But here, however, it possesses a much weaker enzyme activity (Tauschel, 1987b).

Different fixation conditions seem to enable the specific demonstration of

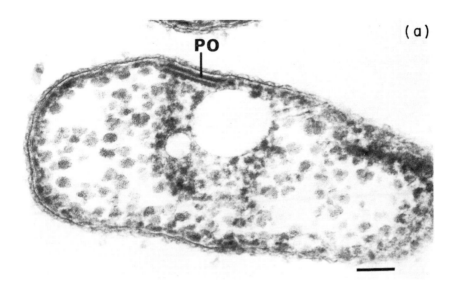

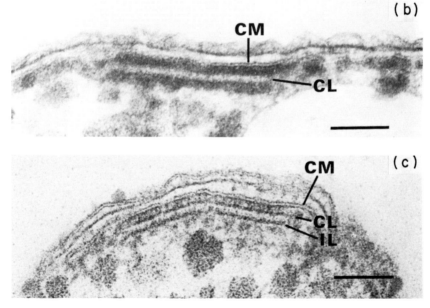

Fig. 4. *Sphaerotilus natans* swarm cells in longitudinal section. (a) Heavy osmiophilic deposits in the spaces of the polar organelle (PO) usually traversed by fine fibrils (spokes) demonstrate the presence of cytochrome oxidase activity. The central layer appears unstained in the subpolarly located PO. Bar, 0.2 μm. (b) Enlarged detail of (a) showing the polar organelle with cytochrome oxidase activity. CL, central layer; CM, cytoplasmic membrane. Bar, 0.1 μm. (c) The detailed structure of the polar organelle is clearly visible in cells treated for cytochrome oxidase in the presence of KCN. CL, central layer; CM, cytoplasmic membrane; IL, inner layer. Bar, 0.1 μm.

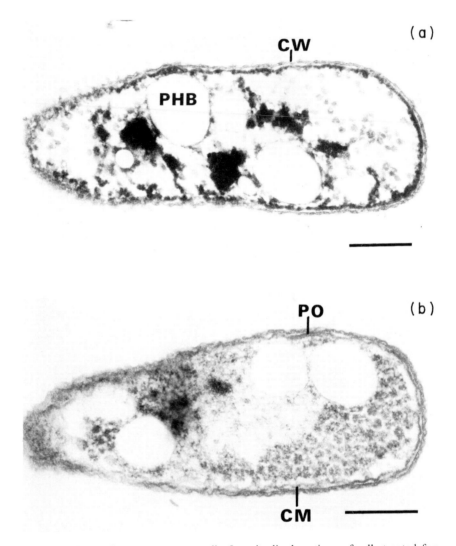

Fig. 5. *Sphaerotilus natans* swarm cells. Longitudinal sections of cells treated for cytochrome oxidase. (a) Cytochrome oxidase characterized by osmiophilic reaction products is associated with the inside of the cytoplasmic membrane. CW, cell wall; PHB, polyhydroxybutyric acid granules. Bar, 0.5 μm. (b) Control cell incubated in the presence of KCN. This shows an electron-transparent space between the cytoplasm and the cytoplasmic membrane (CM). PO, polar organelle. Bar, 0.5 μm.

two differently located cytochrome oxidases. This might be due to different sensitivities of the enzymes to the fixative, in this case to formaldehyde. While the membrane-associated cytochrome oxidase seems to lose its activity very quickly, even under short prefixation (5 min) and in quite low formaldehyde concentrations (2%), the polar organelle-associated enzyme, in contrast, is activated under these conditions.

A search in the literature showed that the membrane-associated oxidoreductases in bacteria (Stevens et al., 1977; Gierczak et al., 1982), cyanobacteria (Lauritis et al., 1975) and the corresponding enzyme system in mitochondria associated with the inner mitochondrial membrane (Seligman et al., 1968; Reith and Schüler, 1972; Anderson et al., 1975; Perotti et al., 1983) all behave very similarly. In contrast to these, the cytochrome oxidase located in the intracristate spaces of the mitochondria, as well as the DAB oxidizing, light-driven oxidoreductase system of the chromatophores of the photosynthetic bacterium *Rhodospirillum rubrum* and of the thylakoids in cyanobacteria, are quite resistant, or at least not so sensitive, to fixation.

However, fixation may not be the only critical and dangerous step for the cytochrome oxidase system of the bacterial cytoplasmic membrane and the inner mitochondrial membrane. Scholes et al. (1971) described a method by which they could easily remove cytochrome c from intact bacterial cells of *Micrococcus denitrificans* just by washing with 0.01 M phosphate buffer, or with the same buffer containing KCl up to about 0.15 mol litre^{-1}. Therefore one has to ask whether the observed loss or reduction in enzyme activity might be due to leakage or loss of cytochrome c from the membranes, which might be washed away by the buffers used before and after prefixation to remove the fixative. Very often phosphate buffers are used for this step, but other buffers might also be harmful in this sense. The loss or leakage of cytochrome c might be aggravated by the prefixation of cells (Beard and Novikoff, 1969; Seligman et al., 1973), making their outer membranes more leaky. As the loss of cytochrome c would also cause a reduction of cytochrome oxidase activity, a reduced oxidation of DAB would be the consequence, with reduced formation of DAB polymer (Seligman et al., 1968, 1970, 1973; Anderson et al., 1975; Frash et al., 1978). But, as DAB or auto-oxidized DAB directly interact with and bind to cytochrome c (Hirai, 1971), a loss of cytochrome c would also lead to the absence of or to a reduced cytochemical reaction.

The pH of washing solutions, prefixation solutions as well as of DAB reaction solutions, is probably not very critical for cytochemical demonstration of the cytochrome oxidases. Solutions within the range of pH 7.0–7.2 (Perotti et al., 1983; Tauschel, 1985) up to pH 7.6 (Stevens et al., 1977; Gierczak et al., 1982) have been used with positive results. *In vitro* studies on cytochrome oxidase membranes, which resemble in many respects natural cytochrome oxidase-bearing membranes, reveal that a variation of the pH between 6.4 and

8.0 is not harmful to the cytochrome oxidase lattice (Vanderkooi, 1974), but the lattice becomes unstable if the pH rises above 8.0. These data correspond very well with DAB oxidation by the cytochrome oxidase-carrying electron-transport particles (ETP) of *Azotobacter vinelandii*, which show their highest oxidation rate at pH 7.0 (Jurtshuk *et al.*, 1979).

IV. Outlook: importance of enzyme systems for flagella rotation

Based on the cytochemical findings for the polar organelle (as presented schematically in Fig. 6a) and the biochemical results known for the respiratory chain in mitochondria, a hypothetical model for the polar organelle is presented in Fig. 6b combining these data. According to Racker (1970), the vectorial topography of the various electron-transport components is essential for their functional coordination and for the coupling of the energy produced during oxidation. As a result, the polar organelle, with its interesting fine structural architecture, would function as a power generator producing ATP and hydrogen ions which might be needed for flagella rotation. This would be in agreement with the hypothesis of Murray and Birch-Andersen (1963) that the polar organelle (polar membrane) is responsible for the liberation of energy for the flagella.

At this point it may be recalled that the bacterial flagella, at least those of *Rhodopseudomonas palustris*, are inserted in spherical basal bodies with, at their very proximal part, a rod-like structure (Tauschel and Drews, 1970). These basal bodies are very fragile and barely discernible as they are not surrounded by a unit membrane structure, and they possess a highly ordered spoke-like system as illustrated in Fig. 1. Assuming that these spherical basal bodies are real structures which are formed by the polar organelle (Tauschel and Drews, 1969b), a new hypothesis might be postulated to explain flagella rotation.

Taking into account the widely accepted view of the "universality of filament sliding" (Rinaldi and Baker, 1969; Donaldson, 1972), this scheme might also be transferred to a rotational sliding movement of the bacterial flagellum, similarly to the rotational movement which has been demonstrated for the actomyosin interaction (Yano *et al.*, 1982). Like the interactions that occur between actin and myosin under involvement of ATP as substrate, the interactions in the bacterial flagella apparatus must occur between the very proximal end of the flagellum and the spoke-like system or subunits of the spherical basal bodies. Energy in the kind of ATP generated either in the basal bodies or at the polar organelle must be put into this system only at the flagella base. However, Mitchell (1984) suggests that ATP might also be used in an ion-pumping system to generate an ion-motive force.

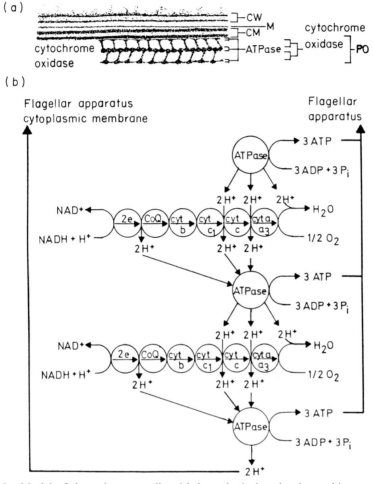

Fig. 6. Model of the polar organelle with hypothetical molecular architecture. (a) Localization of enzyme systems (cytochrome oxidase, ATPase) as observed by electron microscopic cytochemistry. (b) Combination of the electron microscopical findings with known biochemical data from the respiratory chain in mitochondria (see Sengbusch, 1979).

A variation on the above-mentioned hypothesis is the hypothesis of rotational flagella movement presently favoured. According to this model, the flagella are driven by a proton-motive/ion-motive force, presumably generated at the basal discs of the flagellum which are inserted into the cytoplasmic membrane (Macnab, 1979, 1983, 1984; Berg and Khan, 1983; Block and Berg, 1984; Mitchell, 1984).

As the polar organelle has only been observed in flagellated bacteria, and then mostly in close association with the flagella, it might be postulated that it works as a proton-generating and/or proton-translocating pumping station. This would be in good agreement with the recent findings that cytochrome oxidase of *Bacillus subtilis* can generate a proton-motive force in proteoliposomes and membrane vesicles (Vrij *et al.*, 1986).

References

Anderson, W. A., Bara, G. and Seligman, A. M. (1975). *J. Histochem. Cytochem.* **23**, 13–20.

Beard, M. E. and Novikoff, A. B. (1969). *J. Cell Biol.* **43**, 12a.

Berg, H. C. and Khan, S. (1983). In *Mobility and Recognition in Cell Biology* (H. Sund and C. Veeger, eds), pp. 485–497. Walter de Gruyter, Berlin and New York.

Block, S. M. and Berg, H. C. (1984). *Nature* **309**, 470–472.

Brunori, M. and Wilson, M. T. (1982). *Trends Biochem. Sci.* **7**, 295–299.

Burnham, J. C. and Hageage, G. J. Jr (1973). *Can. J. Microbiol.* **19**, 1059–1064.

Cohen-Bazire, G. and London, J. (1967). *J. Bacteriol.* **94**, 458–465.

Donaldson, J. G. (1972). *J. Theoret. Biol.* **37**, 75–91.

Dulbecco, R. and Vogt, M. (1954). *J. Exp. Med.* **99**, 167–182.

Firth, J. A. (1978). *Histochem. J.* **10**, 253–269.

Frash, A. C. C., Itoiz, M. E. and Cabrini, R. L. (1978). *J. Histochem. Cytochem.* **26**, 157–162.

Gierczak, J. S., Stevens, F. J., Pankratz, H. S. and Uffen, R. L. (1982). *J. Histochem. Cytochem.* **30**, 901–907.

Hanker, J. S., Seaman, A. R., Weiss, L. P., Ueno, H., Bergman, R. A. and Seligman, A. M. (1964). *Science* **146**, 1039–1043.

Hanker, J. S., Kasler, F., Bloom, M. G., Copland, J. S. and Seligman, A. M. (1967). *Science* **156**, 1737–1738.

Hayat, M. A. (1972). *Basic Electron Microscopy Techniques.* Van Nostrand Reinhold, New York, Cincinatti, Toronto, London and Melbourne.

Hayat, M. A. (1981). *Fixation for Electron Microscopy.* Academic Press, New York, London, Toronto, Sydney and San Francisco.

Hirai, K.-I. (1971). *J. Histochem. Cytochem.* **19**, 434–442.

Iterson, W. van and Leene, W. (1964a). *J. Cell. Biol.* **20**, 361–375.

Iterson, W. van and Leene, W. (1964b). *J. Cell Biol.* **20**, 377–387.

Jurtshuk, P. Jr, McQuitty, D. N. and Riley, W. H. IV (1979). *Curr. Microbiol.* **2**, 349–354.

Lauritis, J. A., Vigil, E. L., Sherman, L. and Swift, H. (1975). *J. Ultrastruct. Res.* **53**, 331–344.

Lewis, P. R. (1977a). In *Staining Methods for Sectioned Material* (P. R. Lewis and D. P. Knight, eds), pp. 225–287. North-Holland, Amsterdam, New York and Oxford.

Lewis, P. R. (1977b). In *Staining Methods for Sectioned Material* (P. R. Lewis and D. P. Knight, eds), pp. 137–223. North-Holland, Amsterdam, New York and Oxford.

Macnab, R. M. (1979). *Trends Biochem. Sci.* **4**, N10–N13.

Macnab, R. M. (1983). In *Mobility and Recognition in Cell Biology* (H. Sund and C. Veeger, eds), pp. 499–516. Walter de Gruyter, Berlin and New York.

Macnab, R. M. (1984). *Trends Biochem. Sci.* **9**, 185–188.

Mitchell, P. (1984). *FEBS Lett.* **176**, 287–294.

Murray, R. G. E. and Birch-Andersen, A. (1963). *Can. J. Microbiol.* **9**, 393–401.
Perotti, M. E., Anderson, W. A. and Swift, H. (1983). *J. Histochem. Cytochem.* **31**, 351–365.
Racker, E. (1970). In *Essays in Biochemistry* (P. N. Campbell and F. Dickens, eds), Vol. 6, pp. 1–22. Academic Press, London.
Reith, A. and Schüler, B. (1972). *J. Histochem. Cytochem.* **20**, 583–589.
Reynolds, E. S. (1963). *J. Cell Biol.* **17**, 208–213.
Rinaldi, R. A. and Baker, W. R. (1969). *J. Theoret. Biol.* **23**, 463–474.
Ritchie, A. E., Keeler, R. F., Bryner, J. H. and Elmore, J. (1966). *J. Gen. Microbiol.* **43**, 427–438.
Ryter, A., Kellenberger, E., Birch-Andersen, A. and Maaløe, O. (1958). *Z. Naturforsch.* **13b**, 597–605.
Scholes, P. B., McLain, G. and Smith, L. (1971). *Biochemistry* **10**, 2072–2076.
Seligman, A. M., Karnovsky, M. J., Wasserkrug, H. L. and Hanker, J. S. (1968). *J. Cell Biol.* **38**, 1–14.
Seligman, A. M., Seito, T. and Plapinger, R. E. (1970). *Histochemie* **22**, 85–99.
Seligman, A. M., Shannon, W. A. Jr, Hoskino, Y. and Plapinger, R. E. (1973). *J. Histochem. Cytochem.* **21**, 756–758.
Sengbusch, P. V. (1979). *Molekular- und Zellbiologie.* Springer-Verlag, Berlin, Heidelberg and New York.
Stevens, F. J., Pankratz, H. S. and Uffen, R. L. (1977). *J. Histochem. Cytochem.* **25**, 1264–1268.
Tandler, C. J. and Solari, A. J. (1968). *J. Cell Biol.* **39**, 134a.
Tandler, C. J. and Solari, A. J. (1969). *J. Cell Biol.* **41**, 91–108.
Tauschel, H.-D. (1985). *Arch. Microbiol.* **141**, 303–308.
Tauschel, H.-D. (1987a). *Arch. Microbiol.* **148**, 159–161.
Tauschel, H.-D. (1987b). *Arch. Microbiol.* **148**, 155–158.
Tauschel, H.-D. and Drews, G. (1969a). *Arch. Microbiol.* **66**, 166–179.
Tauschel, H.-D. and Drews, G. (1969b). *Arch. Microbiol.* **66**, 180–194.
Tauschel, H.-D. and Drews, G. (1970). *Cytobiologie* **2**, 87–107.
Vaituzis, Z. (1973). *Can. J. Microbiol.* **19**, 1265–1267.
Vaituzis, Z. and Doetsch, R. N. (1969). *J. Bacteriol.* **100**, 512–521.
Vanderkooi, G. (1974). *Biochim. Biophys. Acta* **344**, 307–345.
Vrij, W. de, Driessen, A. J. M., Hellingwerf, K. J. and Konings, W. N. (1986). *Eur. J. Biochem.* **156**, 431–440.
Wachstein, M. and Meisel, E. (1957). *Am. J. Clin. Path.* **27**, 13–23.
Yano, M., Yamamoto, Y. and Shimizu, H. (1982). *Nature* **299**, 557–559.

11

Electron Microscopy of Isolated Microbial Membranes

JOCHEN R. GOLECKI

*Institut für Biologie II, Mikrobiologie, Universität Freiburg,
Federal Republic of Germany*

I. Introduction

Most of the information about the function and chemical nature of microbial membranes has been obtained from highly enriched membrane fractions after disintegration of the cells. Only after the isolation and purification of defined and homogeneous fractions has it been possible to obtain reproducible and exact results concerning their function and chemical composition. Because there is a large number of different bacteria, a large number of disintegration and isolation methods also exists. Therefore only a small selection of methods can be presented here which are representative of the great variety of specific methods available (Salton, 1964, 1974, 1976; Wolk, 1971; Rogers *et al.*, 1980).

The most important consideration in membrane isolation and purification must be to minimize membrane destruction. The state of the isolated membranes should be as close as possible to their native state in intact cells. However, it should be clear that every manipulation during the isolation and

METHODS IN MICROBIOLOGY
VOLUME 20 ISBN 0 12 521520 6

purification procedure is artificial, and hence must influence the state of the isolated membranes. Therefore it seems advantageous to use more than one isolation procedure, and to compare them for their specificity and effectiveness.

In comparison to the great variety of membrane types and systems in eukaryotic cells (plasma membrane, nuclear membrane, endoplasmic reticulum, Golgi apparatus, mitochondrial membranes), the bacteria appear relatively undifferentiated. Nevertheless, several membrane types exist among the bacteria. They exhibit few morphological differences, but are characterized by their different physiological functions (see Section IV and Golecki, 1988).

II. Disintegration of cells: purification and isolation of microbial membranes

For the isolation of microbial membranes it is usually necessary to break up the individual cells because only the outer surface layers and the outer membrane of the bacterial envelope are assailable from outside. The cytoplasmic membrane, which surrounds the protoplast, and the intracytoplasmic membranes like the photosynthetic membranes of photosynthetic bacteria are accessible only after the removal of a more or less rigid cell envelope by one of the mechanical or chemical methods mentioned below. Very often a combination of several methods is necessary in cases where several membrane types can be isolated from the same cell. This applies especially to photosynthetic bacteria and cyanobacteria. For these organisms, not only the isolation of the outer membrane but also the separation of cytoplasmic and photosynthetic membranes, which are localized intracytoplasmatically, require closer attention.

Several environmental factors influence, firstly, the quantity (i.e. the yield) and, secondly, the quality (i.e. the stability and integrity) of the membranes which are recovered by a specific isolation procedure. For instance, in photosynthetic bacteria the amount of photosynthetic membranes depends on the intensity of light. At lower light intensity the cells develop larger amounts of intracytoplasmic membranes than in high light intensities (reviewed by Oelze and Drews, 1981). The state of growth phase may also have an influence on the membrane isolation. Therefore the isolation usually is performed with cells from the logarithmic or early stationary phase of growth. Another important external factor influencing the stability of the isolated membranes is the change of the normal physiological environment of the cells during the disintegration. This can be caused, for example, by an exchange of the culture medium with a buffer of different pH and ionic strength. Therefore a buffer system guaranteeing an optimum in stability of the normal membrane architecture and of the normal enzymatic function of the membranes should be chosen. Several chemical and biochemical markers are well known and are very helpful in establishing optimal conditions (Salton, 1976, Tables 2 and 3). As well as the chemical markers, some morphological indications are also available for evaluating the quality of a membrane preparation. In freeze-fracture preparations, the distribution of intramembrane particles is an indication of whether it is a preparation with inside-out or right-side-out vesicles (see Section IV).

A. Mechanical disintegration methods

The principle of the methods described is the disintegration of bacteria by mechanical processes. The mechanical power can be obtained by different methods: (1) by shaking the cells together with glass beads or aluminium powder in a special apparatus (Mickle, Vibrogen cell mill); or (2) by the use of pressure, for example in a French pressure cell, or by sonication with a sonifier. The breakage of cells by mechanical procedures is dependent on the availability of considerable amounts of energy. The production of energy, however, results in a consequent production of heat. It is necessary to remove this heat by an adequate and efficient cooling method to avoid a denaturation of cellular material.

The bacteria are used as a dense cell suspension in water or a buffer system. Very often the cell suspension contains different enzymes or chemicals to inhibit autolytic cellular enzymes which break free with the cell rupture. Sometimes specific chemicals or enzymes (for example lysozyme) with degradation activity are added to the suspension to support the mechanical disintegration process.

The methods presented in this section work by shaking small glass beads (0.1–0.2 mm diameter) or abrasives like Al_2O_3 at high numbers of oscillations per minute. In the Braun disintegrator (Merkenschlager et al., 1957; manufactured by B. Braun Melsungen Apperatebau, Melsungen, FRG) the cell suspension (20–50 mg dry wt cells ml^{-1}) is contained in a stoppered glass bottle of 65 ml volume. Half of the bottle volume is filled with glass beads. The cell suspension (approximately 30 ml) is added until only a small air space remains at the top of the bottle. In a chamber connected to an eccentric cam the bottle is shaken at between 2000 and 4000 oscillations per minute. During the oscillation, compressed CO_2 streams around the glass bottle and cools the cell suspension. However, freezing of the suspension should be avoided. Because the cell disintegration is very rapid with this apparatus, several (2–3) short periods of shaking lasting 10 s are recommended, after which a morphological examination in the phase contrast microscope should be performed to obtain information about the progress of cell disruption.

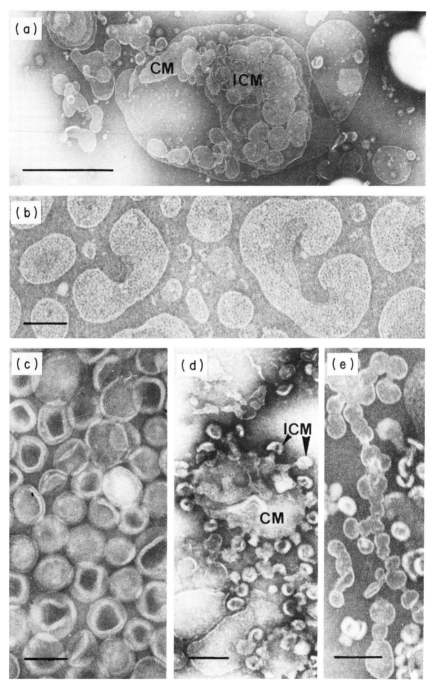

Usually not more than 1–5 min of treatment are required, depending on the species of bacteria. If a longer procedure is necessary, the microscopical control procedure should be repeated from time to time to avoid overextensive cell disruption resulting in very small membrane fragments.

In the same or a similar way using either glass beads or Al_2O_3, bacterial cell suspensions can be prepared in several other machines, including the Mickle disintegrator (Mickle, 1948; Salton and Horne, 1951), the Bühler-Vibrogen cell mill (Drews and Meyer, 1964) or the Sorvall omnimixer (Wolf-Watz et al., 1973).

Other very effective machines for mechanical cell disruption are the French press (Milner et al., 1950) and the Ribi press (Ribi et al., 1959; Wolk, 1971). In this type of machine the cold cell suspension is placed in a precooled pressure chamber provided with a needle valve. The cells are ruptured as they pass through the needle valve under high pressure (up to $25\,000\,lb\,in^{-2}$). For the isolation of chromatophore membranes of photosynthetic bacteria the French press is the most favoured method (Oelze and Drews, 1981). However, as shown by Hurlbert et al. (1974), the isolated chromatophores lose their original interconnections after the French press procedure (Figs 1c–e). This also happens if the bacteria are disrupted by ultra-sonication with a sonifier. Although mechanical disruption is the method of choice for the isolation of cell walls, it has many disadvantages for membrane studies (Salton, 1974).

B. Enzymatic and chemical disintegration methods

A variety of other methods have been developed which avoid using mechanical strength, and instead use lytic enzymes or degrading chemicals for membrane isolation. One of the first used and most powerful lytic enzymes is lysozyme (final concentration $100\,\mu g\,ml^{-1}$ cell suspension), which attacks very specifically the peptidoglycan (= murein) in bacterial cell walls (Weibull, 1953a,b; Repaske, 1956). The treatment of Gram-positive and Gram-negative

---

Fig. 1. Negative staining preparations with phosphotungstic acid. (a) Spheroplast of *Rhodobacter capsulatus* as revealed after treatment with lysozyme and osmotic shock. The cell envelope contains numerous cytoplasmic (CM) and intracytoplasmic (ICM) membrane vesicles. Bar, 0.5 μm. (b) Membrane vesicles after reconstitution with membrane proteins (porins) and lipopolysaccharides from isolated outer membranes of *R. capsulatus*. Bar, 100 nm. (c) Typical chromatophore fraction isolated from French press-treated cells of *Rhodospirillum rubrum*. The ICM vesicles appear to be single flattened spheres of fairly uniform diameter. Bar, 100 nm. (d,e) Chromatophore fraction isolated from lysozyme-induced spheroplasts of *Chromatium vinosum* and separated by sucrose gradient centrifugation. The ICM vesicles are attached to each other and to flat membrane sheets (from Hurlbert et al., 1974). Bar, 100 nm.

bacteria with lysozyme results in the formation of protoplasts and spheroplasts, respectively (Fig. 1a). Both will lyse unless protected by a suitable osmotic environment like isotonic or slightly hypertonic solutions. Osmotic lysis is achieved by a rapid dilution into a system with hypotonic conditions (for example distilled water). To facilitate the effect of lysozyme, in most cases 1 mM EDTA (ethylendiaminetetraacetic acid) is used, together with the lysozyme (Repaske, 1956; Birdsell and Cota-Robles, 1967; Braun et al., 1970; Schnaitman, 1971b; Oelze et al., 1975). The treatment with EDTA as chelating agent permits the entry of lysozyme through the outer membrane by a reaction with Ca^{2+}, which is localized in this layer. The treatment with EDTA, however, has the disadvantage of attacking not only the outer membrane but also the cytoplasmic membrane, resulting in a loss of lipopolysaccharide (Levy and Leive, 1968) and proteins (Nachbar and Salton, 1970).

Approximately the same result—the production of spheroplasts—is obtained by a penicillin treatment (references in Rogers et al., 1980) of growing bacteria suspensions. After their production, the spheroplasts are subjected to various specific disintegration methods to achieve the separation of outer, cytoplasmic and intracytoplasmic membranes (Oelze et al., 1969; Schnaitman, 1971a,b; Osborn et al., 1972a,b; Wolf-Watz et al., 1973; Rogers et al., 1980; Oelze and Drews, 1981). The principle of these methods is the gradual use of different enzymes or chemicals for the specific degradation of membranes or cell wall layers. The outer membrane, for example, can be dissolved by a treatment with hot sodium dodecyl sulphate (Primosigh et al., 1961; Braun and Rehn, 1969; Golecki, 1977; Figs 2b, 3c, 4b, 4c) or an extraction with phenol/water at 68°C (Merkenschlager et al., 1957; Westphal et al., 1961). Other methods for extraction of the outer membrane are the use of phenol–chloroform–petroleum ether (Galanos et al., 1969), proteases (Inouye and Yee, 1972) and 0.9% NaCl (Weckesser et al., 1972).

The cytoplasmic membrane can be removed by treatment with 2% Triton X-100 (Schnaitman, 1971a,b). This is used for the separation of outer and cytoplasmic membranes of Gram-negative bacteria. As mentioned above, the peptidoglycan layer can be dissolved from isolated cell envelopes of Gram-negative bacteria by treatment with lysozyme, resulting in rod-shaped ghosts free of peptidoglycan. The shape of these ghosts is maintained by a unit membrane soluble in sodium dodecyl sulphate (Henning et al., 1973). An enzymatic digestion with proteolytic enzymes (Braun and Rehn, 1969) can also split the connection between the outer membrane and the peptidoglycan layer by dissolving the lipoprotein. It should be mentioned here that the digestion or separation of the murein layer from the outer membrane with lysozyme or trypsin leads to an artefactual reorientation of the lipopoly-saccharide molecules (Mühlradt and Golecki, 1975; Fig. 3d).

C. Purification and separation methods

After the specific procedures of cell disruption and disintegration described in the preceding section, additional steps of enrichment and purification are necessary to obtain homogeneous and defined membrane fractions.

It is necessary first to inactivate autolytic enzymes which are present in the suspension of broken cells. The activity of autolytic enzymes can be reduced by a preparation at ice bath temperature. Another method is a short heat treatment (approximately 10 min) of the suspension at higher temperatures (60–100°C). Treatment with sodium dodecyl sulphate is also used for the inactivation of autolytic enzymes (Shockman et al., 1967; Wolk, 1971), but it should be borne in mind that this dissolves the outer membrane.

Before beginning the specific separation procedure by differential or gradient centrifugation, a short treatment of the broken cell suspension with deoxyribonuclease (20 µg mg^{-1}; 20 min) may be necessary to reduce the viscosity. The specific centrifugation technique depends on the species of bacteria and the different types of membranes which are to be isolated. The most favoured techniques are isopycnic and continuous or discontinuous density gradient centrifugation with saccharose or CsCl, as developed by some laboratories working routinely with the separation of bacterial membranes (Wolk, 1971; Osborn et al., 1972a,b; Hasin et al., 1975; Mizushima and Yamada, 1975; Oelze and Drews, 1981). Sometimes a combination of several techniques must be used for a successful isolation procedure (Oelze et al., 1975).

Each preparation step should be examined by a negative staining preparation with the electron microscope to guarantee the homogeneity and integrity of the isolated fraction.

III. Electron microscopic preparation

After completion of the isolation and enrichment of the different membrane fractions, the specific electron microscopic preparation can begin. Because there is sometimes a long space of time from the beginning of the disruption to the end of the isolation procedure, it should be mentioned that during this time many artefacts can be produced. Therefore the electron microscopic preparation and examination should be performed as soon as possible.

The choice of the specific electron microscopic preparation procedure depends on several factors such as time and equipment available. The dimension of effort, which is undertaken with the preparation technique, depends on the nature and state of the samples and the kind of information required. If only the homogeneity of a membrane fraction is to be determined, then a comparatively simple negative staining procedure is sufficient. However, if information regarding the supramolecular architecture of the membranes is wanted, a complicated and time-consuming freeze-fracture preparation must be performed.

A. Negative staining and shadowing

For both techniques (reviewed by Kay, 1976) only a short preparation time is necessary, and morphological information with the electron microscope is available within a few minutes. Both methods are especially suitable for isolated membranes. Usually they are restricted to demonstrating the outer surface or shape of bacteria and not internal details inside the unbroken cell. The short routine preparation can produce some artefacts like deformation and shrinkage. If the structural details of interest are not impaired by these artefacts, the routine methods are very often used in order to save time.

1. Negative staining procedures

The negative staining technique introduced by Hall (1955) works by surrounding biological objects like membranes, bacteria or viruses with a thin film of the contrast medium. The biological objects consist of elements with a low atomic number and have therefore only a low electron contrast. In the electron microscope the light object is visible against the dark background formed by the electron opaque film of the contrast medium (Figs 1a–e, 4c).

(a) Preparation of the material

For electron microscopic examination it is necessary to place the objects on grids which are usually coated with a thin film of carbon, Formvar or collodion. The quality of the film, i.e. the stability, electron transparency and wettability, is important for the quality of the preparations and the resulting electron micrographs. Therefore, in particular for high resolution, very thin films should be used like perforated films coated with ultra-thin carbon layers. The specimens, for example microbial membranes, have to be suspended in a fluid and placed on a grid. Some time is allowed for the suspended objects to sink onto the grid before the excess fluid is removed and replaced by the staining solution. If the support film is not hydrophilic the membrane suspension, and later on the contrast medium, cannot spread homogeneously on the grid. In this case a more efficient spreading can be achieved by the addition of several wetting agents such as bacitracin (Gregory and Pirie, 1972), 0.1% bovine serum albumin (Valentine, 1961) and octadecanol (Gordon, 1972), or by treatment with a glow discharge unit (Henderson and Griffiths, 1972).

If the membrane fractions are disintegrated by enzymes or detergents and isolated by gradient centrifugation in sucrose or CsCl as described above, these substances may still be present in the suspension (Fig. 4a). Because they influence the quality of the preparation, it is necessary to remove them by dialysis before negative staining. Another way of removing them is to wash the settled membranes on the coated grid with distilled water or buffers several times before continuing the staining procedure. The agar filtration technique (Kellenberger and Arber, 1957; Woldringh *et al.*, 1977) is also an effective method of removing disturbing substances. For this reason the grids are placed on the surface of agar plates before the membrane suspension is dropped onto the supporting film. The soluble substances diffuse into the agar, and only the membranes remain attached to the support film. The result of the agar filtration is a very clean background (Fig. 4b).

Sometimes the samples have to be fixed on the grid because they represent infectious material or have labile structures. After allowing the specimens to attach to the coated grid, the fixation can be performed by transfer to droplets of the fixative for 5–10 min. Glutardialdehyde (1–5%), formaldehyde (1–3%) or osmium tetroxide (0.1–1.0%) are used as fixatives. After removing the fixatives, the grids are rinsed with water and transferred to the staining solution.

As mentioned above, during the described preparations some artefacts may be produced, for example the deformation of the objects caused by surface tension during the evaporation process on the grid. This artefact may not be so important for isolated bacterial membranes as for other objects like virus particles, but in some cases, for example outer membranes with highly ordered structures, it may be better to avoid this artefact by using a special freeze-drying method (Nermut, 1973). In this method the grids with the membranes suspended in the staining solution are quickly frozen in liquid nitrogen ($-196°C$) or propane ($-190°C$) and then transferred to a vacuum chamber of 1×10^{-6} Torr and an object temperature of $-100°C$. Under these conditions sublimation of the ice takes place and the samples are freeze-dried without any deformation.

(b Negative staining substances and techniques

The most frequently used negative staining substances are sodium or potassium salts of phosphotungstic acid (PTA). Caused by the anion complex of 12 tungsten atoms (atomic number 74), PTA represents an intensely electron-opaque medium. It is used at 1% or 2% in aqueous solution adjusted to a pH between 6 and 7 with sodium or potassium hydroxide. Sodium silicotungstate (Valentine *et al.*, 1968) works in the same manner as PTA. Some other routinely used negative stains with properties similar to PTA are

reviewed by Kay (1976) and are listed in Table I. Uranyl acetate (van Bruggen
et al., 1960) should be prepared just before use and should be stored in the
dark. Uranyl formate (Leberman, 1965) can be used alone or in combination,
followed by uranyl acetate. Uranyl oxalate can be prepared with equimolar
(12 mM) solutions of uranyl acetate and oxalic acid adjusted to a pH between
6.5 and 6.8 with dilute NH_4OH (Mellema et al., 1967). Ammonium molybdate
is recommended (Muscatello and Horne, 1968) for contrasting membrane-
bound systems because the tonicity of the stain can be adjusted over a large
range to match the required conditions for stabilizing the particular
membrane structures.

TABLE I
Substances, concentration and pH range of negative staining solutions

Stain	Concentration	pH range
Phosphotungstic acid $H_3PW_{12}O_{40} \cdot x H_2O$	1.0–2.0% aqueous	6.0–7.0 with NaOH or KOH; down to 4.0
Uranyl acetate $UO_2(CH_3COO)_2 \cdot 2 H_2O$	0.2–2.0% aqueous	4.5 without titration; with NH_4^+ to 6.0
Uranyl oxalate $UO_2(COO)_2 \cdot 3 H_2O$	0.5% aqueous	5.0–7.0
Uranyl formate $UO_2(HCOO)_2 \cdot 3 H_2O$	1.0% aqueous	3.5 without titration; with NH_4^+ to 4.5–5.2
Ammonium molybdate $(NH_4)_6Mo_7O_{24} \cdot 4 H_2O$	3.5–10.0% aqueous	5.2–7.5
Silicotungstic acid $H_4SiW_{12}O_{40} \cdot x H_2O$	4.0% aqueous	7.0–7.4 with NaOH

Several techniques are described for transferring the objects and the stains
onto the coated grids (Kay, 1976). The drop method is a very simple and
therefore frequently used procedure. All the operations can be performed on
the surface of the grid, which is held with a pair of forceps. Using a pipette a
drop of the specimen is placed on the grid and left for a period ranging from a
few seconds to several minutes, depending on the density of the suspension
used. A suitable density is 10^6–10^8 particles or membranes per ml, which
allows a comfortable examination of the sample in the electron microscope
without a time-consuming search. The fluid used for the suspension of the
specimens can be water, buffer or the negative staining solution itself. After an
adequate period of time the excess fluid is removed with filter paper. However,
a small film of the fluid should be left on the grid surface to prevent it from
totally drying up. The grid can then be washed with distilled water or buffer,
followed finally by the addition of the stain solution. The incubation time of
the stain ranges from a few seconds to one minute, depending on the object

and the stain. Most of the stain is removed with filter paper and a thin air-dried film of stain is left surrounding the specimens. If too much of the stain is removed, no film of sufficient thickness remains around the specimens and only positive stained objects are the consequence (Fig. 4d). If difficulties with a homogeneous spreading of the objects or stains arise, the above-mentioned wetting agents should be used.

The inverted droplet method works with the same sequence of preparation steps (Nermut, 1972). However, the grids are presented on the top of the suspension and stain droplets, which are placed on a dental wax plate. In the spray method (Brenner and Horne, 1959) a commercial apparatus for nasal spray is used to spray a thin film of a mixture of the specimen and the stain solution on the grid surface. The advantage of this method is an equal distribution of specimens and stain in the air-dried film. Other less economical methods (Kay, 1976) are not described here.

2. Shadowing materials and techniques

The evaporation of high density metals as thin layers to enhance the contrast of the shadowed objects (membranes, bacteria, viruses, etc.) has been used for a long time in electron microscopy (reviewed by Henderson and Griffiths, 1972). The preparation of the specimens is done in the same way as described for negative staining. The specimens have to be suspended in fluid before they are placed on grids to attach to the surface of the support film. Because the same artefacts can arise during this procedure as in negative staining, the same precautions (wetting agents, agar filtration, freeze-drying) should be taken.

The evaporation of the metal onto grids carrying the specimens can be performed in one or two directions (portrait technique) or by rotary shadowing (for references see Henderson and Griffiths, 1972) in a coating unit equipped with a sufficient vacuum system, a cooling trap and a quartz crystal thickness monitor to measure accurately the thickness of the evaporated metal layer. The metal is evaporated with an electron beam source under specific angles depending on the nature of the specimen. For the preparation of isolated microbial membranes usually an angle of 20–45° is used. A variety of high density metals like platinum, platinum/carbon (simultaneously), gold and palladium can be used. The thickness of the deposited metal layer must depend upon the required resolution and the type of specimen examined. If the metal layer is too thick, smaller structure details cannot be resolved and also the contrast is too strong. Shadow casting can be used in combination with the replication technique (Henderson and Griffiths, 1972). However, due to the comparatively low thickness of bacterial membranes, this technique is not necessary; the isolated membranes can be left between the support film and the evaporated metal layer (Figs 4a and b).

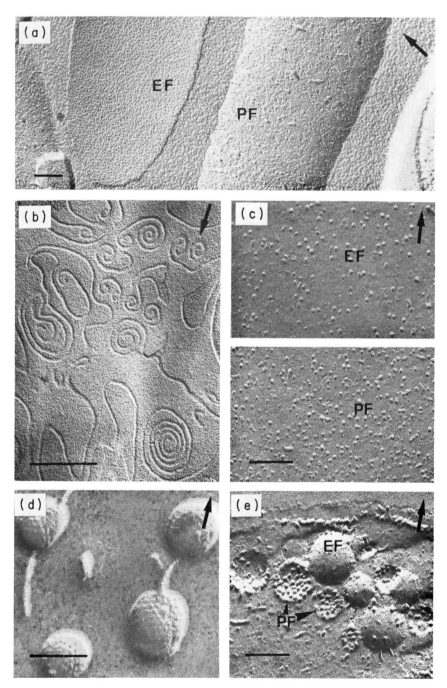

B. Ultra-thin section and freeze-etch techniques

Both techniques require more time and preparation than negative staining and shadow casting. However, they render possible some additional information about the internal organization of bacterial membranes which are not available without these techniques. For example, the supramolecular architecture of the bacterial membranes can be demonstrated only by freeze-fracture methods (Figs 2a, c and e).

1. Embedding and ultra-thin sectioning

The preparation of isolated bacterial membranes for embedding and ultra-thin sectioning follows the technique used for intact bacteria as described elsewhere in this volume. Briefly, before starting the procedure with fixation, dehydration and embedding, it may be useful to examine the sample for homogeneity with negative staining. The fixation may be shortened to 10–30 min because in suspension the fixative can very quickly penetrate the membranes. If only a very small membrane amount is available it is useful to start the whole procedure by embedding in agar (Kellenberger et al., 1958), which ensures no further loss of material during the following steps. This also applies for membrane sediments which can be obtained only by an ultra-centrifugation of long duration. If the sedimented membranes are agar embedded it is not necessary to use the centrifuge for the subsequent steps and much time can be saved. The block-contrasting with uranyl acetate (Kellenberger et al., 1958) should be performed because isolated membranes sometimes do not have sufficient contrast after staining of the ultra-thin sections. Two other possibilities to obtain more contrast in those materials are

Fig. 2. Freeze-fracture and freeze-etch preparation. (a) Freeze-fracture preparation of outer membranes of *Salmonella typhimurium* strain 1195 exposing the typical exoplasmic (EF) and plasmic (PF) fracture faces. Arrow in the upper right-hand corner indicates direction of shadowing as in all following micrographs. Bar, 100 nm. (b) Freeze-fractured peptidoglycan layers of the cyanobacterium *Anacystis nidulans*. The isolation was performed by disintegration in the Bühler apparatus and by treatment with 4% hot sodium dodecyl sulphate and Triton X-100. Bar, 0.5 μm. (c) Both fracture faces of the cytoplasmic membrane of the cyanobacterium *Nostoc muscorum* with lower particle number on the outer membrane leaflet (EF) and numerous intramembrane particles on the inner membrane leaflet (PF). Bar, 100 nm. (d) Freeze-etch preparation of isolated French press-treated chromatophores of *Rhodospirillum rubrum* FR1. Note the regular structures (10 nm centre-to-centre spacing) in the membrane representing photosynthetic units. Bar, 100 nm. (e) Freeze-fractured chromatophore vesicles of *Rhodobacter capsulatus* strain Ala$^+$ exposing the exoplasmic (EF) and plasmic (PF) fracture faces. The plasmic fracture face is densely covered with intramembrane particles representing structural equivalents of the photosynthetic units. Bar, 100 nm.

to use thicker sections or a lower high voltage in the electron microscope, both in connection with a loss of resolution.

2. The freeze-etch preparation

To obtain freeze-etch or freeze-fracture preparations of isolated bacterial membranes (Figs 2a–e) the samples can be frozen in liquid propane and prepared according to the same procedure as used for intact bacteria (Golecki, 1988). Also the same precautions regarding temperature, cooling rate and other preparation conditions should be undertaken to avoid the well-known artefacts.

IV.   Electron microscopic examination

The multiplicity of microbial membrane examinations by electron microscopy admits only a limited number of examples of each described method to be presented. In our laboratory the negative staining technique is preferred for the examination of membrane preparations of photosynthetic bacteria. In photosynthetic bacteria three types of membranes are present (Oelze and Drews, 1981): the outer membrane (OM) (Figs 1a,b; 2a; 3a,b,d), the cytoplasmic membrane (CM) (Figs 1a,e) and the photosynthetic intracytoplasmic membranes (ICM) (Figs 1a,c–e; 2d,e). The photosynthetic membranes are also called chromatophores or thylakoids, especially for cyanobacteria. The outer membrane and the cytoplasmic membrane are separated by the peptidoglycan layer (PG) (Figs 2b; 3a,c; 4a–c). Outer membrane and peptidoglycan layer represent the cell wall.

The different membrane types can be easily distinguished from each other, and also from cell envelopes or spheroplasts which have not disintegrated, by using the negative staining method. Spheroplasts (Fig. 1a) and cell envelopes (Figs 4a–d) can be identified by their characteristic shape and the content of the different membrane types. The outer membrane fraction is represented by large membrane vesicles of different size. From the outer membrane of different Gram-negative bacteria (Steven et al., 1977; Yamada and Mizushima, 1978) a major protein (= porin) was isolated which could be reconstituted in form of vesicles with a regular pattern (Fig. 1b). The porins could be demonstrated as a periodic monolayer with hexagonal lattice on the surface of the peptidoglycan layer if the cell walls were extracted with sodium dodecyl sulphate at non-denaturing temperatures (Rosenbusch, 1974). The cytoplasmic membrane is visible as flat vesicles of irregular shape (Fig. 1e).

The appearance of the intracytoplasmic membranes depends on the preparation procedure. Figure 1c represents the typical chromatophore fraction as revealed from French press-treated cells. The membranes are well dispersed, flattened spheres with a fairly uniform diameter (approximately 60–80 nm). If the membranes are prepared by lysozyme treatment from spheroplasts, the chromatophore vesicles are attached to each other or to flat membrane sheets (Figs 1d,e). The great majority of vesicles exists as clumps or groups, but not as single vesicles as revealed by French press treatment.

Structural equivalents of the photosynthetic units could be demonstrated as regularly arranged particles in intracytoplasmic membrane preparations of several photosynthetic bacteria and cyanobacteria (reviewed by Oelze and Drews, 1981; Golecki and Drews, 1982). The units are visible with negative staining (Giesbrecht and Drews, 1966) and after freeze-etching (Oelze and Golecki, 1975) or freeze-fracturing (Giesbrecht and Drews, 1966; Reed et al., 1975) on the surface or on internal fracture faces of the membranes, respectively (Figs 2d,e).

On the inner side of the cytoplasmic membrane ATPase complexes could be demonstrated for different bacterial species (reviewed by Haddock and Jones, 1977) with negative staining. Uncertainty regarding the localization of the ATPase complexes existed because, depending on the isolation procedure used, membrane vesicles of different orientation, i.e. inside-out and right-side-out vesicles, were compared. The question about the orientation of isolated cytoplasmic and intracytoplasmic membrane vesicles can easily be solved by a freeze-fracture preparation (Figs 2c,e). The distribution of the intramembrane particles is a simple indication for the membrane orientation. Usually on the plasmic fracture face (PF) a higher number of intramembrane particles is visible in right-side-out vesicles than on the exoplasmic fracture face (EF). For inside-out vesicles the opposite particle distribution is presented. The number of intramembrane particles per membrane area is an indication for the biological activity of the membrane (Golecki et al., 1979, 1980; Golecki and Oelze, 1980; Golecki and Drews, 1982).

Isolated fragments of deoxyribonucleic acid–envelope complexes from *Escherichia coli* were demonstrated in ultra-thin sections by Olsen et al. (1974). The fragments were produced by a modified procedure according to Schnaitman (1971a) and subsequent preparative free-flow electrophoresis. The fragments contained adhesion sites of the cytoplasmic membrane to the cell wall which have a special function in phage adsorption (Bayer, 1968) and export of lipopolysaccharides (Mühlradt et al., 1973).

Differentiating between outer and cytoplasmic membrane fractions, which is sometimes not so easy using negative staining or ultra-thin sectioning, is very easy by freeze-fracturing. The ultrastructure of the outer membrane (Fig. 2a) of all Gram-negative bacteria presented by freeze-fracturing is characteristically different from the morphology of the cytoplasmic membrane (Fig. 2c). The outer membrane presents many small particles on the

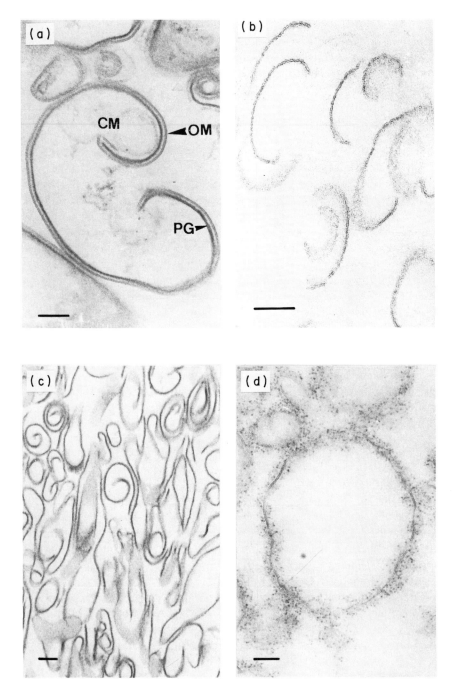

outer leaflet (EF) and pits and only occasional particles of larger size on the inner leaflet (PF). The number of intramembrane particles on the exoplasmic fracture face of the outer membrane varies in mutants of *E. coli* and *Salmonella typhimurium* with the form of lipopolysaccharide (rough or smooth form) and the content of specific outer membrane proteins (Smit *et al.*, 1975; further references in Braun and Hantke, 1981). The particles are interpreted to be structural equivalents of the porins forming transmembrane pores. In ultra-thin sections of isolated outer membranes C-shaped membrane pieces are visible, presenting the characteristic double-track layer (Hasin *et al.*, 1975) (Fig. 3b).

Isolated cytoplasmic membranes present, in ultra-thin sections, the same organization into a double-track layer as the outer membrane, but they are of irregular shape. The fracture faces of the cytoplasmic membrane are characterized by a plasmic fracture face (PF = inner leaflet) densely covered with particles of different size and an exoplasmic fracture face (EF = outer leaflet) sparsely covered with larger intramembrane particles (Fig. 2c; Golecki and Oelze, 1980).

In Gram-negative bacteria the peptidoglycan or murein layer (Fig. 3a) is localized between the outer and cytoplasmic membrane. Weidel and co-workers (Weidel *et al.*, 1960; Martin and Frank, 1962) were the first to isolate this specific layer. They demonstrated the characteristic rigidity of this layer with electron micrographs of isolated sacculi after shadow casting. Shadow casting has been used by many other investigators to show the stepwise degradation of isolated cell walls from a large variety of bacterial species (Figs 4a,b; Martin and Frank, 1962; Drews and Meyer, 1964). Henning *et al.* (1973) demonstrated with this method that ghosts of *E. coli*, that is the isolated outer membrane only, retain their specific rod-like shape even after the loss of the rigid layer by lysozyme treatment. In isolated cell walls of *S. typhimurium* it was shown by Mühlradt and Golecki (1975) that the disturbance of the murein layer can have negative influences on the molecular architecture of the

---

Fig. 3. Ultra-thin section preparation. (a) Isolated cell wall of *Anacystis nidulans* after disintegration in the Bühler mill. Note the thick layer of peptidoglycan (PG) and some remains of the cytoplasmic membrane (CM). The outer membrane (OM) is visible as double-track layer. Bar, 100 nm. (b) C-shaped outer membranes of *Rhodospirillum rubrum*. The peptidoglycan layer is dissolved after the treatment with lysozyme. Cell disintegration by French press. Bar, 100 nm. (c) Isolated rigid layers (= peptidoglycan) of *Rhodomicrobium vannielii* after cell disruption with the French press. Outer and cytoplasmic membranes were dissolved by the treatment with 4% hot sodium dodecyl sulphate. Bar, 100 nm. (d) Isolated outer membrane of *Salmonella typhimurium* labelled with ferritin-conjugated antibodies against lipopolysaccharide. The symmetrical distribution of lipopolysaccharide molecules was caused by their artefactual reorientation at 37°C after the murein was digested with lysozyme (from Mühlradt and Golecki, 1975). Bar, 100 nm.

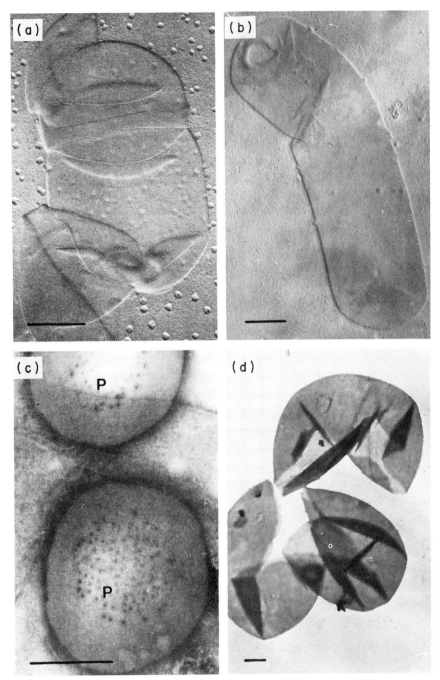

outer membrane. The loss of the peptidoglycan layer results, at physiological temperature, in an artefactual symmetrical reorientation of the lipopoly-saccharide molecules in the outer membrane (Fig. 3d). This artefactual reorientation can be prevented by preparation of the outer membranes at 0°C. This is demonstrated by an asymmetrical distribution of the lipopoly-saccharide molecules in the external half of the outer membrane visualized by specific ferritin-conjugated antibodies directed against the polysaccharide moiety of the lipopolysaccharide.

In cross-walls of filamentous cyanobacteria pores were detected in the peptidoglycan layer (Fig. 4c; references in Drews and Weckesser, 1982). These pores, called microplasmodesmata, ensure a rapid communication between adjacent cells of a filament. In ultra-thin sections isolated peptidoglycan layers have a good electron contrast. This applies especially to the peptidoglycan layers of cyanobacteria (Figs 3a,c), which have a greater thickness (usually 1–10 nm, occasionally up to 200 nm; Drews and Weckesser, 1982) than in Gram-negative bacteria due to the three-dimensional net of peptidoglycan in cyanobacteria.

For the many other structural characteristics of microbial membranes not reported here, the reader is directed to numerous reviews (Braun and Hantke, 1981; Oelze and Drews, 1981; Drews and Weckesser, 1982; Golecki and Drews, 1982).

V. Conclusion

As shown above, many variations of disintegration and electron microscopic methods are available for the preparation and examination of bacterial membranes. Each of these methods has specific advantages and dis-advantages. For the newcomer to electron microscopy it is sometimes difficult to select the most successful method when examining a new object. In this case it is useful to look for published methods which have been successful in the

---

Fig. 4. Shadow casting and negative staining of isolated cell walls. (a) Isolated cell envelopes (outer membrane and peptidoglycan) of *Anacystis nidulans* after disintegration in the Bühler mill and treatment with Triton X-100 to dissolve the cytoplasmic membrane. The dots in the background are caused by salt crystals from the buffer. Shadowing with platinum/carbon. Bar, 0.5 μm. (b) Isolated peptidoglycan layer of *A. nidulans* after treatment with hot sodium dodecyl sulphate and Triton X-100. By the use of the agar filtration technique (Kellenberger and Arber, 1957) no salt crystals are visible on the support film. Bar, 0.5 μm. (c) Negative staining (2% PTA) of isolated peptidoglycan layers of *Anabaena variabilis* with pores (P) in the centre of the cross-walls. The isolated walls were treated with 4% hot sodium dodecyl sulphate. Micrograph: J. R. Golecki in Drews and Weckesser (1982). Bar, 0.5 μm. (d) Isolated cell envelopes of *A. variabilis* positively stained with phosphotungstic acid. Bar, 0.5 μm.

280 J. R. GOLECKI

preparation of similar objects. But beside the proposed procedures, one should apply as many different methods as possible if time and circumstances permit. Thus, it is possible to obtain a general impression of an object from the multiplicity of information gleaned from the different preparation and examination methods. That this assumption is correct has been amply demonstrated in the case of the membrane system in photosynthetic bacteria. The complex morphology and specific architecture of these membranes were elucidated only by the use of many different methods such as freeze-fracturing or freeze-etching, ultra-thin sectioning, negative staining and shadow casting.

References

Bayer, M. E. (1968). *J. Virol.* **2**, 346–356.
Birdsell, D. C. and Cota-Robles, E. H. (1967). *J. Bacteriol.* **93**, 427–437.
Braun, V. and Rehn, K. (1969). *Eur. J. Biochem.* **10**, 426–438.
Braun, V. and Hantke, K. (1981). In *Organization of Prokaryotic Cell Membranes* (B. K. Ghosh, ed.), Vol. II, pp. 1–73. CRC Press, Boca Raton, Florida.
Braun, V., Rehn, K. and Wolff, H. (1970). *Biochemistry* **9**, 5041–5049.
Brenner, S. and Horne, R. W. (1959). *Biochim. Biophys. Acta* **34**, 103–110.
Drews, G. and Meyer, H. (1964). *Arch. Microbiol.* **48**, 259–267.
Drews, G. and Weckesser, J. (1982). In *The Biology of Cyanobacteria* (N. G. Carr and B. A. Whitton, eds), pp. 333–357. Blackwell Scientific, Oxford.
Galanos, C., Lüderitz, O. and Westphal, O. (1969). *Eur. J. Biochem.* **9**, 245–249.
Giesbrecht, P. and Drews, G. (1966). *Arch. Microbiol.* **54**, 297–330.
Golecki, J. R. (1977). *Arch. Microbiol.* **114**, 35–41.
Golecki, J. R. (1988). In *Methods in Microbiology* (F. Mayer, ed.), Vol. 20, pp. 61–77. Academic Press, London and New York.
Golecki, J. R. and Drews, G. (1982). In *The Biology of Cyanobacteria* (N. G. Carr and B. A. Whitton, eds), pp. 125–141. Blackwell Scientific, Oxford.
Golecki, J. R. and Oelze, J. (1980). *J. Bacteriol.* **144**, 781–788.
Golecki, J. R., Drews, G. and Bühler, R. (1979). *Cytobiologie* **18**, 381–389.
Golecki, J. R., Schumacher, A. and Drews, G. (1980). *Eur. J. Cell Biol.* **23**, 1–5.
Gordon, C. N. (1972). *J. Ultrastruct. Res.* **39**, 173–185.
Gregory, D. W. and Pirie, B. J. S. (1972). In *Proc. 5th Eur. Congr. Electron Microscopy* (V. E. Cosslett, ed.), pp. 234–235. The Institute of Physics, London.
Haddock, B. A. and Jones, C. W. (1977). *Bacteriol. Rev.* **41**, 47–99.
Hall, C. E. (1955). *J. Biophys. Biochem. Cytol.* **1**, 1–12.
Hasin, M., Rottem, S. and Razin, S. (1975). *Biochim. Biophys. Acta* **375**, 381–394.
Henderson, W. J. and Griffiths, K. (1972). In *Principles and Techniques of Electron Microscopy* (M. A. Hayat, ed.), Vol. 2, pp. 151–193. Van Nostrand Reinhold, New York.
Henning, U., Höhn, B. and Sonntag, I. (1973). *Eur. J. Biochem.* **39**, 27–36.
Hurlbert, R. E., Golecki, J. R. and Drews, G. (1974). *Arch. Microbiol.* **101**, 169–186.
Inouye, M. and Yee, M.-L. (1972). *J. Bacteriol.* **112**, 585–592.
Kay, D. (1976). In *Methods in Microbiology* (J. R. Norris, ed.), Vol. 9, pp. 177–215. Academic Press, London and New York.
Kellenberger, E. and Arber, W. (1957). *Virology* **3**, 245–255.

Kellenberger, E., Ryter, A. and Séchaud, J. (1958). *J. Biophys. Biochem. Cytol.* **4**, 671–678.

Leberman, R. (1965). *J. Mol. Biol.* **13**, 606.

Levy, S. B. and Leive, L. (1968). *Proc. Natl Acad. Sci. USA* **61**, 1435–1439.

Martin, H. H. and Frank, H. (1962). *Z. Naturforschung.* **17b**, 190–196.

Mellema, J. E., van Bruggen, E. F. J. and Gruber, M. (1967). *Biochim. Biophys. Acta* **140**, 180–182.

Merkenschlager, M., Schlossman, K. and Kurz, W. (1957). *Biochem. Z.* **329**, 332–340.

Mickle, H. (1948). *J. R. Microsc. Soc.* **68**, 10–12.

Milner, H. W., Lawrence, N. S. and French, C. S. (1950). *Science* **111**, 633–634.

Mizushima, S. and Yamada, H. (1975). *Biochim. Biophys. Acta* **375**, 44–53.

Mühlradt, P. F. and Golecki, J. R. (1975). *Eur. J. Biochem.* **51**, 343–352.

Mühlradt, P. F., Menzel, J., Golecki, J. R. and Speth, V. (1973). *Eur. J. Biochem.* **35**, 471–481.

Muscatello, U. and Horne, R. W. (1968). *J. Ultrastruct. Res.* **25**, 73–83.

Nachbar, M. S. and Salton, M. R. J. (1970). *Biochim. Biophys. Acta* **223**, 309–320.

Nermut, M. V. (1972). *J. Microsc.* **16**, 351–362.

Nermut, M. V. (1973). In *Principles and Techniques of Electron Microscopy* (M. A. Hayat, ed.), Vol. 7, pp. 79–117. Van Nostrand Reinhold, New York.

Oelze, J. and Drews, G. (1981). In *Organization of Prokaryotic Cell Membranes* (B. K. Ghosh, ed.), Vol. II, pp. 131–195. CRC Press, Boca Raton, Florida.

Oelze, J. and Golecki, J. R. (1975). *Arch. Microbiol.* **102**, 59–64.

Oelze, J., Biedermann, M. and Drews, G. (1969). *Biochim. Biophys. Acta* **173**, 436–447.

Oelze, J., Golecki, J. R., Kleinig, H. and Weckesser, J. (1975). *Antonie van Leeuwenhoek* **41**, 273–286.

Olsen, W. L., Heidrich, H.-G., Hannig, K. and Hofschneider, P. H. (1974). *J. Bacteriol.* **118**, 646–653.

Osborn, M. J., Gander, J. E., Parisi, E. and Carson, J. (1972a). *J. Biol. Chem.* **247**, 3962–3972.

Osborn, M. J., Gander, J. E. and Parisi, E. (1972b). *J. Biol. Chem.* **247**, 3973–3986.

Primosigh, J., Pelzer, H., Maass, D. and Weidel, W. (1961). *Biochim. Biophys. Acta* **46**, 68–80.

Reed, D. W., Raveed, D. and Reporter, M. (1975). *Biochim. Biophys. Acta* **387**, 368–378.

Repaske, R. (1956). *Biochim. Biophys. Acta* **22**, 189–191.

Ribi, E., Perrine, T., List, R., Brown, W. and Goode, G. (1959). *Proc. Soc. Exp. Biol. Med.* **100**, 647–649.

Rogers, H. J., Perkins, H. R. and Ward, J. B. (1980). In *Microbial Cell Walls and Membranes.* Chapman and Hall, London and New York.

Rosenbusch, J. P. (1974). *J. Biol. Chem.* **249**, 8019–8029.

Salton, M. R. J. (1964). *The Bacterial Cell Wall.* Elsevier, Amsterdam.

Salton, M. R. J. (1974). In *Methods in Enzymology* (S. P. Colowick and N. O. Kaplan, eds), Vol. 31, pp. 653–667. Academic Press, New York.

Salton, M. R. J. (1976). In *Methods in Membrane Biology* (E. D. Korn, ed.), Vol. 6, pp. 101–150. Plenum Press, New York and London.

Salton, M. R. J. and Horne, R. W. (1951). *Biochim. Biophys. Acta* **7**, 177–197.

Schnaitman, C. A. (1971a). *J. Bacteriol.* **108**, 545–552.

Schnaitman, C. A. (1971b). *J. Bacteriol.* **108**, 553–563.

Shockman, G. D., Thompson, J. S. and Conover, M. J. (1967). *Biochemistry* **6**, 1054–1065.

Smit, J., Kamio, Y. and Nikaido, H. (1975). *J. Bacteriol.* **124**, 942–958.

Steven, A. C., ten Heggeler, B., Müller, R., Kistler, J. and Rosenbusch, J. P. (1977). *J. Cell Biol.* **72**, 292–301.

Valentine, R. C. (1961). *Adv. Virus Res.* **8**, 287–318.

Valentine, R. C., Shapiro, B. M. and Stadtman, E. R. (1968). *Biochemistry* **7**, 2143–2152.

van Bruggen, E. F. J., Wiebenga, E. H. and Gruber, M. (1960). *Biochim. Biophys. Acta* **42**, 171–172.

Weckesser, J., Drews, G. and Ladwig, R. (1972). *J. Bacteriol.* **110**, 346–353.

Weibull, C. (1953a). *J. Bacteriol.* **66**, 688–695.

Weibull, C. (1953b). *J. Bacteriol.* **66**, 696–702.

Weidel, W., Frank, H. and Martin, H. H. (1960). *J. Gen. Microbiol.* **22**, 158–166.

Westphal, O., Lüderitz, O. and Bister, F. (1961). *Z. Naturforschung.* **7b**, 148–155.

Woldringh, C. L., de Jong, M. A., van den Berg, W. and Koppes, L. (1977). *J. Bacteriol.* **131**, 270–279.

Wolf-Watz, H., Normark, S. and Bloom, G. D. (1973). *J. Bacteriol.* **115**, 1191–1197.

Wolk, E. (1971). In *Methods in Microbiology* (J. R. Norris and D. W. Ribbons, eds), Vol. 5A, pp. 361–418. Academic Press, London and New York.

Yamada, H. and Mizushima, S. (1978). *J. Bacteriol.* **135**, 1024–1031.

12

Analysis of Dimensions and Structural Organization of Proteoliposomes

FRANK MAYER and MANFRED ROHDE

*Institut für Mikrobiologie der Georg-August-Universität Göttingen,
D-3400 Göttingen, Federal Republic of Germany*

I. Introduction

In the anaerobic bacterium *Veillonella alcalescens* the exergonic decarboxylation of methylmalonyl-CoA is coupled to an active transport of Na^+ through the cytoplasmic membrane (Dimroth, 1982, 1985). The decarboxylation is catalysed by the membrane-bound enzyme methylmalonyl-CoA decarboxylase (EC 4.1.1.41). The enzyme has biotin as the prosthetic group. The biotin-containing subunit sticks out of the membrane into the cytoplasm.

In order to simulate the *in vivo* reaction, "reconstitution" of the decarboxylase into "inside-out" proteoliposomes can be done (Dimroth and Hilpert, 1984; Hilpert and Dimroth, 1984; Rohde *et al.*, 1986). In this inverse system it was found that the decarboxylation of 1 mol methylmalonyl-CoA is reversibly coupled to the translocation of 2 mol Na^+ (Dimroth and Hilpert,

METHODS IN MICROBIOLOGY
VOLUME 20 ISBN 0 12 521520 6

1984). It was also shown that the Na^+ is pumped in an electrogenic fashion, creating a membrane potential in addition to a concentration gradient (Dimroth, 1982, 1985).

The quantitation of these energetic parameters can only be obtained if one knows what percentage of the volume enclosed by the proteoliposomal membranes is available to Na^+ pumping enzyme molecules. This volume can then be used to calculate the concentration gradients which are caused by the Na^+ pumps. In addition, it is important to know how many enzyme molecules incorporated into an average proteoliposome have the biotin prosthetic group exposed to the outside. Only enzyme molecules with this orientation will catalyse decarboxylation of substrate applied from the outside and Na^+ translocation from the outside to the inside of the vesicle.

Decarboxylase molecules can be expected to be integrated into a liposome in two orientations (biotin exposed either at the outside or at the inside). In negatively stained samples, it is impossible to differentiate between these two orientations without the application of an additional procedure especially designed for identification of the orientation of the enzyme molecules.

The following approach can be chosen (Rohde et al., 1986) in order to solve the problems described (technical details are given below). Proteoliposomes are mounted and stained for electron microscopy with a technique which keeps the proteoliposomes in a well-preserved spherical shape. This is necessary because the calculation of the vesicle volumes based on measurement of vesicle diameters has to be as close to the natural state as possible. The numbers of decarboxylase molecules (visible as white dots distributed over the surface of the vesicles) can be counted in negatively stained samples where the proteoliposomes are flattened due to air drying. From these values the number of enzyme molecules integrated into the membrane of an average-sized vesicle can be calculated; also the average number of enzyme molecules per unit area of the vesicle surface can be determined.

The numbers obtained by this procedure represent the total number of decarboxylase molecules integrated in the membrane of an average-sized proteoliposome. They do not, however, reflect the number of interest, i.e. the number of enzyme molecules oriented with the substrate-binding site (the biotin prosthetic group) facing the outside. This number can be determined by affinity labelling of the biotin group with avidin–gold. This type of experiment is performed by incubation of the reconstituted vesicles, together with the marker complex prior to mounting and negative staining for electron microscopy. By doing so all biotin moieties exposed at the vesicle surface can be labelled. In mounting, the vesicles are flattened. Although half of the vesicle surface is then in contact with the surface of the support film, the colloidal gold particles all around the vesicle surface can be monitored due to their high immanent contrast. Negative staining of such a sample is only done in order to see the contours of the individual vesicles.

II. Purification of methylmalonyl-CoA decarboxylase

The enzyme was purified according to Hilpert and Dimroth (1983). *Veillonella alcalescens* (ATCC 17745) was grown anaerobically on 2% sodium lactate (De Vries *et al.*, 1977). Cells were harvested after 24 h in the early stationary growth phase, and washed with 50 mM potassium buffer, pH 7.0.

For the following steps diisopropylfluorophosphate (0.1 M) was added to the buffers used. Membranes were prepared by passing a suspension of 30 g of *V. alcalescens* in 120 ml 50 mM potassium buffer, pH 7.0, containing 5 mM MgCl$_2$ and 3 mg deoxyribonuclease twice through a French pressure cell (137.5 MPa). Unbroken cells were removed by centrifugation at 31 000g for 15 min. Membranes were sedimented by centrifugation at 200 000g for 90 min and washed with 2 mM potassium phosphate buffer, pH 7.0. The pellet was homogenized with 60 ml 2 mM potassium phosphate buffer, pH 7.0, containing 1% Triton X-100. Subsequently the homogenized solution was centrifuged at 230 000g for 30 min. The supernatant of the last centrifugation step was adjusted to 0.3 M KCl and applied to a monomeric avidin–Sepharose column. The affinity column was washed with 10 mM potassium phosphate buffer, pH 7.0, containing 0.3 M KCl and 0.05% Brij 58 at a flow rate of 45 ml h^{-1} until the adsorption at 280 nm was below 0.1. Methylmalonyl-CoA decarboxylase was eluted with elution buffer (10 mM potassium phosphate buffer, pH 7.0, 0.15 M KCl and 1.5 mM biotin). The enzyme solution was concentrated to about 1 ml by ultra-filtration on a PM-10 membrane (Amicon) and stored under liquid nitrogen.

III. Reconstitution of methylmalonyl-CoA into phospholipid vesicles

Reconstitution was done according to Dimroth and Hilpert (1984) and Hilpert and Dimroth (1984). Soya bean phosphatidylcholine (Sigma, type II-S, 80 mg) was vigorously agitated on a vortex mixer for 3–5 min under an N$_2$ atmosphere with 20 ml of reconstitution buffer (30 mM potassium phosphate, pH 7.5, containing 1 mM Na$_2$SO$_4$). Then the solution was sonicated at 80 W for about 1 min in intervals of 20 s. The incubation mixture for the reconstitution of the enzyme into phospholipid vesicles contained in 2.8 ml reconstitution buffer:

  80 mg homogenized phosphatidylcholine
  2.85% octyl glucoside
  methylmalonyl-CoA decarboxylase (0.1 mg, 28 U mg^{-1})

After 5 min the incubation mixture was diluted with 60 ml ice-cold re-constitution buffer, followed by centrifugation at 200 000g for 40 min. The resulting proteoliposomes were resuspended in 4 ml reconstitution buffer.

IV.   Preparation of avidin–gold complexes

A.   Succinoylation of avidin

Egg white avidin is succinoylated as follows (Kishida *et al.*, 1975; Morris and Saelinger, 1984; Rohde *et al.*, 1986):

The lyophilized avidin (5 mg) is dissolved in 5 ml of a saturated solution of sodium succinate, pH 8.35. After cooling to 4°C, 5 mg succinic anhydride are added with gentle stirring for 1 h, followed by incubation overnight at 4°C. The solution is then incubated for 1 h at 23°C and dialysed for 40 h against 5 mM K-phosphate buffer, pH 7.5.

B.   Preparation and stabilization of colloidal gold

Colloidal gold of the diameter range 4–5 nm can be prepared as follows (Slot and Geuze, 1981; Rohde *et al.*, 1986): A mixture of 2.5 ml of 0.6% tetrachloroauric acid and 0.7 ml of 0.2 M K_2CO_3 is poured into 120 ml distilled and filtered water, to which a solution consisting of 0.8 ml ether and 0.2 ml ether saturated with white phosphorus has been added. During a 15-min incubation at room temperature the colour of the solution changes to reddish brown and becomes deep red during the subsequent boiling for 5 min. The colloidal gold is stabilized by adsorption of succinoylated avidin (see above) at pH 2.5; the amounts are determined as follows (Georghegan and Ackerman, 1977): succinoylated avidin (1.25 mg) is added to 5 ml colloidal gold with continuous stirring. After 30 min the avidin–gold complexes are collected by centrifugation at 100 000g (e.g. in an Airfuge from Beckman Instruments). The pellet is washed three times with phosphate-buffered saline (5 mM K-phosphate, pH 7.5, 100 mM NaCl) containing 0.2 mg polyethylene glycol per ml to remove free avidin. The final pellet made up of avidin–colloidal gold complexes is resuspended in 2 ml phosphate-buffered saline.

V.   Mounting and staining of proteoliposomes for electron microscopy

A.   Vesicle shape and size

In order to maintain the three-dimensional organization of the proteo-liposomes, it is necessary to avoid artefacts due to air drying or "flattening" when the vesicles are mounted onto the support film. However, artefacts cannot be completely avoided; they can only be reduced. Even with freeze-etching (Fig. 1) wrong values for the measured vesicle diameters may be obtained because the proteoliposomes "embedded" in ice may not exhibit their largest diameter. In chemically-fixed, resin-embedded and ultra-thin-sectioned samples the visible vesicle diameters may also be wrong. There are a number of reasons why this may occur: change of composition by chemical treatment (fixation, staining, dehydration, embedding); change of the diameter by shrinkage during polymerization of the resin used for embedding; or cutting artefacts. Electron microscopy of proteoliposomes in the "frozen-hydrated" state might be a technique producing a low degree of artificial

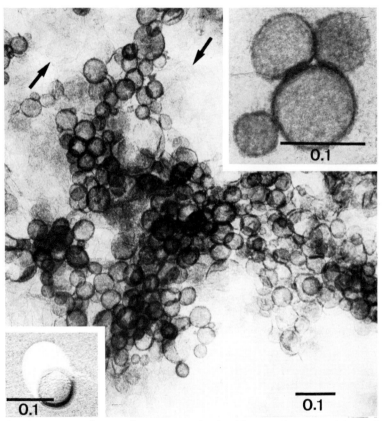

Fig. 1. Aggregate of proteoliposomes stained with uranyl acetate. Most of the vesicles appear to be well preserved (they appear reasonably spherical). Vesicles directly attached to the support film are flattened (arrows). (From Rohde *et al.*, 1986.) Inset at the lower left-hand side: freeze-etched sample of a liposome without inserted protein molecules. Dimensions in µm.

deformation; however, this method is not in common use because in most laboratories the necessary instrumentation is not available.

A simple procedure for the preparation and electron micrography of proteoliposomes with reasonably well-preserved spherical shape is as follows (Rohde *et al.*, 1986): the proteoliposomes are centrifuged after reconstitution. The pellet is removed from the tube and placed into a drop of adequate buffer with saline on a glass slide. The pellet is divided with a needle into small aggregates of different sizes. Care must be taken that the proteoliposomes never dry out during this procedure. Now this suspension is transferred into an Eppendorf tube, and 0.5 ml of adequate buffer (with saline) is added. An aliquot is removed with a Pasteur pipette and put into a small container (less than 0.5 ml volume) in such a way that a convex surface is formed. A carbon support film is floated off a sheet of mica (2×3 mm) onto the surface of the suspension and kept there for 30–60 s, depending on the vesicle concentration. Afterwards the carbon film with the adhering small proteoliposome aggregates is transferred onto the surface of an aqueous staining solution containing 4% uranyl acetate, pH 4.5 (usually this stain is also used for negative staining). After a few seconds the sample is picked up with a copper grid, blotted dry with filter paper, and analysed immediately afterwards in the electron microscope. It has to be made sure that a magnification and an electron dosage as low as possible are used for observation and micrography, and that the imaging is done as quickly as possible. As seen in Fig. 1, the proteoliposomes depicted in this way appear to be positively stained, and very many of them exhibit a sphere-like shape and a "wall" as expected for a typical phospholipid vesicle. The proteins, however, cannot be seen. Those vesicles immediately attached to the carbon film are severely flattened and cannot be evaluated. The others, forming the part of the aggregate not directly in contact with the support, appear to be well preserved. The reason for this might be that they stabilize one another, and that the uranyl acetate present in the "wall" adds further stabilization.

B. Determination of the overall number of integrated enzyme molecules

An estimation of the overall number of enzyme molecules integrated into an average proteoliposome can be achieved by routine negative staining (Valentine *et al.*, 1968; Johannssen *et al.*, 1979; Rohde *et al.*, 1986). This is done by using a diluted proteoliposome sample with well-dispersed vesicles. The procedure is very similar to that described above for the determination of shape and size of proteoliposomes (see Section V.A). The negative staining effect is obtained by incomplete blotting on filter paper. Small amounts of staining solution remain on the support film surrounding the vesicles. Thus, the vesicles appear bright and their surroundings are dark (Fig. 2). Flattening

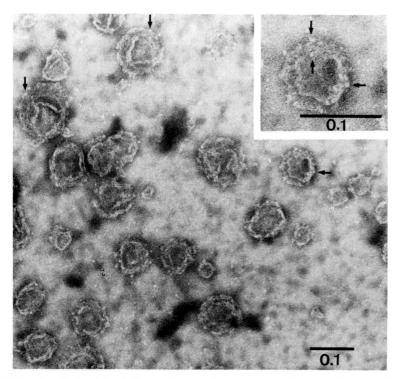

Fig. 2. Negatively stained, flattened proteoliposomes. Projections of incorporated methylmalonyl-CoA decarboxylase molecules are marked (arrows). Dimensions are given in μm. (From Rohde *et al.*, 1986.)

of the vesicles occurs because the concentration of the dispersed vesicles is low; thus aggregate formation is avoided and the vesicles are in contact with the support film and flattened (see Section V.A). As expected, the measurable diameters of the vesicle projections are larger than those determined following the procedure described in the previous section. In addition, white dots are visible, distributed over the surface of the vesicles (Fig. 2). These dots represent the enzyme molecules. This is demonstrated by the fact that phospholipid vesicles devoid of integrated enzyme molecules do not exhibit these white dots (data not shown). When estimating the overall number of integrated enzyme molecules per average vesicle, one should keep in mind that probably not all enzyme molecules are visible; presumably those enzyme molecules integrated in that part of the vesicle directly attached to the support film might not be detected. Intimate contact of the vesicle surface with the support prevents penetration of the negative staining solution between vesicle and surface of the support. Thus, the number of white dots counted per vesicle will probably be too low. However, a good estimate of the overall number of enzyme molecules per proteoliposome can be obtained by measuring the distances between the white dots and calculating the mean distance. By using this mean distance the distribution of enzyme molecules in two dimensions can be calculated, and the average number of enzyme molecules on the surface of an average vesicle can be calculated on the basis of the value which is obtained for this surface (according to Section V.A) from the calculated average vesicle diameter.

C. Determination of the number of enzyme molecules with the biotin prosthetic group exposed at the outside of the proteoliposomes

Samples containing different concentrations of proteoliposomes (with integrated methylmalonyl-CoA decarboxylase), or phospholipid vesicles devoid of enzyme (as a control), are incubated at 30°C for 4 h, under occasional shaking, with a mixture of 40 μl of the suspension of avidin–gold complexes in phosphate-buffered saline (see above) and 0.5 ml of 30 mM K-phosphate buffer, pH 8.9, containing 1 mM Na_2SO_4, 0.5 mM dithioerythrol and 1.5 mM NaN_3 (Rohde et al., 1986). The labelling complex has to be used in such an amount that all biotin molecules are labelled. This amount can be determined by comparing the labelling effect measured in samples containing different concentrations of proteoliposomes (see above). Afterwards the samples are negatively stained as described in Section V.B. In Fig. 3a colloidal gold particles surrounded by avidin are shown. Figure 3b depicts a proteoliposome with two labelling complexes attached. This means that this

Fig. 3. (a) Colloidal gold particles covered by avidin. Negative staining with uranyl acetate. Dimension in nm. (Original micrograph by M. Rohde.) (b) Avidin–gold complexes attached to methylmalonyl-CoA decarboxylase molecules in a proteoliposome. Dimension in μm. (Original micrograph by M. Rohde.)

vesicle contained two methylmalonyl-CoA decarboxylase molecules with their biotin prosthetic group exposed at the outside of the proteoliposomes. As the labelling is done prior to mounting for electron microscopy (i.e. the whole vesicle surface is accessible to the label), and as colloidal gold particles have very high immanent contrast, no allowance has to be made for non-labelled correctly oriented enzyme molecules.

The number of avidin–gold complexes attached per proteoliposome, and the ratio of proteoliposomes without any attached label to vesicles with bound complexes, can now be counted from prints. These values should be determined from several hundred vesicles and from at least two sets of experiments (Rohde et al., 1986). Correction for background labelling has to be made. On the basis of these data, the number of interest, i.e. the number of enzyme molecules integrated in the "correct" orientation, per average proteoliposome can be calculated. One should keep in mind that due to steric factors, and due to some loss of label during the mounting procedure, part of the externally oriented biotin may not be detected.

VI. Conclusion

The electron microscopic procedures described above allowed the determination of the internal volume of average proteoliposomes of given vesicle preparations, the overall number of methylmalonyl-CoA decarboxylase molecules incorporated in an average vesicle, and the number of enzyme molecules integrated into an average proteoliposome with the biotin prosthetic group exposed at the outside (Rohde et al., 1986). These morphological properties and measured and calculated numbers were found to support strongly findings obtained from biochemical analyses regarding parameters of the Na^+ pumping methylmalonyl-CoA decarboxylase system in Veillonella alcalescens (Dimroth, 1982, 1985; Hilpert and Dimroth, 1983, 1984; Dimroth and Hilpert, 1984).

Acknowledgements

We thank Petra Kaufmann-Kolle for the electron micrograph of the freeze-etched sample. Preparation of the manuscript was supported by the Fonds der Chemischen Industrie and the Deutsche Forschungsgemeinschaft.

References

De Vries, W., Riedvald-Struijk, T. R. M. and Stouthamer, A. H. (1977). Antonie Leeuwenhoek J. Microbiol. **43**, 153–167.

Dimroth, P. (1982). *Biosci. Rep.* **2**, 849–860.
Dimroth, P. (1985). *Ann. Rev. NY Acad. Sci.* **447**, 72–85.
Dimroth, P. and Hilpert, W. (1984). *Biochemistry* **23**, 5360–5366.
Geoghegan, W. D. and Ackerman, G. A. (1977). *J. Histochem. Cytochem.* **25**, 1187–1200.
Hilpert, W. and Dimroth, P. (1983). *Eur. J. Biochem.* **132**, 579–587.
Hilpert, W. and Dimroth, P. (1984). *Eur. J. Biochem.* **138**, 579–583.
Johannssen, W., Schütte, H., Mayer, F. and Mayer, H. (1979). *J. Mol. Biol.* **134**, 707–726.
Kishida, Y., Olsen, B. R., Berg, R. A. and Prockop, D. J. (1975). *J. Cell Biol.* **64**, 331–339.
Morris, R. E. and Saelinger, C. B. (1984). *J. Histochem. Cytochem.* **32**, 124–128.
Rohde, M., Däkena, P., Mayer, F. and Dimroth, P. (1986). *FEBS Lett.* **195**, 280–284.
Slot, J. W. and Geuze, H. J. (1981). *J. Cell Biol.* **90**, 533–536.
Valentine, R. C., Shapiro, B. M. and Stadtman, E. R. (1968). *Biochemistry* **7**, 2143–2152.

13

Electron Microscopic Analysis of Nucleic Acids and Nucleic Acid–Protein Complexes

EBERHARDT SPIESS

*Deutsches Krebsforschungszentrum, Institut für Zell -und Tumorbiologie
D-6900 Heidelberg, Federal Republic of Germany*

RUDI LURZ

*Max-Planck-Institut für Molekulare Genetik,
D-1000 Berlin 33, Federal Republic of Germany*

I. Introduction

There was a gap of 90 years between the discovery by Friedrich Miescher in 1859 of nucleic acids as the main constituents of the nuclei of living organisms, and their visualization as individual molecules by electron microscopical techniques. The discovery of Avery and co-workers in 1944 that nucleic acids form the molecular basis of inheritance and have a fundamental role in protein synthesis and regulatory processes in the cell focused great scientific interest on the elucidation of their physical structure. In 1959, Kleinschmidt, Zahn and their collaborators (Kleinschmidt and Zahn, 1959) succeeded in visualizing deoxyribonucleic acid (DNA) molecules by means of a new electron microscopical technique. This achievement, still practised as the classical cytochrome c spreading technique, was the start of a development to visualize nucleic acids in all their aspects of structure and function.

The classical spreading technique provides information about conformation and molecular weight of double-stranded DNA (dsDNA). Davis and

METHODS IN MICROBIOLOGY
VOLUME 20   ISBN 0 12 521520 6

Davidson (1968) and Westmoreland *et al.* (1969) were able to extend the method to single-stranded molecules of DNA (ssDNA) and ribonucleic acids (RNA) by introducing into the spreading process denaturing agents, among which formamide is the most widely used. This opened the fields of denaturation mapping of DNA (Inman, 1974), heteroduplex formation to compare different genomic materials (Davis *et al.*, 1971) and mapping of individual RNA transcript regions on DNA molecules known as the R-loop formation technique (Thomas *et al.*, 1976; White and Hoagness, 1977). Another field is the visualization of transcriptional units by the centrifugation technique introduced by Miller and co-workers (Hamkalo and Miller, 1973).

Technical simplifications are the so-called diffusion methods (Lang *et al.*, 1964; Mayor and Jordan, 1968; Lang and Mitani, 1970), which reduce the impact of mechanical disturbance.

With growing interest in the interactions of proteins with DNA in nucleic acid replication, regulation of DNA transcription, degradation of DNA by endonucleases, packaging of DNA into higher-order complexes and detection of distinct conformational regions like Z-DNA or modified bases by means of specific monoclonal antibodies, the use of cytochrome c as the mediator in the spreading of nucleic acids has become a hindrance. New techniques have been developed: either cytochrome c is substituted by a smaller molecule, i.e. ethidium bromide (Koller *et al.*, 1974), benzyldimethylalkylammonium chloride (BAC) (Vollenweider *et al.*, 1975), polylysine coating of the supporting carbon film (Williams, 1977), or nucleic acids are mounted directly onto the support, mediated by positive charges on carbon films (Dubochet *et al.*, 1971) or bivalent cations, mostly Mg^{2+}, on mica (Koller *et al.*, 1974; Portmann and Koller, 1976).

A considerable number of original and comprehensive articles with detailed recipes for visualization of nucleic acids have been published (i.e. Kleinschmidt, 1968; Davis *et al.*, 1971; Younghusband and Inman, 1974; Davidson, 1978; Ferguson and Davis, 1978; Fisher and Williams, 1979; Brack, 1981; Vollenweider, 1981; Burkhardt and Pühler, 1984; Garon, 1986; Coggins, 1987; Sogo *et al.*, 1987). Therefore it is the aim of this chapter to present an introduction into the field of nucleic acid preparation also suitable for beginners, trying not to repeat all the methods already presented in detail in other reviews. The selected basic techniques are routine methods in our laboratories. They cover a wide range of applications, but do not need sophisticated equipment. In the second part we will focus on examples predominantly useful in the imaging of protein–DNA complexes.

With the exception of exact sequencing, a long-aspired but never-achieved goal, electron microscopy has contributed to all aspects of nucleic acids research. In recent years electron microscopical methods have lost some of their importance to other biophysical or biochemical methods, which are more easily accessible to most workers. Nevertheless, electron microscopy is superior to these biochemical methods concerning the accuracy of analysis of molecule structure or sample composition, and is equivalent in speed when practised in a well-established laboratory by experienced investigators.

II. Methodology

A. General technical requirements

1. Solutions

Water for the solution of reagents and buffers is the most critical material in preparation of nucleic acids. Only water of the highest purity is appropriate, as even traces of detergents can ruin all efforts. Solutions of buffers, additives and stains have to be cleared from all pollutants by chemical purification methods, filtration or high speed centrifugation.

Many of the materials to be used are hazardous in one way or another; they should therefore be handled properly. Uranyl acetate is poisonous and radioactive; solutions should be kept in the dark, but not wrapped in metal foils.

Collodion (= nitrocellulose; see also *Parlodion*). Dissolve solid material to 3.5% in amyl acetate under water-free conditions. Store aliquots at $-20°C$. Traces of water will lead to holes in the films!

Cytochrome c (Type IV or Type V; Sigma, St Louis, USA, or Type III, Serva, Heidelberg, FRG). Prepare solutions in 1 mM EDTA. Additional purification is recommended by filtration through 0.2 μm pore sized filters; even better is chromatography with Sephadex G100. When ssDNA is prepared, a very fine background is desirable. To achieve this, cleave the cytochrome c with cyanobromide and subsequently purify over Sephadex G50 (Brack, 1981). The solutions of cytochrome c are diluted to 1 mg ml^{-1} (A_{410} reading of about 10^1). This solution can be stored at 4°C for several months or small aliquots are kept frozen at $-20°C$ until use.

Ethidiumbromide (Sigma, St Louis, USA). A stock solution with 5 mg ml^{-1} in water is prepared and stored in the dark.

Formaldehyde. Prepared from solid paraformaldehyde (Fluka, Switzerland). Usually a 10% solution is made by dissolving at 60°C 0.1 g of paraformaldehyde in 1 ml of water to which about 5 μl of 1 N NaOH are added; after cooling to room temperature, the solution is cleared by centrifugation. Do not store in the cold!

Formamide (Fluka, Switzerland). This has to be cleared from ammonia and formic acid either by two-fold crystallization at temperatures around $0°C$ or by ion-exchange extraction (1000 ml formamide, 40 g of Amberlite MB1 and 10 g Norit A are shaken for 2 h, filtered twice, extracted three times with ether and finally freed from ether by streaming N_2 gas through the fluid). Small aliquots should be stored at $-20°C$.

Glutaraldehyde (25%; Fluka, Switzerland). Use EM grade quality; dilute to the desired working concentration and always keep cold.

Glyoxal. 30% solution in water (Serva, Heidelberg, FRG) has to be treated with mixed bed resin (AG 501-X8, Bio-Rad, Richmond, USA) before use. Alternatively, prepare from glyoxal trimer dihydrate (Sigma, St Louis, USA) as with paraformaldehyde.

Nucleic acids. DNAs are stored at $-20°C$ in an EDTA-containing buffer such as TE buffer (10 mM Tris-Cl, pH 7.5, 1 mM EDTA) to protect them against the action of DNases. Samples used frequently should be kept at $4°C$ to avoid multiple freezing and thawing.

DNA, which is suitable for most biochemical techniques, is often not clean enough for electron microscopy. High amounts of contaminating RNA produce coarse background which can obscure the DNA. Remove RNA by RNase digestion and microstep exclusion chromatography (Krieger and Tobler, 1977). Proteins still accompanying the DNA after CsCl preparation have to be removed by repeated phenolization and ethanol precipitation. Any phenol or ethanol has to be removed completely before electron microscopical preparations. DNA fragments isolated from gels are often contaminated with traces of gel material which forms clumps with the DNA, especially after fixation with aldehydes. In such cases use low melting agarose to avoid aggregations, or clean the DNA from contaminants by ion-exchange chromatography (e.g. Elutip-d, Schleicher and Schüll, Dassel, FRG). When working with RNAs, observe strictly the general precautions to avoid the degradation of the material by RNases.

Parlodion (Mallinkrodt Inc., St Louis, USA). See *Collodion.*

Sticky grid solution. Dissolve the glue from 10 cm of an adhesive tape in 50 ml chloroform and store in a dark bottle.

Triethanolamine (*TEA*) (Merck, Darmstadt, FRG). A buffer substance which may be used instead of Tris because it has no primary amino groups reacting with formaldehyde, glyoxal or glutaraldehyde.

Uranyl acetate (*UO₂Ac*) solutions for staining (Merck, Darmstadt, FRG).
(1) A stock solution of 5×10^{-3} M UO_2Ac is prepared in 50 mM HCl. Store in the dark. Before use this solution is diluted 1:50 or 1:100 into 90% ethanol.
(2) A stock solution of 5% in water is prepared. Store in a dark bottle. Centrifuge an aliquot before use and dilute to 0.5 or 1% with water.

2. Supports

Sticky grids. Usually 300–600 mesh copper grids are used for supporting carbon films. Place grids on filter paper and float them with sticky grid solution (see Section II.A.1). After evaporation of the solvent, they can be used. The glue attaches carbon films tightly onto the grid.

Carbon films. Carbon provides homogeneous low noise films of high stability under electron bombardment, but they sometimes have uncontrollable surface charges and are brittle. Films are evaporated onto freshly cleaved mica under high vacuum conditions, subsequently floated on a clean water surface and picked up with sticky grids.

Collodion/Parlodion films. Films of this material are less resistant under electron bombardment and have more impurities and also more holes, but are of higher mechanical stability than those made of carbon. Therefore grids with bigger holes can be used, e.g. 200 or 75 × 300 mesh. Stock solution (about 4 µl per 25 cm²) is dropped onto a clean water surface (most conveniently in a Buchner funnel); in a minute the solvent evaporates and a film in silver to pale gold reflection colour remains. It is then lowered onto copper grids and dried. Films should be used within a few days. To improve the stability under electron bombardment, these films should be stabilized by carbon. This should be done in the case of protein-free preparations before mounting the nucleic acids, and in the case of protein spreading, after mounting and shadow casting.

Mica. Cut rectangular pieces (max. 1 × 2 cm) from mica sheets and cleave them with clean razor blades, needles or tweezers. Make sure that "old" and "fresh" surfaces cannot be mixed up, e.g. by marking the outsides with a water-resistant felt-tipped marker. DNA is mounted onto the fresh surface! Films are removed from the mica by immersing the mica under a very low angle slowly into a trough or funnel filled with clean water. It is usually no problem to float off pure carbon films. But when nucleic acids were prepared, it may be that the films do not detach properly from the mica. In such cases, some of the following methods can help:

(1) Use conditioned carbon rods: dip the tips of the carbon rods into 1 M NaCl solution and dry in air before using the rod for film preparation.

(2) Place the mica with film in a humid chamber for some time—several hours may be necessary—and try from time to time to detach the film.

(3) Instead of water, use a solution of 1 N NaOH or KOH at 40°C; pick up the film with the same sheet of mica and transfer it to clean water to wash off the alkali.

(4) Float the covered mica on concentrated HCl for some hours; this will solve even extreme cases without disturbing the quality of the carbon film or the shadow cast.

Samples prepared under very good vacuum conditions, especially in vacuum evaporators which are equipped with a turbo molecular pumping device, often produce very tightly sticking films if vacuum conditions are better than 10^{-3} Pa ($= 10^{-5}$ mbar).

3. Shadow casting

Shadow casting is the predominant contrasting method for nucleic acids. It is achieved by evaporation of metal under high vacuum conditions. Pure platinum (Pt), its alloys with palladium (Pa) or iridium (Ir), both 80:20, or mixed evaporation of Pt and carbon (C), give stable specimens with contrasts of sufficient resolution for most work. In cases of high resolution requirement, tungsten (W) or tantalum–tungsten (Ta–W) alloy has to be used. Castings are done under an angle of between 3 and 10 degrees while rotating the specimen. Optimal contrast conditions have to be elaborated and should then be reproduced by monitoring with a quartz measuring device, or simply by observing the deposit on a piece of white paper close to the specimen.

4. Electron microscopy

In a good preparation, the linear nucleic acid molecules appear as relaxed individuals in a completely random arrangement, neither with intra- nor with interstrand overlapping. Circular molecules, either covalently closed circular DNA (cccDNA), also called supertwisted DNA, or form I DNA and open circular DNA (ocDNA), also called form II DNA (Vollenweider et al., 1976), i.e. circular DNA which is nicked at least in one strand, should also appear as individual molecules. However, intramolecular strand crossings depend on the degree of relaxation.

EM imaging is to be done under normalized conditions at a calibrated magnification. Possible movements of plastic films, which are the consequence of electron bombardments, should be avoided by sufficient carbon enforcement of the supporting plastic film.

5. Quantitations

Length measurements are the basis for molecular weight or base number determinations. Such measurements can be obtained in different ways: (1) directly in the microscope with special optical equipment; (2) from digitized images taken directly from the microscope and processed by computer programs; (3) from photographic images with graphic tablets supported by computer programs; and (4) with a map ruler from highly enlarged positives. To optimize the measurement, the following rules should be observed. Take the measurement from one image; if this is not possible, all images should be photographed under standardized conditions. For sample collection, strictly apply statistical procedures. The length to be defined should read several cm, therefore choose appropriate magnifications. Standards and unknown molecules should be in the same sample. For routine, a minimum of 25 measurements from a certain type of molecule renders a sufficient statistical accuracy; more than 100 measurements do not improve the outcome. For a specimen with n different subpopulations, a minimum of $(25 \times n)$ measurements are necessary. Median and average value with standard deviation must be calculated. Molecules differing in length more than three times from a given standard deviation should be discarded. In a homogeneous population, median and average are almost identical, and the percentage error is around 2%; but up to 5% is tolerable; for ss nucleic acids this error is about two-fold higher. (For more detailed statistics see Ferguson and David, 1978.) In general, large DNA fragments can be measured more accurately than smaller ones. The size of linear DNA pieces shorter than 1 kb may be determined more accurately by gel electrophoresis.

For the definition of the true length or base number an internal standard is recommended. Circular standards are more accurate than linear standards. Standard and unknown molecules should be in the length ratio between 5:1 to 1:5. Some possible standards which are commercially available are given in Table I.

For the rough calculation of the molecular weight or base numbers the following formula is valid:

$$1 \,\mu m \; DNA = 2.07 \,(\pm 0.03) \, MD = 3.14 \, kb$$

or simply $\qquad 1 \,\mu m \; DNA = 2 \, MD = 3 \, kb$

(bp = base pairs; kb = kilobase pairs or kilobases in ds or ss nucleic acids, respectively).

An example for a typical application of length measurements is given in Fig. 1. The length distributions of some unknown plasmids and pBR322 as a standard from an electron microscopic preparation are compared with those from agarose gel electrophoresis of the unknown plasmids.

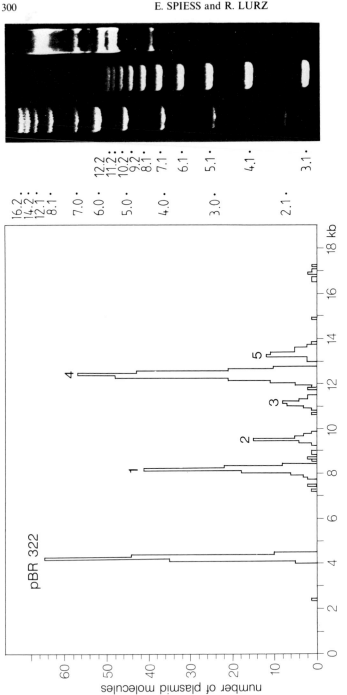

Fig. 1. Size distribution of plasmids isolated from a thermophilic eubacterium. Comparison of lengths measured from electron micrographs with gel electrophoresis data. Plasmid pBR322 was added as an internal standard; the five peaks were calculated to: (1) 8.244 ± 0.104 kb ($n = 97$); (2) 9.553 ± 0.098 kb ($n = 28$); (3) 11.224 ± 0.153 kb ($n = 27$); (4) 12.438 ± 0.166 kb ($n = 120$); (5) 13.442 ± 0.151 kb ($n = 40$). Adsorption to mica was done as described in Section II.B.3(a). Gel electrophoresis shows in the first lane ccc DNA standard, in the second lane standard DNA (BRL, New York, USA), and in the third the isolated plasmid fraction (Ulbrich and Lurz, original preparation).

TABLE I
Some commercially available nucleic acids which are suitable as internal standards for electron microscopical size determination

ssDNA	φX174 phage DNA	1.7 MD	5 386 b^c
	M13mp9 phage DNA	2.39 MD	7 229 b^{a,b}
dsDNA	plasmid pUC12	1.8 MD	2 730 bp^b
	plasmid pBR322	2.84 MD	4 361 bp^{a,b,c}
	plasmid pBR328	3.24 MD	4 907 bp^b
	SV40 virus DNA	3.41 MD	5 423 bp^b
	plasmid ColE1	4.21 MD	6 460 bp^a
	fd109 RF DNA	5.47 MD	8 284 bp^b
	PM2 phage DNA	6.52 MD	10 020 bp^b
ssRNA	MS2 phage RNA	1.18 MD	3 569 b^b
rRNA	16S from *E. coli*	0.44 MD	1 542 b^b

^{a} Amersham, Buckinghamshire, England.
^{b} Boehringer, Mannheim, FRG.
^{c} GIBCO BRL, New York, USA.

B. Preparation methods

For the following standard procedures, shown schematically in Fig. 2, we have compared preparations all made with an identical mixture of two plasmid DNAs and a ss phage DNA. Micrographs are shown in Fig. 3; length measurements of the DNAs in these preparations are compared in Fig. 4 and Table II.

1. Cytochrome c spreading by microdiffusion

Cytochrome c spreading by microdiffusion is a variation of the diffusion method developed by Lang *et al.* (1964). A micro version was described later by Lang and Mitani (1970). In contrast to cytochrome c spreading methods, only one phase is necessary. DNA, buffer and cytochrome c are mixed and small droplets of this solution are formed. Cytochrome c and DNA diffuse to the air–buffer interphase where the protein denatures and forms a monomolecular surface film which entraps the nucleic acid molecule unfolded to two-dimensional structures at the air–water interphase.

The method is suitable mainly for the visualization of double-stranded nucleic acids. Although the DNA diffuses from the solution into the cytochrome c surface film and is not extended by spreading forces, relatively long (more than 150 kb) DNA molecules can be prepared in a suitably unfolded manner by this method.

(a) Droplet diffusion technique

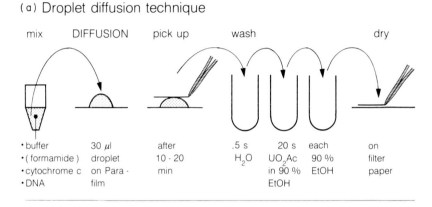

mix	DIFFUSION	pick up	wash			dry

- buffer
- (formamide)
- cytochrome c
- DNA

30 μl droplet on Para- film

after 10 - 20 min

.5 s H$_2$O in 90 % EtOH

20 s UO$_2$Ac in 90 % EtOH

each 90 % EtOH

on filter paper

(b) Carbonate spreading

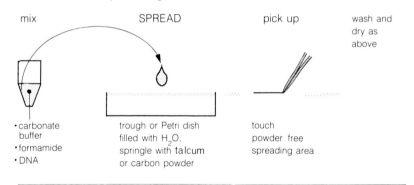

mix	SPREAD	pick up	wash and dry as above

- carbonate buffer
- formamide
- DNA

trough or Petri dish filled with H$_2$O, springle with talcum or carbon powder

touch powder free spreading area

(c) Direct mounting onto mica

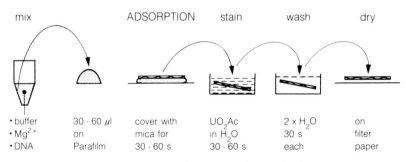

mix	ADSORPTION	stain	wash	dry

- buffer
- Mg^{2+}
- DNA

30 - 60 μl on Parafilm

cover with mica for 30 - 60 s

UO$_2$Ac in H$_2$O 30 - 60 s

2 x H$_2$O 30 s each

on filter paper

Fig. 2. Schemes for preparation methods.

(a) Procedure

(i) *Reagents and buffers*

NH$_4$Ac buffer: 2 M ammonium acetate, 10 mM EDTA, pH 7.0;
1 mg ml^{-1} cytochrome c;
5 × 10^{-5} M UO$_2$Ac in 90% EtOH freshly prepared from UO$_2$Ac stock
solution in 0.1 N HCl.

(ii) *Protocol* (see Fig. 2a)

• Droplet solution: 0.9 ml H$_2$O, 0.1 ml NH$_4$Ac buffer, 3 µl 0.1% cytochrome c.

• 0.5 ml of this droplet solution is carefully mixed with 1 µl DNA (conc. 20 µg ml^{-1}).

• Using a plastic pipette, droplets of 30–40 µl are placed on a hydrophobic surface (Parafilm or clean Petri dish).

• Cover the droplets with a Petri dish and wait for 15 min. Any vibrations or currents of air should be avoided.

• Transfer the surface film of denatured cytochrome c and enclosed DNA onto a grid (covered with carbon or unenforced Parlodion film) by touching the film side of the grid, as parallel as possible, to the surface of a droplet either laterally or at its top.

• Wash 5 s on distilled water.

• Do positive staining with 5 × 10^{-5} M UO$_2$Ac in 90% ethanol and dehydrate for another 10 s in 90% ethanol.

• Dry the grid face down on filter paper.

• Before touching another droplet forceps should be dried carefully, because traces of ethanol vapour would destroy the cytochrome c surface film.

• Do rotary shadowing.

After the protein film has been touched once with a grid, it is possible to repeat this procedure because the protein surface film will reform after a few minutes. This method can be performed with minute quantities of nucleic acids: preparing only one droplet and increasing the time for diffusion to several hours, the amount of DNA may be reduced to 2 × 10^{-5} µg (Lang and Mitani, 1970). Also the amount of cytochrome c has to be reduced, because with longer diffusion time more cytochrome c reaches the surface of the droplet. For two hours diffusion time, 50% of the original cytochrome c concentration is sufficient.

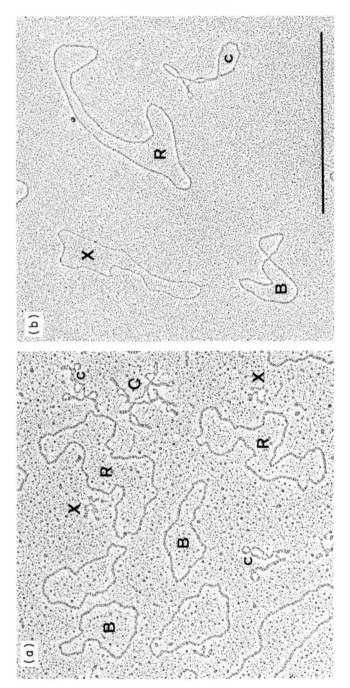

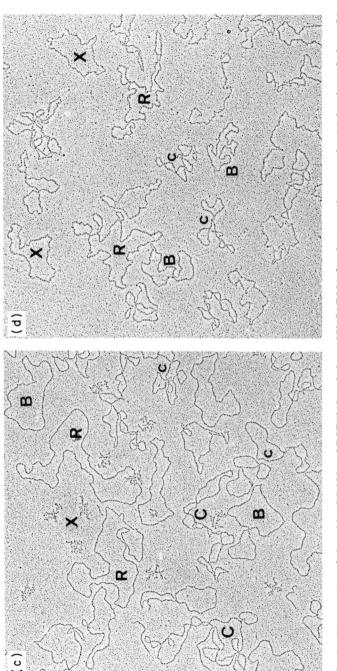

Fig. 3. Preparation of plasmids pBR322 and RSF1010 and phage φX174 DNA for electron microscopy by: (a) droplet technique; (b) carbonate buffer spreading; (c) adsorption to mica; and (d) adsorption to mica under denaturing conditions for ss nucleic acids. B, plasmid pBR322; R, plasmid RSF1010; c, supertwisted molecule of pBR322; C, supertwisted molecule of RSF1010; X, φX174 ss phage DNA. Bar, 1 μm.

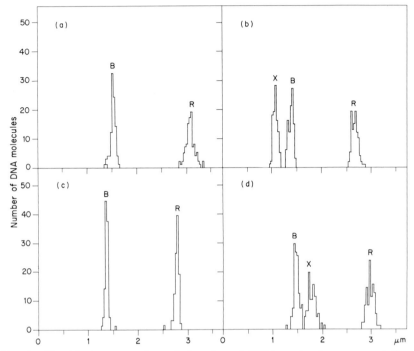

Fig. 4. Graphical presentation of the size of DNA molecules measured from micrographs of the four different DNA preparation methods shown in Fig. 3. Statistical data are given in Table II. Abbreviations as in Fig. 3.

Replacing the water in the droplet by formamide up to 50% final concentration makes this microdiffusion method capable of unfolding ssDNA, but only to a moderate quality.

2. Spreading of ss and ds nucleic acids

Here we describe a method using carbonate buffer with formamide and cytochrome c in the spreading mixture and water for hypophase. This method (Westmoreland et al., 1969; Inman and Schnös, 1970; Paulson and Laemmli, 1977; Rochaix and Malnoe, 1978) can be recommended as a standard procedure for visualizing both ds and ss nucleic acids for problems like evaluation of heteroduplexes, secondary structures (inverted repeats), RNA–DNA hybrids or replication intermediates. Single and double strands can be distinguished easily when spread on a water hypophase, and ss regions have about the same length as the corresponding ds regions (see Figs 3b, 4b, Table II); this facilitates the length evaluations because ss regions do not have to be corrected substantially.

TABLE II
Statistical data of length measurements of different DNA preparations

	Droplet technique	Carbonate buffer spreading	Adsorption to mica	Adsorption to mica for ss
pBR322				
Median	1.550	1.490	1.400	1.395
Mean	1.5496	1.489	1.399	1.394
SD	0.021	0.038	0.024	0.048
Var.co.	2.7%	2.5%	1.7%	3.4%
RSF1010				
Median	3.110	2.990	2.823	2.685
Mean	3.109	2.992	2.823	2.690
SD	0.078	0.065	0.031	0.061
Var.co.	2.5%	2.2%	1.1%	2.3%
φX174				
Median	—	1.785	—	1.095
Mean		1.788		1.091
SD		0.092		0.043
Var.co.		5.2%		3.9%

From the preparations shown in Fig. 3 one hundred individual circular molecules were measured from each type. The graphic representation of these values is shown in Fig. 4. All methods show similar length per kb for dsDNA. Single-stranded nucleic acids cannot be measured using the droplet technique or mica adsorption under non-denaturing conditions. After carbonate buffer spreading the length of ss φX174 DNA is about the same as it would be for dsDNA (within 3%). (Lengths are given in μm; SD = standard deviation of the mean; Var.co. = variation coefficient or percentage error.)

(a) Procedure

(i) *Reagents and buffers*

1 M Na_2CO_3;
0.2 M EDTA;
1 mg ml^{-1} cytochrome c;
purified formamide;
5 × 10^{-5} M UO_2Ac in 90% EtOH freshly prepared from UO_2Ac stock solution in 0.1 N HCl;
graphite or talcum powder.

(ii) *Protocol* (see Fig. 2b)

- Preparation of the carbonate buffer:
  Mix in a plastic tube: 200 μl H_2O, 20 μl 1 M Na_2CO_3, 10 μl 0.2 M EDTA. Cool this buffer in ice; use only fresh buffer.

- Hypophase:
  A clean Teflon trough (or a new Petri dish) is filled with double-distilled water. The spreading area is cleaned thoroughly by pushing a Teflon (or PVC) bar over the water surface. Dust a little bit of graphite (or talcum) powder onto the surface to mark the spreading region.

- Hyperphase:
  Prepare the spreading mixture: 2 µl carbonate buffer, 5 µl formamide, 2 µl nucleic acids solution (conc. 10–$20\,\mu g\,ml^{-1}$), 1 µl cytochrome c.

- Spreading:
  5 µl portions of this mixture are spread immediately by touching the water surface with the drop hanging at the tip of an Eppendorf pipette. Wait for some seconds until the spreading region has been stabilized and transfer the spreading film onto a Parlodion-coated copper grid.

- Wash 5 s in water (optional).

- Do positive staining with uranyl acetate in ethanol.

- Dehydrate 10–20 s in 90% ethanol.

- Dry the grids face down on filter paper (and forceps too!).

- Do rotary shadowing.

When the concentration of nucleic acids is very low, the following modification might be helpful:

- For a final volume of 53 µl formamide-carbonate buffer mix in an Eppendorf tube: 50 µl formamide, 2 µl 1 M Na_2CO_3, 1 µl 0.2 M EDTA.

- Prepare the spreading mixture: 5 µl formamide-carbonate buffer, 5 µl nucleic acid solution, 1 µl cytochrome c.

- Do spreading as described above.

Extremely AT-rich DNA regions may be denatured into the single strands under these formamide concentrations and buffer conditions. In such cases, the amount of formamide and sodium carbonate buffer should be reduced to 50% of the values given.

When spreading is done on pure water hypophase, the spreading region expands greatly, followed by a contraction to less than half the area. Therefore long DNA ($> 30\,kb$) is sometimes parallel compressed. This may be overcome by addition of the detergent octylglucopyranoside (0.005% final concentration; Sigma, St Louis, USA) to the hypophase (Koller et al., 1985). Under these conditions the spreading spot is very small and stable. Another positive effect is the increase of the concentration of nucleic acids per square unit. However, the background is not as fine as when spreading is done without detergent.

3. Preparation of nucleic acids or nucleic acid–protein complexes by adsorption to mica

(a) dsDNA and protein–DNA complexes

This method is very reliable in visualizing nucleic acids and their complexes with proteins. It was originally applied in the presence of ethidium bromide for localization of RNA polymerase binding sites on T7 DNA (Koller *et al.*, 1974) and subsequently modified using divalent cations (Portmann and Koller, 1976). Further simplifications were elaborated in our own laboratories. Length measurements of isolated DNAs have shown (Portmann *et al.*, 1974; Fig. 4d, Table II) that this method gives the most accurate values compared to all other electron microscopical nucleic acid preparation methods.

(i) *Reagents and buffers*

The adsorption buffer should contain a minimum of 5 mM Mg^{2+} (DNA preparations without bound proteins were made with 5 mM Tris-acetate, pH 7.5, 5 mM magnesium acetate); 1% UO_2Ac.

(ii) *Protocol* (see Fig. 2c)

• Prepare mica sheets as described in Section II.A.2 (*Mica*).

• Add DNA (RNA) to the buffer to a final concentration of 0.5–1 µg ml^{-1}.

• 15–20 µl droplets of nucleic acid-containing solution are placed on parafilm.

• Cover the droplets with freshly cleaved mica. The solution should wet the whole area of the mica piece.

• Adsorption time is 1 min.

• Remove excess of solutions by pressing the edge of the mica sheet on a Kleenex tissue or by immersion into water. Subsequent washing and staining can be done either by transferring the mica onto droplets of stain and water or by immersion into stain and water in small dishes.

• Transfer for 1 min on/in 1% aqueous uranyl acetate for positive staining.

• Transfer 2–3 × for 1 min on/in distilled water.

• Dry face down on filter paper.

- Do rotary shadowing.

- Cover the mica from top by an additional carbon film.

- Transfer to grids: cut small pieces (about 2.5 × 2.5 mm) from the mica and
  float the carbon film off on a water surface; pick up the carbon from top
  using sticky 300–600 mesh copper grids and dry specimen up on filter paper.
  Alternative: float off the film completely onto water in a Buchner funnel and
  lower it onto sticky grids on the bottom of the funnel; pick up the grids and
  dry on filter paper.

Ways of detaching tightly adhering carbon films from the mica are given in
Section II.A.2 (*Mica*).

The high accuracy and the relatively easy preparation procedure re-
commend this method as a standard procedure for many nucleic acid and
nucleic acid–protein complexes preparation problems. In addition, this
method may be used at pH values between 3 and 10; it tolerates higher
quantities of salts (up to 0.5 M) and other additives (aldehydes and even traces
of Triton X-100). With this technique the three-dimensional coil of nucleic
acids in solution cannot be unfolded to such an extent as with cytochrome c or
BAC spreading techniques. Therefore problems arise with molecules bigger
than 30 kb.

By addition of ethidium bromide the supertwist torsions in ccc molecules
can be influenced: concentrations of about $2 \mu g \, ml^{-1}$ are sufficient to
compensate the supertwist and all molecules appear relaxed; at higher
concentrations positive supertwists are formed, while at $50 \mu g \, ml^{-1}$ or higher
the ccc molecules appear as rods slightly thicker than normal linear molecules
and are sometimes branched. The intercalating agent ethidium bromide
increases the length of the DNA (Butour *et al.*, 1978) and it should be taken
into account. The addition of ethidium bromide ($50–250 \mu g \, ml^{-1}$ final
concentration) also increases the rigidity of the DNA; it should be applied in
the case of longer molecules.

Furthermore, the diameter of the nucleic acid strand is not enlarged by a
coat of protein when using protein-free DNA preparation methods, allowing
unambiguous tracing of rather entangled molecules.

(b) ss Nucleic acids

The following modification of the mica adsorption technique allows visualization of ssDNA and ssRNA, and proteins bound to them. It is an
alternative to the BAC spreading method (Koller *et al.*, 1974), which is
described in a separate chapter of this volume (Johannssen). The characteristic
backfolding caused by unspecific base pairing of the single strands (see Figs 3a

and c) is prevented in this variation by treatment with formaldehyde, glyoxal and increased temperature. Formaldehyde reacts with amino groups of the bases and prevents base pairing (Haselkorn and Doty, 1961). Glyoxal reacts with amino groups but forms, especially with G residues, a very stable covalent compound (Johnson, 1975). Compared to cytochrome c spreading with carbonate buffer, the measured contour lengths of single strands are considerably shorter, but comparable in their standard deviations (see Figs 3b,d and Table II).

(i) *Reagents and buffers*

TEA-E: 100 mM TEA, 50 mM $MgCl_2$, no pH adjustment;
1 M KCl;
formaldehyde (FA): 10% formaldehyde freshly prepared from para-formaldehyde;
glyoxal (Go): 30% glyoxal in water;
4% UO_2Ac in water.

(ii) *Protocol*

- For a total volume of 50 μl mix in an Eppendorf tube: 26 μl H_2O, 5 μl TEA-E, 4 μl 1 M KCl, 10 μl FA, 3 μl Go, 2 μl nucleic acid (about 20 μg ml^{-1}).

- Incubate for 10 min at 50°C.

- Put 25 μl droplets on a Teflon-coated aluminium block (diameter about 10 cm) which is kept at 50°C (e.g. in a water bath) and cover them immediately with mica because evaporation is very rapid at this temperature.

- Do adsorption to mica on the aluminium block as described for ds nucleic acids, except that positive staining is done for 3 min on 4% aqueous UO_2Ac.

Incubation time and temperature must be varied depending on the sample. If denaturation is too weak, ss nucleic acids will still form secondary structures. If the conditions are too strong, dsDNA will also be denatured. The denaturing conditions given are sufficient to unfold ssDNA (Fig. 3b). The higher standard deviation of the ds plasmid DNAs (Table II) is due to the start of denaturation into single strands which are barely visible in the micrographs. Less denaturing conditions give standard deviations comparable to non-denaturing mica adsorption. For RNA, which forms very stable secondary structures like ribosomal RNAs, the incubation should be extended up to 30 min at 50°C.

C. DNA–protein complexes

Formation of DNA–protein complexes and their visualization follows a three-step protocol: (1) formation of the complexes; (2) removal of unbound protein after complexation by chromatographic methods; (3) mounting and shadow-casting for EM. Only the first step is specific for each DNA–protein complex.

To obtain clear results, it is recommended that unbound proteins are removed prior to preparation for EM. This step is critical when the reaction leads to an equilibrium between the concentration of the reaction partners and the product. Therefore the reaction product has to be stabilized before the separation. Such a stabilization is achievable by cross-linking with glutaraldehyde. The cross-linking effect improves with increasing mass and complexity of the polypeptides, because the cross-links in the polypeptide mass predominantly contribute to the stabilization of the complexes. It is possible to avoid this situation by working at low temperature (4°C) and as fast as possible. Suitable matrices for the separation are Sephacryl S1000, Sepharose 4B or DE52 ion exchanger. The column beds should be between 0.3 and 2 ml. For housing the matrix use a small commercial column, but siliconized Pasteur pipettes or larger pipette tips both plugged with siliconized glass fibres are very suitable.

The following protocols describe examples for a cooperatively binding protein, a DNA modifying enzyme protein (DNA-cytosine-5-methyltransferase), 16S rRNA bound to ribosomal proteins and antibodies which are used to map special bases or particular conformational regions.

1. Protein 10b from Sulfolobus acidocaldarius

(i) *Reagents and buffers*

> *Binding buffer*, pH 7.5: 20 mM TEA-Cl, 100 mM KCl, 5 mM dithioerythritol, 1 mM EDTA;
> *Triethanolamine-acetate buffer (TEA-Ac)*, pH 7.0: 10 mM TEA-acetate, 5 mM Mg-acetate;
> *Glutaraldehyde (Ga)*: 0.6% (dilute from stock solution);
> *DNA* (RSF 1010 plasmid DNA, Haring *et al.*, 1985): 30 μg ml^{-1}; protein 10b: 60 μg ml^{-1} in H$_2$O (Grote *et al.*, 1986).

(ii) *Protocol*

- 3 μl binding buffer are mixed with 1 μl DNA and 3 μl protein 10b and incubated for 15 min at 37°C. The protein–DNA ratio in this case is 4:1.

- After addition of 1 μl Ga, the assay is fixed for another 15 min at 37°C.

• Removal of unbound proteins is not necessary because protein 10b is very small (10 kD) and individual unbound protein molecules are not visible in the preparation. Only at very high protein concentrations does free protein 10b cause rough background.

• The fixed complexes are prepared by adsorption to mica as described in Section II.B.3(a) after addition of 24 μl TEA-Ac.

Protein 10b from *Sulfolobus acidocaldarius* binds cooperatively to dsDNA. The linear complexes formed contain two strands of dsDNA. In Fig. 5 three plasmid molecules are shown which have bound different amounts of protein 10b. The left molecule shows two small complexed regions; the free DNA is still supertwisted. The molecule in the middle has more than 50% of its DNA bound within the complex structure. In such cases, supertwists are not visible in free DNA because superhelical turns are compensated within the complex structure. Therefore it cannot be decided whether this is an open circular or a supertwisted molecule. In the right plasmid molecule, no free DNA is visible any more. This is a typical ccc molecule; in open circular DNA, the protein covers only one dsDNA strand above saturation conditions. The protein complexed ocDNA molecules have about the same morphology as uncomplexed ones. The strands are only somewhat thicker due to bound protein 10b (Lurz *et al.*, 1986).

2. DNA-cytosine-5-methyltransferase

(i) *Reagents and buffers*

TED *buffer*, pH 7.5: 200 mM Tris, 50 nM EDTA, 0.5 mM dithioerythritol;
Adenosylmethionine (*SAM*): 125 mM;
Triethanolamine buffer (*TEA*), pH 7.5: 100 mM TEA;
Triethanolamine-EDTA buffer (*TEA-E*), pH 7.5: 10 mM TEA, 10 mM EDTA;
MgCl: 100 mM $MgCl_2$, adjusted to pH 7.5 with TEA;
Glutaraldehyde (*Ga*): 1% Ga, dilute from stock solutions and adjust pH with alkali to 7.5;
Sephacryl S1000 or Sepharose 4B column: 2 ml gel bed equilibrated in 0.1 × TEA-E buffer;
DNA: 20 μg ml^{-1};
Enzyme: 10 μg ml^{-1} in 0.1 × TED buffer.

(ii) *Protocol*

Binding assay. A typical assay contains about 250 ng of DNA and variable concentrations of the enzyme. For a 100 μl assay mix in a test tube: 50 μl H_2O,

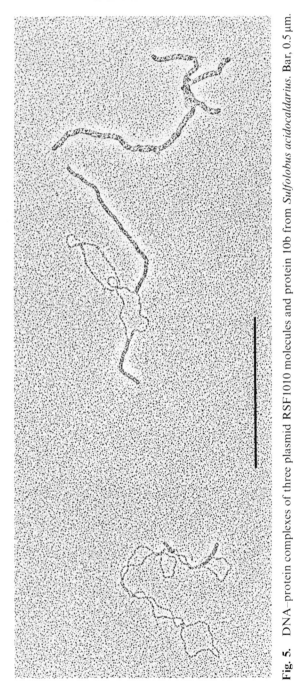

Fig. 5. DNA–protein complexes of three plasmid RSF1010 molecules and protein 10b from *Sulfolobus acidocaldarius*. Bar, 0.5 μm.

10 μl TED, 10 μl SAM, 20 μl DNA, 10 μl enzyme. Incubate at 37°C for various time intervals. Add 12.5 μl TEA and mix in very rapidly 12.5 μl GA-1. Incubate at 37°C for 15 to 30 min and remove unreacted enzyme.

Removal of unreacted enzyme. Apply sample onto the column and elute with TEA-E buffer, collect 1-drop fractions, proceed immediately to mounting.

Mounting for EM. A typical mounting mix contains 50 to 100 ng of DNA. The DNA concentration in the eluate after gel filtration is about 1/3 to 1/5 of the original concentration. Mix in a test tube: 30 μl H_2O, 6 μl TEA-E buffer, 6 μl MgCl and 18 μl of the eluate; mount onto mica as described in Section II.B.3(a).

The enzyme attaches at undefined sites and moves along the DNA molecule; obviously ssDNA sequences are preferred for binding. Figure 6 shows an example with enzyme isolated from mouse cells and reacted with plasmid pBR322 DNA (Pfeifer *et al.*, 1985).

3. Use of antibodies for mapping of special sites on DNA molecules

Monoclonal antibodies may become sticky during incubation and further processing. The result is a considerable loss of material by the gel filtration process. An alternative method to gel filtration is ion-exchange chromatography on DE52 cellulose. This method provides satisfactory yields of DNA–protein complexes. In addition, it works faster than gel filtration; a separation is completed within a few minutes.

For mounting onto mica a higher concentration of Mg^{2+} than in the cases described above is advisable. Mg^{2+} is supposed to improve specificity and stability of antibody–antigen complexes. However, the image of the DNA appears sharper under these conditions.

(i) *Reagents and buffers*

Triethanolamine-EDTA (TEA-E) buffer, pH 7.2: 100 mM TEA, 10 mM EDTA;

TEA-E-1500 mM NaCl (TEA-E-1500S) buffer, pH 7.2: 1500 mM NaCl in TEA-E buffer;

TEA-E-150 mM NaCl (TEA-E-150S) buffer, pH 7.2: 150 mM NaCl in 0.1 × TEA-E buffer;

TEA-E-200 mM NaCl (TEA-E-200S) buffer, pH 7.2: 200 mM NaCl in 0.1 × TEA-E buffer;

TEA-E-1000 mM NaCl (TEA-E-1000S) buffer, pH 7.2: 1000 mM NaCl in 0.1 × TEA-E buffer;

MgCl: 500 mM $MgCl_2$, adjust pH with TEA to 7.2;

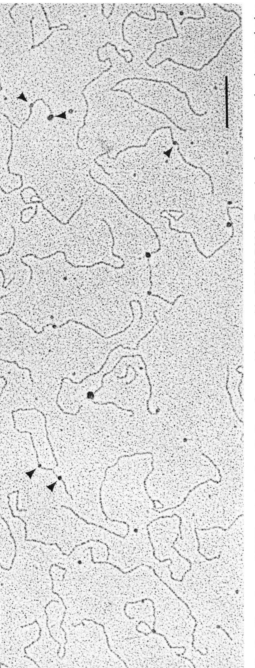

Fig. 6. Complexes of DNA–cytosine-5-methyltransferase with plasmid pBR322 DNA. Examples of enzyme molecules attached to the DNA are indicated by arrows. Bar, 1 kb.

Glutaraldehyde (GA): 1% GA (GA-1), dilute from stock solutions and adjust pH with alkali to 7.2;
Sephacryl S1000 or Sepharose 4B column: about 2 ml gel bed equilibrated in TEA-E-150S buffer.

Alternatively, *DE52 cellulose column*: equilibrate DE52 in TEA-E-200S buffer. Prepare 250–300 µl bed in a 1 ml siliconized pipette tip plugged with glass wool.

(ii) *Protocol*

Formation of complexes. The assay should contain 50 µg ml^{-1} of DNA and an excess of antibodies (AB) above complete saturation of possible antibody binding sites. If this concentration is unknown, make several assays with increasing AB concentrations. For a final volume of 50 µl mix in a test tube: $50 - (5 + x + y)$ µl H_2O, 5 µl TEA-E-1500S buffer, x µl DNA and y µl AB. Incubate at 37°C for 30 to 60 min or at 4°C for several hours (overnight). Rapidly mix into the assay 5 µl of GA-1, incubate at 37°C for 15 to 30 min and proceed with removal of unbound AB.

Removal of unbound AB. Gel filtration method: put the sample onto the gel column, elute with TEA-E-150S buffer and fractionate into 1-drop fractions. Combine the volumes of the first two-thirds of the peak.

DE52 method: put the sample onto the column, wash the column carefully with the five- to six-fold bed volume using TEA-E-200S buffer, elute the DNA with 1×50 µl and 4×100 µl of TEA-E-1000S buffer. The fractions two, three and four contain most of the DNA. Washing and elution is done with light pressure produced with a pipette. Immediately after chromatography, proceed with mounting.

Mounting for EM. A typical mounting mix contains 50–100 ng of DNA. The DNA concentration in the eluate after gel filtration is about 1/3 to 1/5, and after DE52 chromatography about 1/2 of the original concentration. Mg^{2+} may contribute positively to the stability and specificity of AB–DNA complexes. Therefore the molarity of Mg^{2+} in these mounting assays is elevated above usual concentrations up to 50 mM. In the case of DE52 eluates the final dilution of the eluate should be about five-fold to reduce NaCl concentrations. Mix in a test tube: $60 - (12 + x)$ µl H_2O, 6 µl TEA-E buffer, 6 µl $MgCl$ and x µl of eluate and mount onto mica as described in Section II.B.3(a).

Examples of visualized complexes are shown in Fig. 7. Figure 7a shows the mapping of O^6-ethyldesoxyguanosine (O^6-edG) in plasmid DNA. Linearized DNA of the plasmid pUC8 was ethylated *in vitro* by ethylnitrosourea and, in addition, ends were tailed by O^6-edG using terminal desoxynucleotidyl

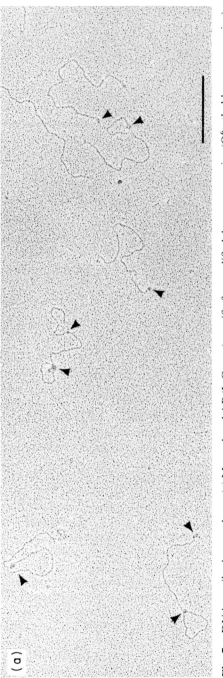

Fig. 7. DNA–antibody complexes. (a) Monoclonal AB (IgG type) specific for modified desoxyguanosine (O⁶-ethyldesoxyguanosine). Original preparation by Spiess and Nehls. (b) Monoclonal AB AK30-10 (IgM type) specific for ds and ss DNA bound to DNA from murine Ehrlich ascites tumour cells. The inset shows the pure antibody prepared for electron microscopy under identical conditions. Original preparation by Spiess and Werner. (c) Monoclonal AB specific for DNA in Z-conformation bound to plasmid pAF6 DNA. Original preparation by Spiess and Nehls. Bar, 1 kb. (Figures 7b and c on facing page.)

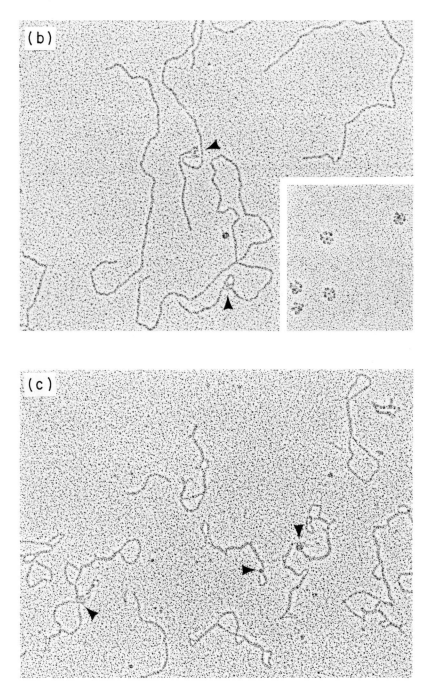

transferase. Reaction with a monoclonal antibody specific for O^6-edG (Nehls et al., 1984) was performed as described above. Unbound ABs were removed with the DE52 method. The tails bear clusters of ABs due to their richness in O^6-edG; single AB markers indicate the modified bases along the DNA molecules.

Figure 7b shows a complex of DNA with an AB of the IgM type. Because of their size, this type of AB allows easy and unambiguous detection of the binding sites, but usually they have lower binding strength than ABs of the IgG type. The AB used (AK30-10) detects ds and ssDNA (Scheer et al., 1987). The reaction was performed as described. Unbound AB were removed by gel filtration (Sephacryl S1000). The DNA is characteristically bent around the AB (Fig. 7b).

AB used to detect Z-DNA conformation is shown in Fig. 7c. Plasmid pAF6 DNA (Alonso et al., 1983) was reacted with ABs as described; unbound ABs were removed by DE52 chromatography. Covalently closed (ccc) DNA molecules are complexed with numerous ABs, while open circle (oc) and linear molecules cannot be recognized. The binding areas of the ABs can be mapped by restriction enzyme analysis (Di Capua et al., 1983).

4. Mapping of ribosomal proteins bound to 16S rRNA

(i) Reagents and buffers

  Protein binding buffer: 10 mM TEA-Cl, pH 7.8, 60 mM KCl, 10 mM MgAc; 1% formaldehyde (prepared freshly from paraformaldehyde).

(ii) Protocol

Formation of the complexes. The ribosomal proteins are bound to the 16S RNA in the protein binding buffer as described (Wiener and Brimacombe, 1987). Unreacted materials and complexes are separated by sucrose gradient centrifugation. After centrifugation, the fraction containing the RNA–protein complexes (0.3 OD) is fixed for 10 min at 37°C by addition of 5 µl 1% formaldehyde to 20 µl sample and stored on ice.

Mounting for EM. 1 µl of the fixed complexes is added to the incubation mixture for ss nucleic acids as described in Section II.B.3(b). Incubation time is 25 min at 50°C and adsorption to mica is carried out as described.

An example showing the complex of head proteins S7, 9, 19, 3, 10, 14 with 16S rRNA, all from the small ribosomal subunit of E. coli, is given in Fig. 8. The proteins are visible as compact dots near one end on the linear RNA molecules.

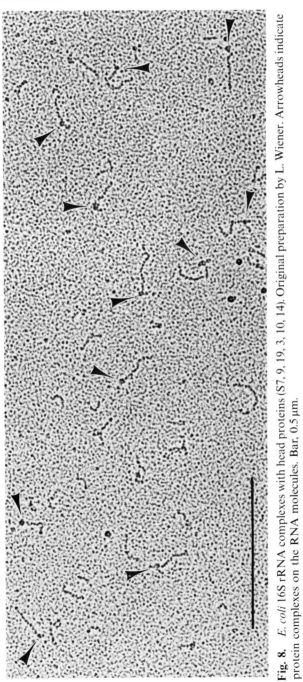

Fig. 8.   *E. coli* 16S rRNA complexes with head proteins (S7, 9, 19, 3, 10, 14). Original preparation by L. Wiener. Arrowheads indicate protein complexes on the RNA molecules. Bar, 0.5 μm.

References

Alonso, A., Kühn, B. and Fischer, J. (1983). *Gene* **26**, 303–306.

Brack, Ch. (1981). *CRC Crit. Rev. Biochem.* **10**, 113–169.

Burkardt, H. J. and Pühler, A. (1984). In *Methods in Microbiology* (P. M. Benett and J. Grinster, eds), Vol. 17, pp. 133–155. Academic Press, London.

Butour, J. L., Delain, E., Coulaud, D., Le Pecq, J. B., Barbet, J. and Roques, B. P. (1978). *Biopolymers* **17**, 873–886.

Coggins, L. W. (1987). In *Electron Microscopy in Molecular Biology* (J. Sommerville and U. Scheer, eds), pp. 1–29. IRL Press, Oxford.

Davidson, N. (1978). In *Ninth International Congress on Electron Microscopy*, Vol. 3: Organisation of Genetic Material, pp. 587–594. Toronto.

Davis, R. W. and Davidson, N. (1968). *Proc. Natl Acad. Sci. USA* **60**, 243–250.

Davis, R. W., Simon, R. M. and Davidson, N. (1971). In *Methods in Enzymology* (L. Grossman and K. Moldave, eds), Vol. 21, pp. 413–428. Academic Press, New York and London.

Di Capua, E., Stasiak, A., Koller, T., Brahms, S., Thomae, R. and Pohl, F. M. (1983). *EMBO J.* **2**, 1531–1535.

Dubochet, J., Ducommun, M., Zollinger, M. and Kellenberger, E. (1971). *J. Ultrastruct. Res.* **35**, 147–167.

Ferguson, J. and Davis, W. (1978). In *Advanced Techniques in Biological Electron Microscopy II* (J. K. Koehler, ed.), pp. 123–171. Springer-Verlag, Berlin.

Fisher, H. W. and Williams, R. C. (1979). *Ann. Rev. Biochem.* **48**, 649–679.

Garon, C. F. (1986). In *Ultrastructure Techniques for Microorganisms* (H. C. Aldrich and W. J. Todd, eds), pp. 161–181. Plenum Press, New York and London.

Grote, M., Dijk, J. and Reinhard, R. (1986). *Biochem. Biophys. Acta* **873**, 405–413.

Hamkalo, B. and Miller, O. L. (1973). *Ann. Rev. Biochem.* **42**, 379–396.

Haring, V., Scholz, P., Scherzinger, E., Frey, J., Derbyshire, K., Hatfull, G., Willetts, N. S. and Bagdasarian, M. (1985). *Proc. Natl Acad. Sci. USA* **82**, 6090–6094.

Haselkorn, R. and Doty, P. (1961). *J. Biol. Chem.* **236**, 2738–2745.

Inman, R. B. (1974). *J. Mol. Biol.* **18**, 464–476.

Inman, R. B. and Schnös, M. (1970). *J. Mol. Biol.* **49**, 93–98.

Johnson, D. (1975). *Nucl. Acids Res.* **2**, 2049–2054.

Kleinschmidt, A. K. (1968). In *Methods in Enzymology* (L. Grossman and K. Moldave, eds), Vol. XII, Part B, pp. 361–377. Academic Press, New York and London.

Kleinschmidt, A. K. and Zahn, R. K. (1959). *Z. Naturforsch., Teil B* **14**, 770–775.

Koller, B., Clarke, J. and Delius, H. (1985). *EMBO J.* **4**, 2445–2450.

Koller, Th., Sogo, J. M. and Bujard, H. (1974). *Biopolymers* **13**, 995–1009.

Krieger, M. and Tobler, J. (1977). *Anal. Biochem.* **81**, 450–453.

Lang, D. and Mitani, M. (1970). *Biopolymers* **9**, 373–379.

Lang, D., Kleinschmidt, A. K., Zahn, R. K. and Hellmann, W. (1964). *Biochim. Biophys. Acta* **88**, 142–154.

Lurz, R., Grote, M., Dijk, J., Reinhardt, R. and Dobrinski, B. (1986). *EMBO J.* **5**, 3715–3721.

Mayor, H. D. and Jordan, L. E. (1968). *Science* **161**, 1246–1247.

Nehls, P., Rajewsky, M. F., Spiess, E. and Werner, D. (1984). *EMBO J.* **3**, 327–332.

Paulson, J. R. and Laemmli, U. K. (1977). *Cell* **12**, 817–828.

Pfeifer, G. P., Spiess, E., Grünwald, S., Boehm, T. L. J. and Drahovsky, D. (1985). *EMBO J.* **4**, 2879–2884.

Portmann, R. and Koller, Th. (1976). In *Sixth European Congress on Electron Microscopy, Jerusalem* (Y. Ben-Shaul, ed.), Vol. 2, pp. 546–548. Tal, Israel.
Portmann, R., Sogo, J. M., Koller, Th. and Zillig, W. (1974). *FEBS Lett.* **45**, 64–67.
Rochaix, J. D. and Malnoe, P. (1978). *Cell* **15**, 661–670.
Scheer, U., Messner, K., Hazan, R., Raska, I., Hansmann, P., Falk, H., Spiess, E. and Franke, W. W. (1987). *Eur. J. Cell Biol.* **43**, 358–371.
Sogo, J., Stasiak, A., De Bernardin, W., Losa, R. and Koller, Th. (1987). In *Electron Microscopy in Molecular Biology* (J. Sommerville and U. Scheer, eds), pp. 61–79. IRL Press, Oxford.
Thomas, M., White, R. L. and Davis, R. W. (1976). *Proc. Natl Acad. Sci. USA* **73**, 2294–2298.
Vollenweider, H. J. (1981). *Meth. Biochem. Analysis* **28**, 201–265.
Vollenweider, H. J., Sogo, J. M. and Koller, T. (1975). *Proc. Natl Acad. Sci. USA* **72**, 83–87.
Vollenweider, H. J., Koller, T., Parello, J. and Sogo, J. M. (1976). *Proc. Natl Acad. Sci. USA* **73**, 4125–4129.
Westmoreland, B., Szybalski, W. and Ris, H. (1969). *Science* **163**, 1343–1348.
White, R. L. and Hoagness, D. S. (1977). *Cell* **10**, 177–192.
Wiener, L. and Brimacombe, R. (1987). *Nucl. Acids Res.* **9**, 3653–3670.
Williams, R. C. (1977). *Proc. Natl Acad. Sci. USA* **74**, 2311–2315.
Younghusband, H. B. and Inman, R. B. (1974). *Ann. Rev. Biochem.* **43**, 605–619.

14

Interaction of Restriction Endonucleases with DNA as Revealed by Electron Microscopy

WALTHER JOHANNSSEN

Reagents Division, E. Merck, D-6100 Darmstadt, Federal Republic of Germany

I. Introduction

The electron microscopical investigation of nucleic acids started in 1948 when Scott attempted to visualize isolated double strands of calf thymus DNA. However, the methods he used produced many artefacts such as lateral aggregation and unpredictable stretching of the DNA double strands. In 1959, Kleinschmidt and Zahn followed by other workers and later co-workers introduced the basic protein film technique. This simple, reproducible method allowed nucleic acids to be studied quantitatively under the electron microscope for the first time ever. These methods were subsequently developed in a variety of ways (reviewed in Ferguson and Davis, 1978). As methods were developed and improved, interest in the electron microscopical investigation of DNA grew in the fields of molecular genetics and molecular and cell biology. A multitude of techniques are now available for analysing the structural and functional relationships existing in the genetic material of the

METHODS IN MICROBIOLOGY
VOLUME 20 ISBN 0 12 521520 6

Copyright © 1988 by Academic Press Limited
All rights of reproduction in any form reserved

cell (e.g. molecular weight determination by length measurement, hetero-duplex analysis, D-loop and R-loop mapping, secondary structure in single strands, visualization of protein–DNA complexes, etc.).

Particular interest has been shown in the elucidation of protein–nucleic acid interactions over the past few years, and has accelerated the development of preparative methods for the visualization of DNA–protein complexes in the electron microscope. The sites at which proteins bind to DNA can be determined over a large area by means of electron microscopy; the proteins and protein components involved in complex formation can also be identified. The most frequently studied DNA–protein complexes consist of polymerases and various phage nucleic acids. They are usually visualized in the electron microscope by spreading with BAC (BAC = benzyldimethylalkylammonium chloride) (Vollenweider *et al.*, 1975), with ethidium bromide (Koller *et al.*, 1974) or by adsorption onto a film of activated carbon (Dubochet *et al.*, 1971; Bordier and Dubochet, 1974). Other studies have been performed on DNA–protein complexes which contained lac-repressors (Hirsch and Schlief, 1976; Zingsheim *et al.*, 1977), Qβ-replicase (Vollenweider *et al.*, 1976), SV40 proteins (Kasamatsu and Wu, 1976), membrane proteins (Alberding *et al.*, 1977) or histones. These and other DNA–protein complexes and the relevant spreading methods have been reviewed by Griffith and Christiansen (1978).

The reaction mechanisms of restriction endonucleases are still being intensively studied, and these enzymes are becoming increasingly important for molecular genetic research (Kessler and Höltke, 1986). They are divided into three main classes according to their recognition and cleavage mechanisms. Class I comprises the restriction endonucleases that combine to form an enzyme complex consisting of subunits with different sizes. Mg^{2+}, ATP and S-adenosylmethionine are required as cofactors. These restriction endo-nucleases recognize a specific nucleotide sequence, but they cleave the double-stranded DNA molecule non-specifically outside this recognition sequence. Class II consists of restriction nucleases that are composed of identical subunits; they recognize a specific nucleotide sequence and cleave the DNA within this region specifically using Mg^{2+} as a cofactor. Modification of the specific recognition sequence is catalysed by another enzyme. Class III is composed of restriction-modification enzyme complexes that consist of subunits of different sizes. Mg^{2+}, ATP and S-adenosylmethionine are required as cofactors. These restriction enzymes recognize a specific nucleotide sequence and specifically cleave the DNA molecule outside this sequence.

The most popular enzyme used in studies on DNA–restriction endo-nuclease complexes is the Class II restriction endonuclease EcoRI. This enzyme is a well-characterized protein with a molecular weight of 31.065 daltons and contains 276 amino acids (Greene *et al.*, 1981; Newman *et al.*, 1981). In solution it forms catalytically active dimers, as well as tetramers at higher protein concentrations (Modrich and Zabel, 1976; Jen-Jacobson *et al.*, 1983). Under physiological conditions, EcoRI cleaves the following canonical sequence consisting of six base pairs:

$$5'\ldots \text{GAATTC} \ldots 3' \qquad 3'\ldots \text{CTTAAG} \ldots 5'$$

to form "sticky ends". Under non-physiological conditions (high pH value, low ionic strength, addition of glycerol or ethylene glycol, replacement of Mg^{2+} by Mn^{2+}), the recognition sequence is changed at one or more base pairs (termed EcoRI* sites) (Polisky *et al.*, 1975; Woodhead *et al.*, 1981). As well as binding specifically to the recognition sequence, the enzyme also binds non-specifically to the DNA even in the presence of Mg^{2+} (Modrich, 1979, 1982; Halford and Johnson, 1980; Woodhead and Malcolm, 1980; Johannssen *et al.*, 1984). It has been postulated that the non-specific binding of EcoRI permits the formation of a specific DNA–EcoRI complex by facilitating diffusion along the DNA (Terry *et al.*, 1983; Ehbrecht *et al.*, 1985).

X-ray structural studies on DNA–EcoRI complexes have been performed at a resolution of 0.3 nm (McClarin *et al.*, 1986) and have revealed the molecular structure of the complex. Specific protein regions, so-called arms, of the EcoRI enzyme are responsible for making DNA adopt the correct conformation. Deletion of these arms does not change the enzyme's specificity, but does reduce its stability (Jen-Jacobson *et al.*, 1986). Pingoud and co-workers have moreover used protein engineering to characterize the roles played by a number of amino acids in the hydrolysis reaction catalysed by EcoRI endonuclease (Scholtissek *et al.*, 1986; Wolfes *et al.*, 1986).

These studies have so far only been carried out on oligonucleotide–EcoRI complexes. If, however, analysis of the molecular structure of the complex is extended, modern electron microscopy, with its sophisticated techniques, will be able to make a valuable contribution to the elucidation of the structural and functional relationships in DNA–endonuclease interactions over wide stretches of DNA.

Studies on DNA–protein complexes have also been conducted with the restriction endonucleases EcoK (Brack *et al.*, 1976), EcoB (Rosamond *et al.*, 1979), EcoP15 (Yuan and Hamilton, 1980), HinHI (Miwa *et al.*, 1979), SalGI (Johannssen, 1983) and EcoRI (Johannssen *et al.*, 1984).

II. Methodology

In order to obtain high-quality electron micrographs of DNA–protein complexes, the individual components of these complexes must be available in a very pure form. All buffers and solutions must therefore be autoclaved prior to use and subsequently filtered through a membrane filter with a pore size of

0.01 μm. General detailed information about the purity requirements of solutions is to be found in Chapter 13 of this volume.

A. DNA isolation

Plasmid DNA provides a highly suitable source of DNA because it has been well characterized and is available in relatively large quantities. Small circular or linear DNA molecules with known absolute molecular weights and defined nucleotide sequences are preferred, e.g.

pBR322	4 363 base pairs
SV40	5 226 base pairs
φX174	5 386 base pairs
fd	6 408 base pairs
λ	48 502 base pairs

The method described by Bøvre and Szybalsky (1971) has been successfully used to isolate DNA. Goebel (1970) has reported a method which has proved to be especially suitable for isolating plasmid DNA (e.g. pBR322). Other methods of purification are to be found in the laboratory manual written by Maniatis et al. (1982). All of the buffers used in these methods are autoclaved before use to ensure that they do not contain any traces of nucleases. The DNA obtained is, however, not sufficiently pure. It also has to be subjected to a special type of phenol extraction to remove any protein residues present.

Procedure

Isolated DNA obtained by one of the methods described above is precipitated with ethanol (Meyers et al., 1976) and resuspended in a buffer solution (10 mM Tris-HCl, pH 7.8, 5 mM EDTA, 0.5% (w/v) SDS) to give a final DNA concentration of 500 μg ml^{-1}. Proteinase K is then added (50 μg lyophilisate ml^{-1} of DNA solution) and the resulting solution is incubated for up to 12 h at 37°C. This step almost completely hydrolyses any proteins present. Protein residues and proteinase K are then removed by adding NaCl to give a final concentration of 1.2 M, followed by a two-fold volume of a mixture consisting of freshly distilled phenol–isoamylalcohol–chloroform (50:1:49) buffered with 0.1 M Tris-HCl, pH 7.5, containing 1.2 M NaCl. The combined solutions are then carefully mixed for 5 min and centrifuged to separate the organic and aqueous phases. The upper organic phase is removed with a sterile glass capillary. After repeating this extraction procedure four times, the aqueous phase is overlaid with a layer of diethyl ether to remove any residual traces of the organic solvent mixture. The ether is then removed, any remaining traces being driven off under sterile conditions with nitrogen. The purified DNA solution is finally dialysed against several changes of 50 mM Tris-HCl buffer, pH 7.2, for 12 h at 4°C; it is stable for several months when stored at -20 or -70°C.

B. Linearization of circular DNA with restriction endonucleases

When interactions between DNA and restriction endonucleases are investigated under the electron microscope, defined linear DNA species or DNA fragments often have to be used which are obtained by using Class II restriction endonucleases.

Procedure

Plasmid DNA (usually supercoiled) is linearized and defined linear DNA fragments are prepared by treating 50 µg DNA ml^{-1} of buffer with 10–100 units of an appropriate restriction endonuclease for 1–12 h at 37°C. The degree of DNA cleavage can be determined by means of electrophoreses on agarose gel as described, for example by Johannssen *et al.* (1979). In some cases the DNA should only be partially digested. The restriction endonuclease(s) is (are) subsequently removed by deproteinization (see above).

C. Isolation of restriction endonucleases

The restriction endonucleases must, like the DNA, be extremely pure (homogeneous). The isolation of such enzymes has recently been reviewed by Kessler and Höltke (1986). The methods described in their article generally yield protein preparations that are devoid of non-specific nucleases, but very few of them contain the desired restriction endonuclease in a homogeneous form. Homogeneity is, however, essential for the successful electron microscopical investigation of DNA–restriction endonuclease complexes. The preparations must therefore be additionally purified by affinity chromatography on DNA-cellulose (Litman, 1968). This method often leads to considerable losses, but usually affords a homogeneous enzyme preparation with protein particles that are capable of binding (Modrich and Zabel, 1976). In some cases restriction endonucleases must be purified in the presence of detergents, which subsequently have to be removed because they would otherwise interfere with the preparative methods used in electron microscopy (e.g. spreading of the DNA; see Chapter 13, this volume). The detergents can be removed by chromatography on Bio-Beads™ SM2 or an equivalent material (Holloway, 1973; Furth, 1980).

D. Preparation of supports for mounting DNA–restriction endonuclease complexes

Many methods have been described for preparing support films suitable for the electron microscopical study of DNA and DNA–protein complexes. The method used often depends on the technique employed for preparing the DNA–protein complex. General methods are described in Chapter 13 as well as in several review articles (Younghusband and Inman, 1974; Ferguson and Davis, 1978; Fisher and Williams, 1979; Brack, 1981). Special mention should, however, be made of a method of support preparation that provides reproducible results in the investigation of DNA–restriction endonuclease complexes (Johannssen, 1983; Johannssen et al., 1984). It employs modified collodion support films that are sufficiently stable, have a homogeneously hydrophilic surface and satisfy the general requirements demanded of electron microscopical specimen supports.

Procedure

Collodion support films are prepared according to the method of Ferguson and Davis (1978) on copper grids (300 mesh). The grids should be as planar as possible and cleaned in acetone prior to use. The preparation of these support films is described in detail in Chapter 13. The surfaces of these films often display areas of varying hydrophobicity which interfere with the DNA preparations deposited on them (Sogo et al., 1975; Coetzee and Pretorius, 1979; Arcidiacono et al., 1980). The surface must be treated using a modification of the method described by Koller et al. (1974) and Arcidiacono et al. (1980) to increase and standardize its hydrophilicity. An ethidium bromide solution ($5\,mg/H_2O$, double-quartz-distilled) is centrifuged for 12 h at $160\,000g$ in a fixed angle A100 aluminium rotor in an Airfuge™ (Beckman Instruments) to remove any interfering particles. The supernatant is carefully removed and adjusted with double-quartz-distilled water to give a final ethidium bromide concentration of $50\,\mu g\,ml^{-1}$. Purification of this solution is extremely important and can also be achieved by microfiltration through a membrane filter with a pore size of $0.01\,\mu m$. Omission of this purification step can readily produce artefacts when studying DNA–restriction endonuclease complexes. Contaminating particles are scattered over the surface of the film but preferentially adhere to the DNA and resemble DNA–restriction endonuclease complexes. Forty microlitre droplets of the purified ethidium bromide solution are applied onto the clean side of a piece of Parafilm[R] or onto clean glass slides. The carbon-strengthened supports are placed on the droplets with their face down and left to stand for 10 min at room temperature in the dark. They are then washed three times for 5 s in double-quartz-distilled water to remove the excess ethidium bromide solution. The grids are then dried on filter paper. They should be used within a few hours; longer storage periods are not recommended.

Comments

DNA–restriction endonuclease complexes can also be studied under the electron microscope on other types of support films. Collodion or Formvar[R] films can be pretreated with poly-L-lysine to increase their hydrophilicity and thus the adsorption of nucleic acids and enzymes (Williams, 1977; Miwa *et al.*, 1979; Rosamond *et al.*, 1979). Collodion or Formvar[R] films can also be treated with an aqueous 0.01 to 0.05% (w/w) solution of anthrabis (1,4-bis-[3-(benzyldimethylamino)-propylamino]-9,10-anthraquinone dichloride) (Thomas *et al.*, 1978) or 0·001% (w/w) solution of Alcian blue (Labhart and Koller, 1981). All of these solutions must be extremely pure. Dubochet and co-workers (1971) developed a different method for increasing the hydrophilicity of support films in which the surface of a carbon film is positively charged by glow discharge in pentylamine vapour *in vacuo*. This technique is more complex than the methods described above and has a number of limitations for the following preparation steps (Brack, 1981). Nevertheless, DNA–restriction endonuclease complexes have been visualized on specimen supports prepared by this technique (Brack *et al.*, 1976; Bickle *et al.*, 1978; Yuan *et al.*, 1980).

E. DNA–restriction endonuclease complexes

In the preparation of DNA–protein complexes for electron microscopy, a distinction must be made between complexes in which the protein is covalently bound to the DNA and those in which protein binding is non-covalent. Covalently bound DNA–protein complexes generally do not require further fixation, but some non-covalently bound complexes, including DNA–restriction endonuclease complexes, have to be stabilized by chemical fixation to prevent them from dissociating (Koller *et al.*, 1974, 1978; Portmann *et al.*, 1974). The necessity for fixation and the suitability of the fixative (glutaraldehyde, formaldehyde) must be carefully established for the problem in question. DNA–EcoK complexes (Brack *et al.*, 1976) and DNA–EcoP15 complexes (Yuan *et al.*, 1980) have been successfully fixed with 0.1% (w/v) glutaraldehyde, but this method proved to be unsuitable for DNA–EcoRI complexes (Johannssen *et al.*, 1984) and DNA–SalGI complexes (Johannssen, 1983). Fixatives produced artefacts that were presumably related to the oligomeric structure of these restriction endonucleases (multiple DNA binding sites on the oligomeric enzyme particle). Furthermore, glutaraldehyde also reacts with amino groups. If glutaraldehyde is used for fixation, buffers that contain amino groups (e.g. Tris-HCl or ammonium acetate) are therefore unsuitable. It should also be remembered that glutaraldehyde fixation can be reversed by the addition of compounds containing amino groups or by high ionic strength.

According to the author's experience, non-fixed DNA–restriction endonuclease complexes (DNA–EcoRI, DNA–SalGI) that are prepared under physiological conditions prove to be sufficiently stable for electron microscopy if they are subjected to DNA spreading with BAC. Non-bound restriction endonuclease particles can be removed from the preparation by using the gel permeation method described in Chapter 13. However, the complex should then first be fixed to prevent possible uncontrolled, concentration-dependent dissociation.

Procedure

DNA–EcoRI complexes can be prepared as follows: DNA (e.g. plasmid DNA pBR322) is linearized with an appropriate restriction endonuclease and then deproteinized as described above. Varying quantities of EcoRI (0.05 to 400 units) are added to 0.01 µg of the resulting DNA. The reaction mixtures should be kept at a constant volume of 10 µl and contain 50 mM Tris-HCl buffer, pH 7.5, either with or without 10 mM $MgCl_2$. Mixtures are incubated for 5 or 60 min at 37°C, the reaction being stopped by freezing in liquid nitrogen. The mixtures are stored at $-20°C$ until they can be mounted for electron microscopy (see below).

F. Visualization of DNA–restriction endonuclease complexes

A number of preparative methods are available for visualizing DNA–restriction endonuclease complexes, but this article will only describe a modification of the original BAC technique (Vollenweider *et al.*, 1975). In this method the DNA and the DNA–protein complexes are prepared by the diffusion and spontaneous surface adsorption of nucleic acids from droplet solutions.

Procedure

Aliquots (120 µl) of the BAC solution are added to 5 µl aliquots of the freshly thawed mixture (see above) containing 0.005 µg DNA and an appropriate amount of EcoRI or another restriction endonuclease (0.025 to 200 units; complexing reaction terminated). The BAC solution is prepared as follows.

Fifteen microlitres of 2 M ammonium acetate (pH 7.0 containing 0.01 M EDTA) and 3 µl of BAC stock solution (0.2% (w/v) BAC (40% benzyl-

dimethyldodecylammonium   chloride,   60%   benzyldimethyltetradecyl-
ammonium chloride) in formamide purified by crystallization—see Chapter
13) are added to a clean test tube containing 2.982 ml double-distilled water.
Forty-microlitre aliquots of the 125 µl containing the above-mentioned
amounts of DNA and EcoRI are removed and deposited as droplets onto a
sheet of Parafilm®. After 15 min, collodion grids prepared as described are
briefly touched onto the surface of the droplets (one grid per droplet). The
grids are washed by touching them for 10 s onto 90% (v/v) ethanol, and finally
blotted dry on filter paper. For contrasting with high resolution, the DNA and
the DNA–restriction endonuclease complexes are subjected to electron beam
evaporation with platinum–iridium (80:20%) or tungsten–tantalum alloys or
a combination of alloys and carbon. This is performed in a vacuum-generating
system that is equipped with an electron beam evaporation device. The
preparations are subjected to shadowing at a pressure of 0.1 mPa on a rotating
holder at an angle of 6° with respect to the evaporation source. The chamber of
the vacuum-generating system should be thoroughly cleaned and free of oil
residues from the vacuum pump because they may evaporate also and
contaminate the preparation. The vacuum pressure should not exceed 0.5–
1 mPa during electron beam evaporation; it is therefore an advantage to have
the vacuum-generating system equipped with a liquid nitrogen cooling trap.
The instruction manuals supplied by the manufacturers should be consulted
for further details.

When combined with the BAC spreading method, the above contrasting
method results in a DNA diameter of around 5 nm. This is a prerequisite for
the visualization of DNA–restriction endonuclease complexes under the
electron microscope. Contrasting with platinum–iridium or platinum–carbon
is preferable to that with tungsten–tantalum because contrast and stability are
higher and the alloys evaporate more readily. The resolution obtained with
tungsten–tantalum is, however, higher.

Comments

Figures 1 and 2 show typical examples of DNA–restriction endonuclease
complexes that have been prepared by the methods described above. The
DNA–EcoRI complexes can clearly be seen as pearl necklace-like aggregates.
The diameters of most of the bound EcoRI molecules are 14–16 nm, which is
the expected diameter for bound, metal-shadowed, tetrameric EcoRI
particles. EcoRI subunits (Fig. 1a; small arrows) and smaller EcoRI particles
assumed to be dimers (Fig. 1a,b; arrowheads) are visible.

Spreading of the DNA and DNA–restriction endonuclease complexes with
BAC or by other protein-free spreading methods does not produce the high
degree of reproducibility and simplicity obtained with the cytochrome c

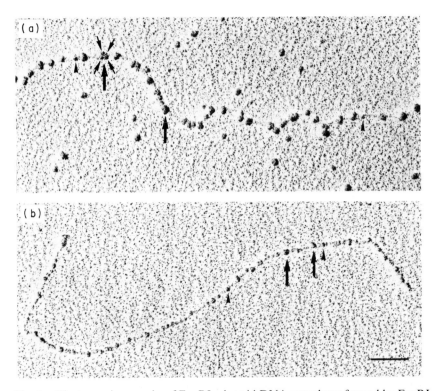

Fig. 1. Electron micrographs of EcoRI–plasmid DNA complexes formed by EcoRI and SalGI-digested pGW10 DNA, prepared from solutions containing Mg^{2+}. The contrast enhancement was done by electron beam evaporation of platinum (angle 6°, rotary shadowing). The scale bar represents 100 nm. (a) Sample prepared from solution containing 0.01 μg DNA and 200 units EcoRI. The EcoRI particles are predominantly bound in the tetrameric state (large arrows point to two examples). The small arrows indicate projections of EcoRI subunits. The arrowheads point to occasionally occurring small bound EcoRI particles. (b) Sample prepared from solution containing 0.01 μg DNA and 10 units EcoRI. Most of the bound EcoRI particles are of small diameter (arrowheads point to two examples). Some larger bound EcoRI particles are visible (arrows). [Reprinted with permission from Johannssen *et al.* (1984). *Arch. Microbiol.* **140**, 265–270.]

spreading technique (Fisher and Williams, 1979). Artefacts are to be expected but can largely be avoided by using the methods described in this article. Figure 3 shows typical artefacts obtained if the preparative solutions are not carefully purified or the support films are not pretreated:

(1)   Most of the DNA is incompletely spread. It is either laterally aggregated and partially coiled or displays sharp bends with straight,

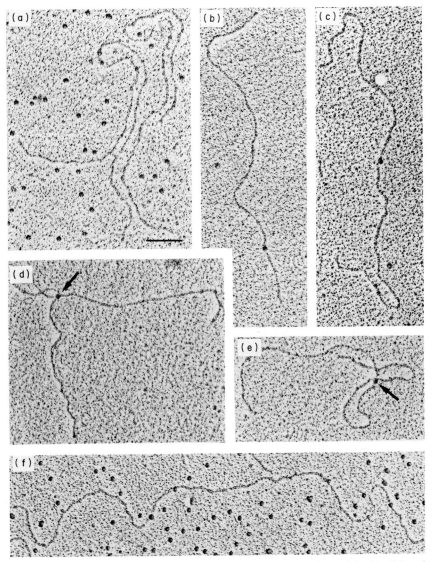

Fig. 2. Electron micrographs of EcoRI–pGW10 complexes formed by EcoRI and SalGI-digested pGW10 DNA (a–e) and of an EcoRI–pBR322 DNA (SalGI-digested) complex (f) prepared and mounted in the absence of Mg^{2+}. Contrast enhancement and scale bar as in Fig. 1. (a) Sample prepared from a solution containing 0.01 μg DNA and 200 units EcoRI. (b) and (c) Samples prepared from a solution containing 0.01 μg DNA and 10 units EcoRI. (d) and (e) Samples prepared from a solution containing 0.01 μg DNA and 45 units EcoRI. The DNA loops indicate the existence of two binding sites on each of the EcoRI particles positioned at the DNA cross-over points. (f) Sample prepared from a solution containing 0.01 μg pBR322 DNA and 200 units EcoRI. As in (a), only few EcoRI particles are bound to the DNA. [Reprinted with permission from Johannssen et al. (1984). Arch. Microbiol. **140**, 265–270.]

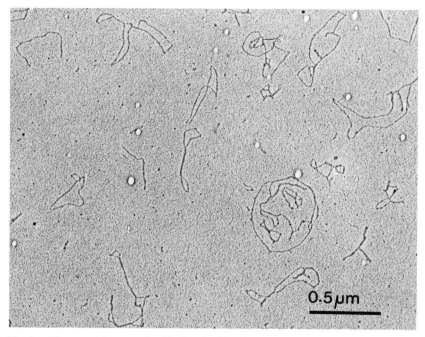

Fig. 3. Electron micrograph of isolated pGW10 DNA after spreading with BAC as described; in this case, however, without using purified solutions and without pretreatment of the support film with ethidium bromide. The contrast enhancement was done by electron beam evaporation of platinum–carbon. For further comments see text.

elongated stretches of DNA. These artefacts prevent reliable measurement of the length of the DNA contour.

(2)   The DNA diameters vary widely from 6 to 13 nm.

(3)   Contaminating particles can be seen on the background; some of them are located on the DNA and resemble DNA–restriction endonuclease complexes.

G.   Evaluation of electron micrographs of DNA–restriction endonuclease complexes

The basic technique for taking electron micrographs of DNA–restriction endonuclease complexes, and methods for their evaluation, are described in Chapter 13. For length measurements of DNA and evaluation of the positions of the binding sites of restriction endonuclease particles, the negatives are projected 5–10 times enlarged on an opalescent glass plate. The length

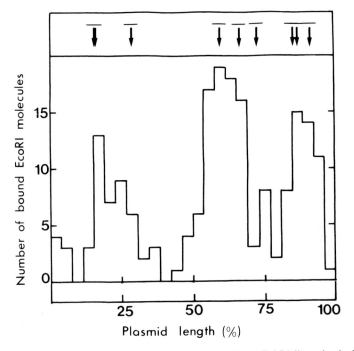

Fig. 4. Distribution of the binding positions of EcoRI on SalGI-linearized pBR322 DNA. Data taken from micrographs of EcoRI–DNA complexes prepared in the absence of Mg^{2+}. The arrowheads at the top of the figure indicate the positions of the EcoRI site and the sites 5′…AATT…3′ along the pBR322 DNA. Each of the scale bars above the arrowheads has a length of 5% of the pBR322 contour length, which corresponds to the standard deviation obtained in the experiments when positions of bound EcoRI particles along the plasmid DNA were evaluated. [Modified reprint, with permission from Johannssen *et al.* (1984). *Arch. Microbiol.* **140**, 265–270.]

measurements are then carried out on this plate. The data thus obtained can be processed using a computer system with a program according to Burkardt *et al.* (1978), as demonstrated by Lurz and co-workers (1981).

As an example, Fig. 4 shows the evaluation of the positions of the binding sites of EcoRI along the plasmid DNA pBR322. At the top of Fig. 4 are indicated the positions of the EcoRI recognition sequence (5′…GAATTC…3′) and of the EcoRI* recognition sequences (5′…AATT…3′) along the pBR322 DNA, as derived from published sequence data. It is shown that the sites (5′…AATT…3′) may be favoured as binding sites for EcoRI over other sites on pBR322 DNA, independent of whether they are part of the EcoRI recognition sequence or not (Johannssen *et al.*, 1984).

References

Alberding, M., Griffith, J. and Attardi, G. (1977). *Proc. Natl Acad. Sci. USA* **74**, 1348–1352.

Arcidiacono, A., Stasiak, A. and Koller, Th. (1980). *Proc. 7th Eur. Congress on EM*, The Hague, Vol. 2, pp. 516–523.

Bickle, T. A., Brack, C. and Yuan, R. (1978). *Proc. Natl Acad. Sci. USA* **75**, 3099–3103.

Bordier, C. and Dubochet, J. (1974). *Eur. J. Biochem.* **44**, 617–624.

Bøvre, K. and Szybalsky, W. (1971). In *Methods in Enzymology* (L. Grossman and K. Moldave, eds), Vol. 21, Part D, pp. 383–413. Academic Press, New York and London.

Brack, C. (1981). *CRC Crit. Rev. Biochem.* **10**, 113–169.

Brack, C., Eberle, H., Bickle, T. A. and Yuan, R. (1976). *J. Mol. Biol.* **108**, 583–593.

Burkardt, H. J., Mattes, R., Pühler, A. and Heumann, W. (1978). *J. Gen. Microbiol.* **105**, 51–62.

Coetzee, W. F. and Pretorius, G. H. J. (1979). *J. Ultrastruct. Res.* **67**, 33–39.

Dubochet, J., Docummun, M., Zollinger, M. and Kellenberger, E. (1971). *J. Ultrastruct. Res.* **35**, 147–167.

Ehbrecht, H.-J., Pingoud, A., Urbanke, C., Maass, G. and Gualerzi, C. (1985). *J. Biol. Chem.* **260**, 6160–6166.

Ferguson, J. and Davis, R. W. (1978). In *Advanced Techniques in Biological Electron Microscopy* (J. K. Koehler, ed.), Part II, pp. 123–171. Springer-Verlag, Berlin, Heidelberg and New York.

Fisher, H. W. and Williams, R. C. (1979). *Ann. Rev. Biochem.* **48**, 649–679.

Furth, A. J. (1980). *Anal. Biochem.* **109**, 207–215.

Goebel, W. (1970). *Eur. J. Biochem.* **15**, 311–320.

Greene, P. J., Gupta, M., Boyer, H. W., Brown, W. E. and Rosenberg, J. M. (1981). *J. Biol. Chem.* **256**, 2143–2153.

Griffith, J. D. and Christiansen, G. (1978). *Ann. Rev. Biophys. Bioeng.* **7**, 19–35.

Halford, S. E. and Johnson, N. P. (1980). *Biochem. J.* **191**, 593–604.

Hirsch, J. and Schleif, R. (1976). *J. Mol. Biol.* **108**, 471–490.

Holloway, P. W. (1973). *Anal. Biochem.* **53**, 303–308.

Jen-Jacobson, L., Kurpiewski, M., Lesser, D., Grable, J., Boyer, H. W., Rosenberg, J. M. and Greene, P. J. (1983). *J. Biol. Chem.* **258**, 14638–14646.

Jen-Jacobson, L., Lesser, D. and Kurpiewski, M. (1986). *Cell* **45**, 619–629.

Johannssen, W. (1983). *Z. Allgm. Mikrobiol.* **23**, 197–201.

Johannssen, W., Schütte, H., Mayer, F. and Mayer, H. (1979). *J. Mol. Biol.* **134**, 707–726.

Johannssen, W., Schütte, H., Mayer, H. and Mayer, F. (1984). *Arch. Microbiol.* **140**, 265–270.

Kasamatsu, H. and Wu, M. (1976). *Biochem. Biophys. Res. Commun.* **68**, 927–936.

Kessler, C. and Höltke, H.-J. (1986). *Gene* **47**, 1–153.

Kleinschmidt, A.-K. and Zahn, R. K. (1959). *Z. Naturforsch., Teil B* **14**, 770–779.

Koller, Th., Sogo, J. M. and Bujard, H. (1974). *Biopolymers* **13**, 995–1009.

Koller, Th., Kübler, O., Portmann, R. and Sogo, J. M. (1978). *J. Mol. Biol.* **120**, 121–131.

Labhart, P. and Koller, Th. (1981). *Eur. J. Cell Biol.* **24**, 309–316.

Litman, R. M. (1968). *J. Biol. Chem.* **243**, 6222–6233.

Lurz, R., Danbara, H., Rückert, B. and Timmis, K. N. (1981). *Mol. Gen. Genet.* **183**, 490–496.

Maniatis, T., Fritsch, E. and Sambrook, J. (1982). *Molecular Cloning*. Cold Spring Harbor Laboratory, Cold Spring Harbor.

McClarin, J. A., Frederick, C. A., Wang, B.-C., Greene, P. J., Boyer, H. W., Grable, J. and Rosenberg, M. (1986). *Science* **234**, 1526–1541.

Meyers, J. A., Sanches, D., Elwell, L. P. and Falkow, S. (1976). *J. Bacteriol.* **127**, 1529–1537.

Miwa, T., Takanami, M. and Yamagishi, H. (1979). *Gene* **6**, 319–330.

Modrich, P. (1979). *Q. Rev. Biophys.* **12**, 315–369.

Modrich, P. (1982). *CRC Crit. Rev. Biochem.* **13**, 287–323.

Modrich, P. and Zabel, D. (1976). *J. Biol. Chem.* **251**, 5866–5874.

Newman, A. K., Rubin, R. A., Kim, S. H. and Modrich, P. (1981). *J. Biol. Chem.* **256**, 2131–2139.

Polisky, B., Greene, P. J., Grafin, D. E., McCarthy, B. J., Goodman, H. M. and Boyer, H. W. (1975). *Proc. Natl Acad. Sci. USA* **72**, 3310–3314.

Portmann, R., Sogo, J. M., Koller, Th. and Zillig, W. (1974). *FEBS Lett.* **45**, 64–67.

Rosamond, J., Endlich, B. and Linn, S. (1979). *J. Mol. Biol.* **129**, 619–635.

Scholtissek, S., Pingoud, A., Maass, G. and Zabeau, M. (1986). *J. Biol. Chem.* **261**, 2228–2234.

Scott, J. F. (1984). *Biochim. Biophys. Acta* **2**, 1–5.

Sogo, J. M., Portmann, R., Kaufmann, P. and Koller, Th. (1975). *J. Microsc.* **104**, 187–198.

Terry, B. J., Jack, W. E., Rubin, R. A. and Modrich, P. (1983). *J. Biol. Chem.* **258**, 9820–9825.

Thomas, J. D., Sternberg, N. and Weisberg, R. (1978). *J. Mol. Biol.* **123**, 149–176.

Vollenweider, H. J., Sogo, J. M. and Koller, Th. (1975). *Proc. Natl Acad. Sci. USA* **72**, 83–87.

Vollenweider, H. J., Koller, Th., Weber, H. and Weissmann, C. (1976). *J. Mol. Biol.* **101**, 367–377.

Williams, R. C. (1977). *Proc. Natl Acad. Sci. USA* **74**, 2311–2315.

Wolfes, H., Alves, J., Fliess, A., Geiger, R. and Pingoud, A. (1986). *Nucl. Acids Res.* **14**, 9063–9080.

Woodhead, J. L. and Malcolm, A. D. B. (1980). *Nucl. Acids Res.* **8**, 389–402.

Woodhead, J. L., Bhave, N. and Malcolm, A. D. B. (1981). *Eur. J. Biochem.* **115**, 293–296.

Yuan, R. and Hamilton, D. L. (1980). *J. Mol. Biol.* **144**, 501–519.

Yuan, R., Hamilton, D. L. and Burckhardt, J. (1980). *Cell* **20**, 237–244.

Younghusband, H. B. and Inman, R. B. (1974). *Ann. Rev. Biochem.* **43**, 605–619.

Zingsheim, H. P., Geisler, N., Mayer, F. and Weber, K. (1977). *J. Mol. Biol.* **115**, 565–570.

15

Preparation of Two-dimensional Arrays of Soluble Proteins as Demonstrated for Bacterial D-Ribulose-1,5-bisphosphate Carboxylase/Oxygenase

ANDREAS HOLZENBURG*

*Maurice E. Müller-Institut für hochauflösende Elektronenmikroskopie am
Biozentrum der Universität Basel, Basel, Switzerland*

I. Introduction

Two-dimensional (2-D) crystals, i.e. crystalline monolayers, are generally very welcome in electron microscopical studies of biomacromolecules since they contain information that can easily be averaged. All molecules within the 2-D lattice are aligned along the vertical axis so that those problems related to the structural analysis of single molecules, the major one of which is caused by different projections (e.g. Ottensmeyer *et al.*, 1977; Frank *et al.*, 1978; Van Heel and Frank, 1981; Frank, 1984; Van Heel, 1984, 1986; Giersig *et al.*, 1986; Kunath *et al.*, 1986), can be reliably overcome (see e.g. Unwin and Henderson,

*On leave from Institut für Mikrobiologie der Universität Göttingen, Göttingen, Federal Republic of Germany.

METHODS IN MICROBIOLOGY
VOLUME 20 ISBN 0-12-521520-6

1975; Aebi, 1978; Saxton and Baumeister, 1982; Aebi *et al.*, 1984; Frank, 1984; on the subject treated herein see Holzenburg *et al.*, 1987). The multiprojection problem is especially true for molecules that consist of more than four subunits and possess a more or less globular shape, like the soluble enzyme D-ribulose-1,5-bisphosphate carboxylase/oxygenase (abbreviated to RuBisCO), the CO_2-fixing key enzyme of the reductive pentose phosphate cycle (Calvin cycle) and O_2-fixing key enzyme in photorespiration (reviewed by Miziorko and Lorimer, 1983). Like the plant RuBisCOs, the enzyme isolated from the chemolithoautotrophic, H_2-oxidizing bacterium *Alcaligenes eutrophus H16* is a particle consisting of eight large and eight small subunits (L_8S_8) with a total molecular mass of 534 kD (Bowien *et al.*, 1976; Bowien and Mayer, 1978).

Using the *A. eutrophus* RuBisCO as a well standardized example, this chapter describes the main requirements and strategies for crystallizing a soluble protein in two dimensions.

II.   Methodology

A.   Theoretical and practical principles

Crystallizing a protein in three dimensions (3-D) requires us to influence the very sensitive thermodynamic equilibrium between the two competitive processes, precipitation and crystallization, in such a way that the latter one is favoured (for an excellent review on 3-D crystallization see McPherson, 1976). In 2-D crystallization experiments the same kind of approach is involved, but in addition the fundamental difference between 2-D and 3-D crystallization events has to be taken into account: 2-D crystals are formed in the presence of a suitable heterogeneous nucleation surface (HNS) only, whereas the growth of 3-D crystals is mediated by a homogeneous nucleation process (cf. Kleber, 1983; Holzenburg, 1984; Feher and Kam, 1985).

A HNS is a surface that consists of anything else but protein molecules. In contrast to homogeneous nucleation, where the nucleus is formed exclusively by protein–protein interactions within the solution, 2-D nucleation involves additional interactions between the protein and a heterogeneous surface. Hence the 2-D crystallization process includes one more parameter to be kept under control—the properties of the HNS. Moreover, the samples have to be prepared for examination in the transmission electron microscope (TEM), so that another influencing factor must be taken into account.

Two-dimensional crystallization and subsequent electron microscopical investigation thus include a total of four main parameters, two of which (1 and

4) apply to 2-D crystallization only, while (2) and (3) are important for both 2-D and 3-D experiments:

(1) properties of the HNS (see Section II.C);

(2) crystallization conditions (see Sections II.D and II.E);

(3) crystallization method (see Section II.E); and

(4) post-treatment(s) of the crystallization set-ups (see Section II.F).

Each of these main parameters, as well as related (sub)parameters, have to be optimized in order to achieve the desired results. This is done by answering the following questions:

(1) Which parameters are absolutely essential for 2-D crystallization and therefore require exact reproducibility?

(2) Which parameters may be *a priori* subjected to statistical fluctuations without disturbing the crystallization process?

(3) Which parameters can be reproduced quite easily?

(4) How do the parameters have to be constituted to support the crystallization process?

In order to fulfil this tedious task (cf. Carter and Carter, 1979), a step-by-step procedure has been used, especially designed to answer the above questions in an effective and simple way. On the basis of a visual examination of the crystallization specimens (i.e. post-treated crystallization set-ups) in the TEM, and the experimental verification that led to those specimens, each parameter, or parameter variation, has been assessed according to a binary (x/y) evaluation where:

$x = +$ stands for good standardizable experimental realization;

$x = -$ for poor standardizable experimental realization;

$x = \pm$ for an average degree of standardization;

$y = +$ for a reproducible positive influence on the crystallization process;

$y = -$ for a reproducible negative influence on the crystallization process;

$y = \pm$ for a non-reproducible positive or negative influence on the crystallization process; and

$y = + +$ for reproducible 2-D crystals of high quality.

While the estimation of the x-values will always remain subjective, the y-values can be assessed in a much more objective way. Parameter specifications that exert a positive influence on the crystallization $(y = +)$ can be recognized in the TEM immediately because the corresponding specimens exhibit ordered arrays (Figs 1b–d) that cannot be found in $y = -$ specimens (Fig. 1a).

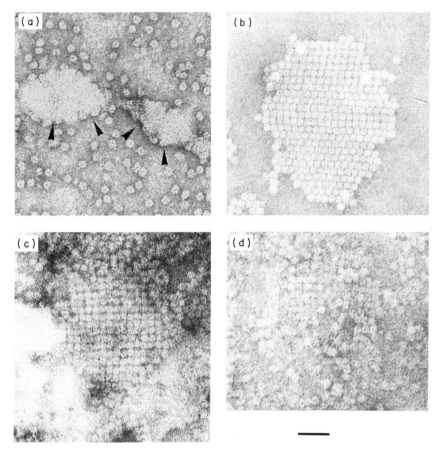

Fig. 1. Examples for the appearance of $y = -$ (a) and $y = +$ (b–d) rated crystallization specimens in the TEM. The specimens were negatively stained with uranyl acetate according to the procedure described in Section II.F. The arrows in (a) are pointing at a precipitate. Bar, 50 nm.

The quality of the arrays can easily be quantified using optical diffraction techniques (see e.g. Beeston *et al.*, 1972; Misell, 1978). Electron micrographs of some $y = + +$ specimens are shown in Fig. 2.

From the experimental point of view, the stepwise optimization is started by gradually varying different parameters, with only one parameter varied per crystallization set-up at a time, so that the x/y-values correspond to a single parameter. Hence a single variation can be judged to be either advantageous or disadvantageous. Prior to any modification, it is necessary to have a standard parameter configuration serving as a starting point. In the case of *A*.

eutrophus RuBisCO, the starting point was a parameter configuration described for a 2-D approach to the spinach RuBisCO (Barcena and Shaw, 1985). However, in general any crystallization recipe for a given protein known from the literature should do for this purpose. When a 3-D recipe is employed one has to keep in mind that it is probably necessary to adopt certain 2-D-specific modifications before defining the starting conditions (see following paragraphs and Section III).

To reach the desired optimization the binary evaluation served as a helpful tool when applied in the following way: the first evaluation was determined relative to a specimen prepared according to the starting parameter constellation. All $x/y = +$ variations were maintained and taken as a basis for further variations pointing in the same direction (e.g. Variation 1: 3 mg protein ml^{-1} instead of 2 mg ml^{-1} → Variation 2: 4 mg ml^{-1} → Variation 3: 5 mg ml^{-1}, etc.), whereas variations with $x/y = -$ were discarded. Variations rated $\pm$ were further modified in a given direction till a $+$ or $-$ was found unambiguously, or they were discarded also. The latter possibility was preferred when further modifications in the direction already envisaged did not look promising in advance.

The optimization procedure is deemed to have been successful when at least one parameter is given $+/++$ and all others $+/+$. Until this condition is satisfied, reproducibility of the whole crystallization process cannot be achieved. Exact reports concerning each crystallization set-up and its result are indispensable, not only for sake of reproducibility. As experience has shown, the use of forms for this purpose can be recommended.

B. Protein solutions used

As in 3-D crystallization, the protein to be crystallized is the most important prerequisite and should always be as homogeneous as possible in order to avoid any unspecific events that *per se* reduce reproducibility. RuBisCO from *Alcaligenes eutrophus H16* (Wilde, 1962; ATCC 17699, DSM 428) was purified according to the procedure described by Bowien *et al.* (1976), slightly modified by employing three sucrose density gradient runs instead of one (Holzenburg, 1987). Homogeneity of RuBisCO was about 99%, as determined by analytical SDS polyacrylamide gel electrophoresis (Weber *et al.*, 1972). Furthermore, only fresh and not proteolytically affected material with a high specific activity (see below) was used. It is also recommended that the whole optimization is carried out with enzyme from one batch (purification), so that the protein solutions are of constant quality.

With the *A. eutrophus* RuBisCO no problems occurred in this respect. The enzyme was stored in small portions of approximately 1 mg protein as ammonium sulphate (AS) pellets at $-20°C$; protein solutions were brought to

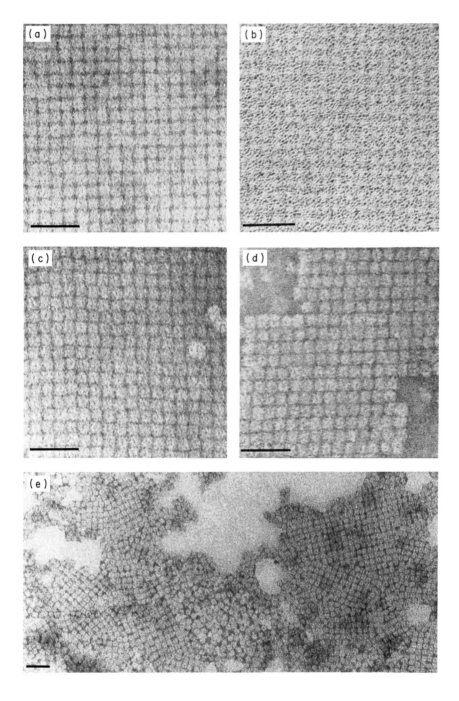

45% saturation using a 100% saturated AS solution in double distilled water adjusted with ammonia to a pH of 7.8 at room temperature. Under these storage conditions no activity loss could be detected, even after a period longer than 8 months. A specific activity of 1.78 U (mg protein)$^{-1}$ (U = μmol $^{14}CO_2$ fixed min^{-1}) was determined for both freshly purified (values between 1.56 and 1.70 U (mg protein)$^{-1}$ have been reported by Bowien $et\ al.$, 1976, 1980; Bowien and Mayer, 1978; and Bowien and Gottschalk, 1982) and stored enzyme after subsequent dialysis. The carboxylase activity was measured by the incorporation of $^{14}CO_2$ into an acid-stable product at 30°C (Bowien $et\ al.$, 1976). All protein concentrations were consistently determined by the method of Lowry $et\ al.$ (1951), using bovine serum albumin as a standard.

Ready-for-use enzyme solutions were prepared from the frozen AS pellets by a modified flow-dialysis method originally described by Colowick and Womack (1969). One AS pellet was thawed gently at 4°C, dissolved in 30 μl of buffer (the particular enzyme buffer as specified below), transferred to the hemispherical sample chamber (chamber volume 120 μl) of a double flow-dialysis chamber, and dialysed against 1 litre of the buffer with a flow rate of 1–1.5 ml min^{-1} at 4°C. The dialysis membrane was situated between the sample chamber and the buffer chamber (diameter 6 mm, height 1.2 mm) which had been connected to the buffer reservoir. The dialysate then was subjected to centrifugation for 15 min at 18 000 × g (4°C), the resulting pellet discarded, and the supernatant stored on ice for not longer than 7 days, or 48 h for the enzyme complexed with CABP (see below). During this period no loss of activity could be observed.

A visual control of the enzyme solutions to be crystallized was achieved by electron microscopical examinations. For this purpose the enzyme was prepared according to the negative staining procedure described by Valentine $et\ al.$ (1968) without a fixation step (Fig. 3).

By means of flow-dialysis about 40 μl of a dialysate with a protein concentration around 22 mg ml^{-1} were obtained from a 1-mg AS pellet. Thus all necessary protein concentrations (usually lower than 10 mg ml^{-1}) could easily be prepared by simple dilution with an appropriate buffer.

RuBisCO was always purified under activating conditions, i.e. in the presence of Mg^{2+} and HCO_3^- ions (Bowien and Gottschalk, 1982). The following buffer (enzyme buffer, abbreviation EB) was employed (Bowien $et\ al.$,

---

Fig. 2. Electron micrographs depicting 2-D arrays of high quality ($y = + +$). (a) Ea, negatively stained with an aqueous solution of uranyl acetate as described in Section II.F. (b) Ea, rotationally Pt/C-shadowed using electron beam evaporation (elevation angle 20°); the protein appears dark. (c) Eia, negatively stained as in (a). (d) E-$CABP$, negatively stained as in (a). (e) = (a), but instead of AS, PEG 8000 was employed as precipitant. Regarding the above abbreviations, please see Section II.B. Bars, 50 nm.

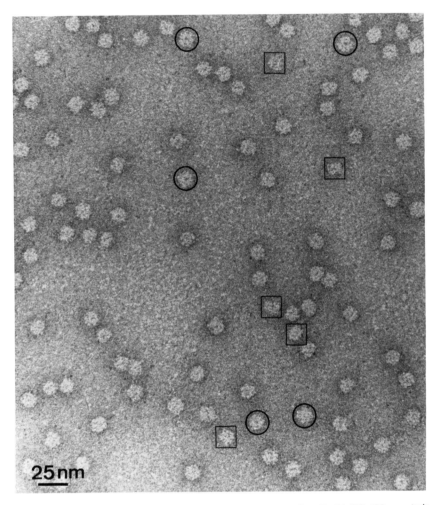

Fig. 3. Electron micrograph of homogeneous *A. eutrophus* RuBisCO ($80 \, \mu g \, ml^{-1}$) prepared according to Valentine *et al.* (1968) using a 4% (w/v) aqueous solution of uranyl acetate (pH 4.5). The encircled molecules represent typical face-on views. Some side-on views are marked by black squares (cf. Bowien *et al.*, 1976).

1976): 20 mM Tris/HCl, 10 mM $MgCl_2$, 50 mM $NaHCO_3$, 1 mM EDTA and 1 mM DTE in double distilled water (pH 7.8 at room temperature). The buffer was stored at 4°C. DTE was added immediately before use.

Absence of Mg^{2+} and HCO_3^- is correlated with an inactivation of the enzyme (Bowien and Gottschalk, 1982, and literature given therein). A residual activity of 0.07 U (mg protein)$^{-1}$ was determined for the inactivated enzyme (abbreviated to *Eia*, with *Ea* standing for the activated enzyme).

Furthermore, the enzyme could be locked in the transition state of the carboxylase reaction upon incubation of the *Ea* with the transition state analogue 2-carboxyarabinitol-1,5-bisphosphate (CABP) in a two-fold stoichiometric excess (on the basis of eight active sites per RuBisCO molecule, this gives a molar ratio of RuBisCO:CABP = 1:16) at 30°C, resulting in an *Ea* completely saturated with CABP (abbreviation *E-CABP*; Wishnick *et al.*, 1970; Siegel and Lane, 1972; Pierce *et al.*, 1980; Johal *et al.*, 1985; Pal *et al.*, 1985). *Eia* was prepared by thorough dialysis (flow-dialysis as described above) against EB lacking Mg^{2+} and HCO_3^- (abbreviation EBia; procedure after Bowien and Gottschalk, 1982).

A. eutrophus RuBisCO, in each of its three different functional states (as *Ea*, *Eia* and *E-CABP*), could be reproducibly crystallized into a monolayer directly on the support films used for electron microscopy (Fig. 2). How to proceed further with these different enzyme solutions in order to obtain 2-D crystals will be described below.

C. The HNS

The properties of the HNS (heterogeneous nucleation surface) are determined by the properties of the support film, which in turn are dependent on the chemical composition of the film in connection with the type of grid used. Best results (see Fig. 2) were obtained with carbon-coated copper grids of type G200F1 (Science Services GmbH, Munich, Federal Republic of Germany). The grid windows of G200F1 grids were partially marked by letter–number combinations (A1, A2, A3, ..., B1, ..., etc.) which allow excellent orientation. In addition, they offer a highly differential grid-morphology (different sized carbon-coated areas, different plane/concave patterns, etc.) which helps 2-D crystallization by increasing the probability of heterogeneous nucleation events taking place.

Pure carbon films were preferred, as they possess the following advantageous properties, and were therefore superior to all other film materials tested so far (Collodion, Formvar, Pioloform, in each case carbon- or not carbon-coated):

- sufficient mechanical stability;
- no essential changes of surface properties occurred during storage;
- no tendency to form holes was observed;
- good reproducibility in fabrication; and
- optimal affinity for RuBisCO molecules in order to attract and bind the molecules but, on the other hand, allow those statistical fluctuations (gliding of the molecules on the HNS) to take place that are essential for the formation of 2-D-specific nuclei (cf. Kleber, 1983; Holzenburg, 1984).

Carbon films 10–20 nm thick were produced by indirect (see Robinson *et al.*, 1987) resistance evaporation of carbon onto freshly cleaved mica (approximately 2.5 × 7.5 cm) as proposed by Bradley (1954). The carbon film was floated on a clean triple distilled water surface and subsequently covered with grids (smooth sides facing the carbon film). About 10 min later the grids were separated, liberated from overhanging carbon, taken off the water surface, blotted on filter paper, and stored in Petri dishes for not longer than six weeks in the dark at room temperature.

D. Crystallization buffers

The two criteria for a suitable crystallization buffer were:

(1) the buffer had to produce at least $y = +$ specimens; and

(2) should compare well with the respective enzyme buffer (EB for the *Ea* and *E-CABP*; EBia for the *Eia*).

The only additional component allowed was the precipitant (AS or PEG 8000), and NaCl in the case of *Eia* in order to compensate for the missing ionic strength caused by the Mg^{2+} and HCO_3^- deficiency of the EBia compared to the EB.

The following crystallization buffers (abbreviation CB) were found to be optimal:

$$CB_{Ea} = CB_{E-CABP} = EB \text{ without DTE but supplemented with AS up to}$$
7.77% saturation (pH 7.8);

$$CB_{Eia} = EBia \text{ without DTE but supplemented with 6.8% (w/v)}$$
PEG 8000 (Sigma Chemie GmbH, Munich, Federal Republic of Germany) and 70 mM NaCl (pH 7.8).

CB_{Ea} with PEG 8000 (6.8% w/v) instead of AS was found to form less perfect crystals with slightly bigger unit cell dimensions in the case of *Ea* (see Fig. 2e), whereas it was essential in the case of *Eia*. PEG 8000 had already been mentioned for the 2-D crystallization of spinach RuBisCO (Barcena and Shaw, 1985). The substitution of 6.8% (w/v) PEG 8000 by 7.77% saturated AS was carried out after Johal *et al.* (1980), where 0.375 M AS was substituted for 8% PEG 6000. DTE was omitted in all CBs because it was found to quench the contrast of specimens that had been stained with uranyl acetate.

E. Crystallization method

The crystallization method is designed to mediate between the protein solution to be crystallized and the crystallization conditions such as incubation time, incubation temperature, pH value, ionic milieu, type and concentration of the precipitant, etc. It should always be as simple and versatile as possible. Versatile in this case means allowing all kinds of modifications and variations to be carried out in an easy and reproducible way. In general, it is the crystallization method that leads to a moderate—otherwise precipitation would be favoured—increase of the effective protein concentration, i.e. the density of the molecules, and subsequently to crystals (cf. McPherson, 1976).

The method (after Barcena and Shaw, 1985) described here fulfils all desired properties, and thus was the only one employed. The use of only one method reduces the total number of crystallization set-ups to be tested by a factor of at least 10, so that the experimenter can concentrate on the crystallization conditions. In Fig. 4 the procedure used is shown in detail for the *Ea*. For the *Eia* an initial protein concentration of $7.5 \, mg \, ml^{-1}$ was necessary. The optimum incubation time at $25°C$ was $18 \pm 2 \, min$. Two-dimensional crystals of the *E-CABP* were grown at $25°C$ using a slightly shorter incubation time $(16 \pm 2 \, min)$ and an initial protein concentration of $4.2 \, mg \, ml^{-1}$. The incubation temperature of $25°C$ was found to be about the lowest temperature required for reproducibly growing 2-D crystals of high quality. Thus temperatures from $25–28°C$ were superior to temperatures lower than $25°C$. At $4°C$, for instance, long incubation times ($\geq 40 \, min$) had to be applied to achieve moderately good results ($y = \pm$) only.

There is experimental evidence that the crystallization process as described here consists of two phases, one taking place in the droplet (see Fig. 4) during incubation, the other directly on the HNS. Both are essential for growing the 2-D crystals (for details see Holzenburg, 1987).

F. Post-treatments

The last step in the procedure cannot be regarded as a crystallization step, but it is here that it is decided whether 2-D crystals that might have formed on the grid are stabilized and stained, or are destroyed (dissolved and/or washed off). This gives rise to another problem: how to evaluate a single crystallization set-up, if one cannot be sure that crystals which have been formed, will necessarily be visible in the TEM? Crystals that cannot be imaged will lead to wrong conclusions about the respective optimization steps, and hence make an optimization impossible when based on TEM observations. Consequently an absolutely reliable preparation procedure is indispensable.

The standard procedure used during the optimization was as follows: 2–3 12-µl droplets of an aqueous uranyl acetate solution (4% w/v in triple distilled water (H 4.5, filtered) were applied onto the grid. In between the applications, the negative staining solution was allowed to act upon the specimen for 5–30 s.

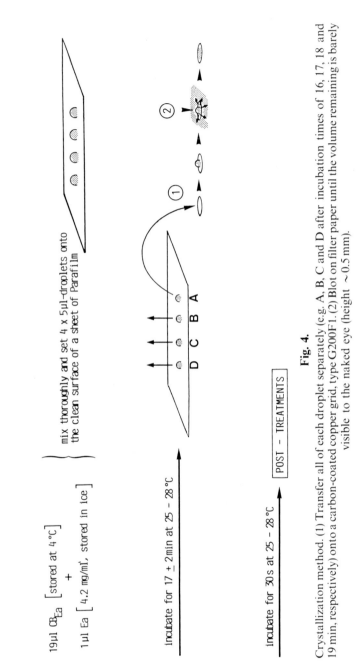

Fig. 4.

Crystallization method. (1) Transfer all of each droplet separately (e.g. A, B, C and D after incubation times of 16, 17, 18 and 19 min, respectively) onto a carbon-coated copper grid, type G200F1. (2) Blot on filter paper until the volume remaining is barely visible to the naked eye (height ~0.5 mm).

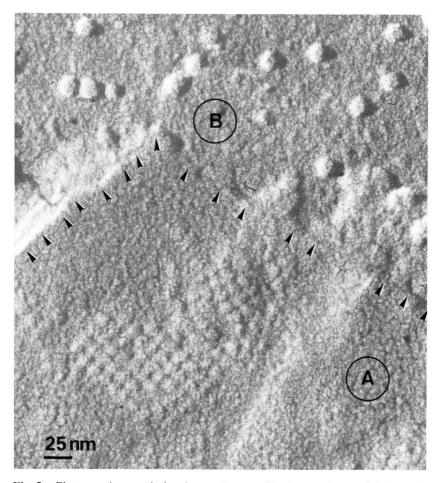

Fig. 5. Electron micrograph showing an *Ea*-crystallization specimen quick-frozen in liquid nitrogen (using a plunge-cooling device after Robards and Crosby, 1983), completely freeze-dried (2 h at −80°C; cf. Kistler *et al.*, 1977; Umrath, 1983; Robards and Sleytr, 1985), and unidirectionally shadowed with W/Ta at 30°C (elevation angle 45°) using electron beam evaporation (for further details regarding the preparation, see Holzenburg, 1987). The protein appears light. In area A the integrity of the crystal is maintained by the surrounding crystallization buffer (CB), whereas in B the stabilizing CB has been washed off. Consequently crystals disappear in these areas, and only single molecules are visible. However, a problem connected with the CB-embedded crystals is the smear effect: structural details are covered by the CB, and are thus not accessible to the shadowing procedure. This loss of information does not apply to the molecules in area B because the disturbing CB is not present. The boundary between A and B is marked by arrows.

Then 50–90% of the total volume on the grid was removed by pressing the grid briefly against filter paper. After the last application, the solution was blotted off immediately and completely.

The same procedure, but using phosphotungstic acid (3% w/v in triple distilled water or EB without DTE and adjusted to pH values of 4.7, 5.8, 7.0 and 7.8) instead of uranyl acetate, did not yield any positive results since no ordered aggregates of the Ea and Eia could be observed in the TEM. Negative results were also obtained when using glutaraldehyde solutions. Furthermore, the 2-D crystals were very sensitive to quick-freezing procedures (see Fig. 5). Best results were obtained (Fig. 2) by the negative staining method with uranyl acetate, and by simple air drying with subsequent shadowing (Fig. 2b).

The reason why phosphotungstic acid probably dissolved the 2-D crystals, whereas uranyl acetate did not, has not yet been elucidated. Different viscosities and abilities to stabilize protein aggregates may play a subtle role. Even if uranyl acetate exhibits excellent properties regarding staining and stabilizing the 2-D crystals of the A. eutrophus RuBisCO, the situation can be totally different for another protein. This is especially true for proteins that undergo precipitation in response to the low pH of uranyl acetate solutions.

III.   Concluding remarks

When using the methodology discussed in this chapter for 2-D crystallization of a soluble protein, the optimization procedure should include the search for:

(1) a reliable specimen preparation method in order to obtain good imaging conditions in the TEM;

(2) a suitable HNS and precipitant;

(3) the optimum initial protein concentration as well as the precipitant concentration; and

(4) the optimum incubation time at possibly $\geq 25°C$.

An appropriate precipitant can often be taken from 3-D crystallization recipes, if available, or it has to be discovered by trial and error: precipitants that cause microcrystalline "precipitates" as observed under a binocular microscope deserve special attention in this respect. Regarding the initial protein concentration, about 1/3 of the concentration recommended for 3-D crystallization should do for a 2-D approach. The initial precipitant concentration should be sufficient with around 1/6 of the concentration required for precipitating the protein. If nothing helps, it may also be worth trying di- or multivalent cations of different Van der Waals' radii (screening necessary).

Finally, one should be aware of the fact that, even in the case of optimized crystallization parameters, there are always statistical fluctuations giving rise to a certain variability of the crystal quality. For further reading on 2-D crystallization experiments the reader is referred to Van Bruggen *et al.* (1986).

Acknowledgement

The preparation of the manuscript was supported by a grant for Förderung der wissenschaftlichen Forschung in Niedersachsen.

References

Aebi, U. (1978). *9th Int Conf. Electron Microsc.* **3**, 81–86.
Aebi, U., Fowler, W. E., Buhle, E. L. and Smith, P. R. (1984). *J. Ultrastruct. Res.* **88**, 143–176.
Barcena, J. A. and Shaw, P. J. (1985). *Planta* **163**, 141–144.
Beeston, B. E. P., Horne, R. W. and Markham, R. (1972). In *Practical Methods in Electron Microscopy* (A. M. Glauert, ed.), Vol. 1, pp. 185–444. Elsevier/North-Holland, Amsterdam.
Bowien, B. and Gottschalk, E.-M. (1982). *J. Biol. Chem.* **257**, 11845–11847.
Bowien, B. and Mayer, F. (1978). *Eur. J. Biochem.* **88**, 97–107.
Bowien, B., Mayer, F., Codd, G. A. and Schlegel, H. G. (1976). *Arch. Microbiol.* **110**, 157–166.
Bowien, B., Mayer, F., Spiess, E., Pähler, A., Englisch, U. and Saenger, W. (1980). *Eur. J. Biochem.* **106**, 405–410.
Bradley, D. E. (1954). *Br. J. Appl. Phys.* **5**, 65–67.
Carter, Jr, C. W. and Carter, C. W. (1979). *J. Biol. Chem.* **254**, 12219–12223.
Colowick, S. P. and Womack, F. C. (1969). *J. Biol. Chem.* **244**, 774–776.
Feher, G. and Kam, Z. (1985). In *Methods in Enzymology* (H. W. Wyckoff, C. H. W. Hirs and S. N. Timasheff, eds), Vol. 114, pp. 77–112. Academic Press, Orlando.
Frank, J. (1984). *8th Eur. Congr. Electron Microsc.* **2**, 1307–1316.
Frank, J., Goldfarb, W., Eisenberg, D. and Baker, T. S. (1978). *Ultramicroscopy* **3**, 283–290.
Giersig, M., Kunath, W., Sack-Kongehl, H. and Hucho, F. (1986). In *Nicotinic Acetylcholine Receptor* (A. Maelicke, ed.), NATO ASI Series, Vol. H3. Springer-Verlag, Berlin.
Holzenburg, A. (1984). Versuche zur Herstellung und elektronenmikroskopischen Analyse zweidimensionaler Kristalle des isolierten Enzyms D-Ribulose-1,5-bis-phosphat-Carboxylase. Diploma, Universität Göttingen.
Holzenburg, A. (1987). Untersuchungen zu Struktur und funktionsabhängigen Konfigurationsänderung der D-Ribulose-1,5-bis-phosphat-Carboxylase/Oxy-genase aus *Alcaligenes eutrophus H16*. Thesis, Universität Göttingen.
Holzenburg, A., Mayer, F., Harauz, G., Van Heel, M., Tokuoka, R., Ishida, T., Harata, K., Pal, G. P. and Saenger, W. (1987). *Nature* **325**, 730–732.
Johal, S., Bourque, D. P., Smith, W. W., Suh, S. W. and Eisenberg, D. (1980). *J. Biol. Chem.* **255**, 8873–8880.

Johal, S., Partridge, B. E. and Chollet, R. (1985). *J. Biol. Chem.* **260**, 9894–9904.

Kistler, J., Aebi, U. and Kellenberger, E. (1977). *J. Ultrastruct. Res.* **59**, 76–86.

Kleber, W. (1983). "Einführung in die Kristallographie". VEB Technik, Berlin.

Kunath, W., Giersig, M., Sack-Kongehl, H. and Van Heel, M. (1986). *10th Int. Congr. Electron Microsc.* **1**, 489–490.

Lowry, O. H., Rosebrough, N. J., Farr, A. L. and Randall, R. J. (1951). *J. Biol. Chem.* **193**, 265–275.

McPherson, A. (1976). *Meth. Biochem. Anal.* **23**, 249–345.

Misell, D. L. (1978). In *Practical Methods in Electron Microscopy* (A. M. Glauert, ed.), Vol. 7. Elsevier/North-Holland, Amsterdam.

Miziorko, H. M. and Lorimer, G. (1983). *Ann. Rev. Biochem.* **52**, 507–535.

Ottensmeyer, F. P., Andrew, J. W., Bazett-Jones, D. P., Chan, A. S. K. and Hewitt, J. (1977). *J. Microsc.* **109**, 259–268.

Pal, G. P., Jakob, R., Hahn, U., Bowien, B. and Saenger, W. (1985). *J. Biol. Chem.* **260**, 10768–10770.

Pierce, J., Tolbert, N. E. and Barker, R. (1980). *Biochemistry* **19**, 934–942.

Robards, A. W. and Crosby, P. (1983). *Cryo-Lett.* **4**, 23–32.

Robards, A. W. and Sleytr, U. B. (1985). In *Practical Methods in Electron Microscopy* (A. M. Glauert, ed.), Vol. 10. Elsevier/North-Holland, Amsterdam.

Robinson, D. G., Ehlers, U., Herken, R., Herrmann, B., Mayer, F. and Schürmann, F.-W. (1987). *Methods of Preparation for Electron Microscopy*. Springer-Verlag, Berlin and New York (in press).

Saxton, W. O. and Baumeister, W. (1982). *J. Microsc.* **127**, 127–138.

Siegel, M. I. and Lane, M. D. (1972). *Biochem. Biophys. Res. Commun.* **48**, 508–516.

Umrath, W. (1983). *Mikroskopie* **40**, 9–37.

Unwin, P. N. T. and Henderson, R. (1975). *J. Mol. Biol.* **94**, 425–440.

Valentine, R. C., Shapiro, B. M. and Stadtman, E. R. (1968). *Biochemistry* **7**, 2143–2152.

Van Bruggen, E. F. J., Van Breemen, J. F. L., Keegstra, W., Boekema, E. J. and Van Heel, M. (1986). *J. Microsc.* **141**, 11–20.

Van Heel, M. (1984). *8th Eur. Congr. Electron Microsc.* **2**, 1317–1325.

Van Heel, M. (1986). In *Pattern Recognition in Practice: II* (E. S. Gelsema and I. Kanal, eds), pp. 291–299. North-Holland, Amsterdam.

Van Heel, M. and Frank, J. (1981). *Ultramicroscopy* **6**, 187–194.

Weber, K., Pringle, J. R. and Osborn, M. (1972). In *Methods in Enzymology* (W. Tirs and S. N. Timasheff, eds), Vol. 26, pp. 3–27. Academic Press, New York and London.

Wilde, E. (1962). *Arch. Mikrobiol.* **43**, 109–137.

Wishnick, M., Lane, M. D. and Scrutton, M. C. (1970). *J. Biol. Chem.* **245**, 4939–4947.

16

Correlation Averaging and 3-D Reconstruction of 2-D Crystalline Membranes and Macromolecules

HARALD ENGELHARDT

Max-Planck-Institut für Biochemie, D-8033 Martinsried,
Federal Republic of Germany

METHODS IN MICROBIOLOGY
VOLUME 20 ISBN 0-12-521520-6

I. Introduction

Why image processing? Though the resolution power of modern electron microscopes is theoretically sufficient to obtain almost atomic images of macromolecules, the directly usable information content of the electron micrographs is disappointingly small. Superimposed structures, variabilities introduced by preparation, noise of any source and, particularly with well-focused projections or unstained specimens, low image contrast, all reduce the perceptibility of the structure of interest. Digital image processing techniques allow an enhancement of the signal through the averaging of many equivalent projections of the molecule under scrutiny and the reconstruction of the spatial structure through the combination of different projections. Using negatively stained preparations, one may obtain structural information to a resolution of about 1.5 nm, corresponding to globular protein domains (embedded in stain) of 10 to 20 amino acids. Higher resolutions are expected with hydrated preparations, embedded in glucose or aurothioglucose, or even completely unstained specimens, investigated by cryoelectron microscopy. Extremely radiation-sensitive unstained biological material requires low dose techniques, and the resulting low contrast micrographs require even more highly sophisticated image processing methods. Image processing has become an indispensable step in modern electron microscopy of biological macromolecules and macromolecular assemblies.

Why two-dimensional crystals? Averaging requires projections of molecules more or less identically oriented on the specimen support. If projections of asymmetric, irregularly oriented single molecules are to be processed, they have to be analysed and sorted for orientational consistency. The molecules in 2-D crystals are ordered *a priori* and render immediate averaging possible. This is a valuable advantage with low contrast micrographs in particular, where the analysis of the orientation becomes a problem with respect to the accuracy required. High resolution reconstructions (<2 nm) are at present restricted to 2-D crystalline specimens. Another advantage is that in 2-D crystals only one domain of the polypeptide is adsorbed onto the specimen support; flattening phenomena are therefore locally restricted. Regular specimens are not as rare in nature as it may seem: the purple membrane, bacterial surface layers, some outer and photosynthetic membranes are examples, as well as helical and other assemblies like flagella, fimbria, phages, viruses, etc., and a few (membrane) proteins that have been artifically reconstituted to form ordered arrays: cytochrome c oxidase, ATPase, insulin receptor, antibodies, haemocyanin, ribulose-1,5-bisphosphate carboxylase, 5s rRNA, etc., suggesting that this approach might be successful for other proteins, too.

Why correlation methods? If a 2-D crystal is "ideal", i.e. the positions of the repetitive units are exactly known from crystallographic considerations, averaging as well as 3-D reconstruction can be performed by straightforward Fourier techniques relying on perfect crystallinity. However, most 2-D crystals are more or less distorted and possess lateral and occasionally rotational disorder, significantly limiting the usable resolution. Correlation methods do not assume a certain arrangement of the molecules, but *analyse* their positions individually and (partly) correct for the lattice distortions. The benefit of using 2-D crystals is that, unlike with single molecules, the positioning can be improved by noise reduction. Furthermore, selective averaging of well-preserved molecules is now possible, excluding distorted and untypical ones that would otherwise reduce the resolution achievable. Correlation averaging has recently been combined with 3-D reconstruction; both approaches are described here.

TABLE I

Review articles and books on image processing in biological electron microscopy

Topic	References
Image analysis, digital image processing	Frank, 1973; Misell, 1978; Saxton, 1978; Hawkes, 1980
Bibliographic compilation	Baker, 1981
Correlation methods, averaging and 3-D reconstruction of single particles	Hoppe and Hegerl, 1980; Frank, 1982; van Heel, 1984
Averaging and 3-D reconstruction of 2-D crystals, principles and techniques	Vainshtein, 1978; Mellema, 1980; Amos *et al.*, 1982; Vainshtein *et al.*, 1982; Engel and Massalski, 1984; Saxton *et al.*, 1984
Principles and applications	Fuller, 1981; Aebi *et al.*, 1982; Glaeser, 1985; Hovmöller, 1986; Stewart, 1986

This chapter is not primarily addressed to specialists in the field, but to scientists who want to become familiar with the principles, possibilities and requirements of digital image analysis in order to obtain appropriate micrographs for subsequent image processing with some assistance in adequately equipped laboratories. Thus, in the following it is not the algorithms, but the strategies, of averaging and 3-D reconstruction of non-ideal 2-D crystals that are outlined, together with the definition of some important operations. Since 2-D crystals may not always be available, a brief survey of the routes of 3-D reconstruction of both single particles and 2-D crystals is included. In Table I some recent reviews and comprehensive articles on image processing techniques are compiled.

II. Image processing tools

A. Image processing methods

A variety of macromolecules have been reconstructed ranging from aperiodic randomly oriented single particles to highly ordered 2-D crystals or even thin sections of embedded 3-D crystals. There is no universal reconstruction procedure which could be applied in a straightforward manner; but a number of methods have been developed in the recent years designed for the particular requirements of the various macromolecular specimens. Figure 1 and the following section briefly outline the characteristic steps of a 3-D reconstruction of single molecules and 2-D crystals.

Obtaining 3-D data.   The reconstruction of one single molecule, 1-D and 2-D crystals requires a series of projections of the tilted specimen to collect information of all dimensions in space (*tilt series*; see Section IV.A). Preparations of single molecules that are randomly oriented on the specimen support do not need to be tilted since the 0° projection of a sufficiently large number of molecules already provides various projected views. If the number of molecules or different projections is limiting, one or a few further tilted views may complete the required information. Particles displaying only one preferred orientation (e.g. adsorption on the flat side of a disc-shaped molecule like glutamine synthetase, for example) can be reconstructed from one tilted view. This provides different projections of the molecule since the particles are randomly oriented in the plane, i.e. each particle is "rotated" around an axis perpendicular to the specimen support ("conical tilting"; Hoppe and Hegerl, 1980; Frank *et al.*, 1986). The 0° projection of the same field furnishes the azimuthal orientation of each molecule.

Determination of parameters.   Several parameters concerning the tilting of the specimen in the microscope and/or the orientation of the molecules must be determined.

For 2-D crystals:
- actual tilt angles
- azimuthal orientation of the tilt axis in plane
- (inclination of the specimen around an axis normal to the tilt axis)
- lattice base vectors
- lateral positions of the molecules

For single particles showing one preferred orientation:
- tilt angle
- lateral positions of molecules
- in-plane orientation of each molecule (azimuth)

For single particles that are randomly oriented:
- lateral positions of molecules
- the three Eulerian angles which describe the orientation of each single particle in the orthogonal coordinate system in 3-D space (the angles include information on any tilting)

In ideal 2-D crystals all unit cells, each containing one or a few molecules arranged in a regular fashion (see Section III), are identically oriented, and it is therefore sufficient to determine three parameters of each projection, as well as the tilt parameters: the two lattice base vectors and the origin of the lattice with respect to the origin of the coordinate system. Normally the lattices are not ideal but show some distortions (short-range and/or long-range disorder) and the lateral positions of the unit cells can no longer be exactly described by the lattice parameters; here the actual positions of the unit cells have to be individually determined by correlation analysis.

Randomly oriented particles do not provide any *a priori* information on the relative orientations. The precise *a posteriori* determination has been one difficulty of the 3-D reconstruction of single molecules. Recently a universal correlation method was introduced capable of determining the three Eulerian angles of relative orientation for each molecule under scrutiny (*angular reconstitution*; van Heel, 1987a). Originally designed for randomly oriented particles, it can, of course, also be applied to molecules with few stable orientations. This method, as well as the 3-D reconstruction itself, assumes the projections to be correctly centred. For aperiodic, randomly oriented molecules the best method appears to be the determination of the centre of gravity (centre of mass) for each particle (van Heel, 1987a). For molecules displaying one stable orientation on the specimen support (if more than one exists, particles with identical orientations have to be selected) only one of the three angles is required, i.e. the azimuthal orientation. The latter can be evaluated by the well-established method of rotational and lateral correlation alignment using the 0° projections as described in Frank *et al.* (1981a).

Refinement of input data for 3-D reconstruction. As well as the exact determination of the tilt and orientational parameters, and the common origin of the projections, the selection of well-preserved molecules which show little variation in stain distribution or position of flexible features, for example, is an important step in improving the resolution. The various criteria for *selective averaging* of 2-D periodic specimens are described in detail below (Section V). The selection criterion for single particles is the degree of similarity between one molecule and a set of related, i.e. more or less identically oriented, particles or the corresponding average, respectively, judged by means of the correlation coefficient or the position of the molecule (its "distance") in the vectorial space, calculated by *correspondence analysis*. Particles which differ greatly from

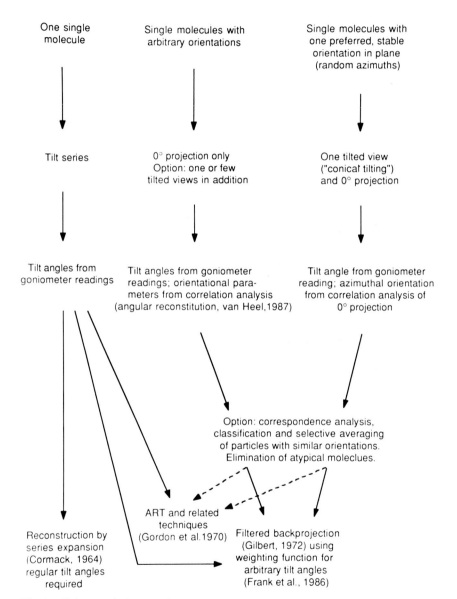

Fig. 1. Scheme of characteristic steps in various routes of 3-D reconstruction of single molecules and 2-D crystals.

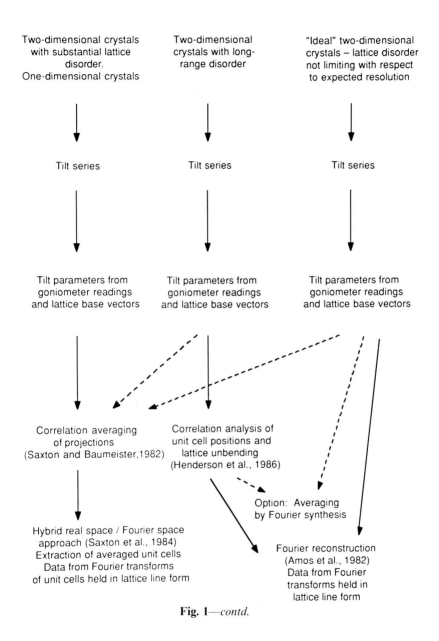

Fig. 1—*contd.*

identically oriented molecules may be excluded from reconstruction. For a detailed description, see Frank *et al.* (1986) and van Heel (1984).

3-D reconstruction.   In the course of a 3-D reconstruction the various projections are combined according to their relative orientations in order to recover the density distribution in three-dimensional space. Using *back-projection* techniques, this may be done in real space by algebraic approaches like ART (Gordon *et al.*, 1970), as well as in reciprocal space via filtered backprojection algorithms (Gilbert, 1972). Two-dimensional crystals are usually reconstructed by crystallographic approaches, accumulating and interpolating the measured data in reciprocal space, and inverse trans-formation (Section VI).

ART is an approximative method; the backprojected data are summed, compared with the original projections, and iteratively corrected so as to match the original projections (Hoppe and Hegerl, 1980). ART and related techniques (Harauz and Ottensmeyer, 1984) are not frequently applied and it has been pointed out that the iterative approach may become prohibitively complicated with a large number of projections (Frank *et al.*, 1986). Backprojected data may also be combined by Fourier techniques, since each projection represents a central section through the reciprocal space (Hoppe and Hegerl, 1980). The problem is to adequately weight the projections. The filter function depends on the projection geometry and has to be individually calculated for arbitrarily oriented particles. General solutions have been developed recently (Frank *et al.*, 1986; Harauz and van Heel, 1986), while the methods currently available are analytical and require regular tilt angles or other analytical distributions of data in Fourier space (Gilbert, 1972; Smith *et al.*, 1973; Cormack, 1964).

If tilting is incomplete ($< 90°$), and that is inevitably the case, or "gaps" in the angular distribution of the projections of single particles do exist, more or less large data gaps will remain in the reciprocal space which in turn create characteristic reconstruction artefacts (see Section VII.A). Hence a data set as complete as possible is an important prerequisite for an interpretable and qualitatively convincing 3-D reconstruction.

The reconstruction methods outlined (Fig. 1) normally expect (thin) isolated particles or macromolecular assemblies; but there are also approaches to reconstruct thin sections of embedded biological material three-dimensionally. Frank *et al.* (1987) applied the tilt series approach and filtered backprojection to projections obtained in a high voltage electron microscope. 3-D reconstructions from one single projection of an *oblique* section through an embedded 3-D crystal can be calculated by the method of Crowther (1984) and Crowther and Luther (1984). The resolution of those reconstructions, however, is clearly limited because of the harsh treatment of the specimen.

B. Image processing systems

Several image processing systems have been developed in various laboratories involved in digital image restoration and reconstruction of electron micrographs. Single computer programs for image handling and processing may have been developed variously; probably the largest collection of computer programs exists at the MRC, Cambridge. But image processing *systems* are more than just a collection of programs. Here the source (Fortran) programs (Fast Fourier Transformation, for example) are packed together in a modular system with a particularly designed command language which allows the user to call for basic operations, to define variables, and to write procedures consisting of a set of commands needed, for example, for a complete averaging job. There are some image processing systems (Table II) whose command language and syntax is convenient, flexible and simple enough to be learnt even by non-specialists in the field. All the systems provide the possibility for normal file and data handling, basic image transformations, display of images on a screen, definition of procedures, interactive and batch mode work, etc. They differ, however, with respect to the major categories of biological macromolecules (single particles or 2-D crystals) for which they were initially designed. The EM, SPIDER and IMAGIC systems were developed for single particle averaging and reconstruction; MDPP and SEMPER, as well as most of the programs of the MRC collection, for 2-D crystal averaging and 3-D reconstruction. However, all the systems have been

TABLE II

Image processing systems for image restoration, averaging and three-dimensional reconstruction in biological electron microscopy

System	Reference	2-D averaging		3-D reconstruction			Correspondence analysis
		Single molecules	2-D crystals	Single molecules	2-D crystals	Helical structures	
EM	Hegerl and Altbauer, 1982	+	+	+	$-^a$	$+^a$	+
SEMPER	Saxton *et al.*, 1979	+	+	+	+	−	−
SPIDER	Frank *et al.*, 1981b	+	+	+	−	−	+
IMAGIC	van Heel and Keegstra, 1981	+	+	+	−	−	+
MDPP	Smith, 1978	−	+	−	+	+	−
"MRC"[b]		−	+	−	+	+	−

[a] Not implemented, but treatable on procedure level.
[b] Program collection of the MRC, Molecular Biology Laboratory, Cambridge (no published account).

developed further and are still in a state of expansion. They are now comparably powerful in many aspects, but not yet universally usable for all applications and processing variants. (Correlation) averaging of 2-D crystals, but not yet 3-D reconstruction, is a standard procedure of all the systems listed here. For a detailed description of the systems the reader is referred to the references in Table II.

III. Two-dimensional crystals

A.   Fourier transforms of two-dimensional crystals

The Fourier transformation of an image (transformation into *reciprocal space*) is equivalent to a two-dimensional spatial frequency analysis (spectral analysis), describing the structures by means of a set of sine and cosine functions of varying spatial frequencies individually weighted by factors (Fourier coefficients). *Large* features are mainly composed of *low* frequency waves (long wavelengths), *small* details (little dots, sharp edges, etc.) of *high* frequency components (short wavelengths). Low frequency information (Fourier coefficients) is located close to the centre of the Fourier transform, while high frequency components are situated further from the centre. The distance r from the centre to a certain spatial frequency in *reciprocal* space is equivalent to structural information in the image at a resolution level of $1/r$. The spatial frequency range in the Fourier transform of a *digitized* image of size s with a pixel size p is given by $1/s$ (the lowest frequency contained in the image), its Fourier coefficient represented by the value of the first pixel from the centre, and the largest interpretable frequency by $1/2p$, the resolution limit of the image (imposed by densitometry, see Section IV.C, and the sampling theorem; Misell, 1978), represented by pixels at the edges of the Fourier transform. Fourier coefficients are complex numbers and, therefore, Fourier transforms are complex pictures consisting of both a *real* and an *imaginary* part or (in the goniometrical form) described by the *moduli* (amplitudes) and the *phases* (arguments) of the Fourier coefficients. The modulus weights the spatial frequencies; the phase comprises information on the position of the sine functions and, thus, on the position of structural detail. (For a comprehensive description of the Fourier transform the reader is referred to Misell, 1978.)

Power spectrum.   In optical diffractograms and computer-calculated quasi-optical diffractograms (power spectra), the phase information is lost and only the *squared* moduli (intensities) of the Fourier coefficients are obtained. (The power spectrum is the product of the Fourier transform multiplied by its

complex conjugate, zeroing the imaginary part.) Thus diffractograms may be
used to identify the spatial frequencies that are heavily weighted or largely
suppressed in an image, but they cannot be used to reconstruct the image
correctly. Lateral translation means a phase shift in reciprocal space and that
is why power spectra are translationally invariant—the diffractogram of a 2-D
crystal does not change if it is shifted (but not rotated), provided that the
structure of the crystal does not change. That is also the reason for why the
structural information of superimposed identical crystals cannot be separated
if the crystals are only shifted with respect to each other.

Lattice constants. The structural information of 2-D crystals is represented
by a set of spatial frequencies related to (multiples of) the reciprocal lattice
constants, i.e. the characteristic distances between neighbouring unit cells. In
fact, the complete structural information from an ideal 2-D crystal is
contained in the diffraction spots arranged on a lattice in reciprocal space. The
analysis of the regular arrangement of the diffraction pattern provides
information on the crystal symmetry, the lattice constant(s) and the lattice
base vectors, defining the shape and orientation of the lattice in the original
image. The indexation of the diffraction spots, the first step necessary in the
course of a 2-D crystal analysis, is illustrated in Fig. 2.

The *lattice spacings a* and *b* (the centre-to-centre distance between
neighbouring unit cells in the real image) are related to the reciprocal *lattice
constants a** and *b** by

$$a = 1/(a^* \sin \gamma^*) \quad \text{and} \quad b = 1/(b^* \sin \gamma^*) \tag{1}$$

where γ^* denotes the angle between the reciprocal lattice base vectors $\mathbf{u}^*$ and
$\mathbf{v}^*$. For $p3$, $p4$ and $p6$ symmetries $b^* = a^*$ applies (see Section III.B).

The lattice constants may be obtained from optical diffractograms or by
means of power spectra calculated from digitized images in the computer.
Starting with optical diffractograms, apply

$$a_{\text{specimen}} = (f \cdot \lambda/M) \cdot 1/(a^*_{\text{diffr.}} \cdot \sin \gamma^*) \tag{2}$$

where f denotes the focus length of the diffractometer, γ the wavelength of the
laser source (632.8 nm for He–Ne lasers), M the magnification factor of the
electron micrograph, and $a^*_{\text{diffr.}}$ the lattice constant as measured in the actual
diffractogram. The diffractogram may also be internally calibrated by means
of the diffraction pattern of a lattice with known lattice parameters.

To calculate the lattice base vectors $\mathbf{u}$ and $\mathbf{v}$, starting with calculated power
spectra (using pixel units or the actual values throughout), apply

$$\begin{cases} \mathbf{u} = d \cdot (v_y^*, -v_x^*) \quad \text{and} \quad \mathbf{v} = d \cdot (-u_y^*, u_x^*), \quad \text{where} \\ \mathbf{u}^* = (u_x^*, u_y^*) \quad \text{and} \quad \mathbf{v}^* = (v_x^*, v_y^*) \quad \text{and} \\ d = s/(|u_x^* \cdot v_y^* - v_x^* \cdot u_y^*|) \end{cases} \tag{3}$$

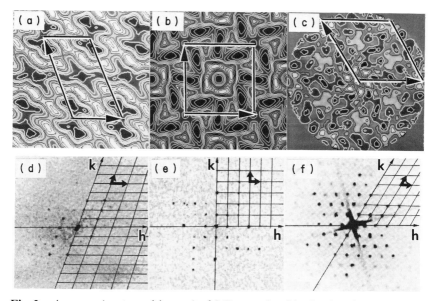

Fig. 2. Averages (contoured images) of 2-D crystals with $p2$, $p4$ and $p6$ symmetry, respectively, (a–c) and the power spectra (d–f) of the corresponding source images. The unit cells (lattice base vectors) and the reciprocal vectors are indicated; the lattices with indices h and k show one set of independent reflections for each crystal. (The averages are from bacterial surface layers: (a) *Aquaspirillum dispar*, (b) *Sporosarcina ureae* and (c) *Thermoproteus tenax* (the latter courtesy of I. Wildhaber, Martinsried).)

s being the size of the power spectrum. Starting with real space vectors, the reciprocal vectors are calculated by exchanging $\mathbf{u}^*$ and $\mathbf{u}$, $\mathbf{v}^*$ and $\mathbf{v}$, u_x^* and u_x, etc. The lattice constant a^* is obtained from the modulus of $\mathbf{u}^*$ by

$$a^* = |\mathbf{u}^*| = \sqrt{(u_x^{*2} + u_y^{*2})} \tag{4}$$

the values for b^*, a and b are determined correspondingly. For a *qualitative* analysis of power spectra, see Section IV.B.

Three-dimensional Fourier transform. The 3-D Fourier transform of a 2-D crystal is periodic in the x^* and y^* directions but aperiodic along the z^*-axis since the crystal is only one unit cell thick. The Fourier components are therefore continuously distributed along lines located in the diffraction spots of, for example, the x^*, y^* plane (*lattice lines*; Fig. 3). A projection through a three-dimensional object tilted by an angle of ψ is equivalent to a central section at ψ degrees in the 3-D reciprocal space. The diffraction spots in the power spectrum (Fourier transform) of the projections appear where the lattice lines are intersected (Fig. 3). The smaller the tilt increments are, the

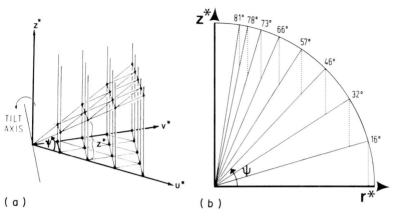

Fig. 3. (a) Data organization of a 2-D crystal in reciprocal space. The (lattice) lines indicate the distribution of the Fourier components in the z^*-direction. The two planes, i.e. central sections through the lattice lines, represent the transforms of an untilted and a tilted (angle ψ) projection. Scheme adapted from Amos *et al.* (1982). (b) Scheme of tilt angles for a tilt series with regularly sampled lattice line data. The resolution to thickness ratio according to equation (5) is 2/7. Scheme according to Saxton *et al.* (1984).

closer will the sections in Fourier space be, and therefore the more complete is the sampling of lattice line data (see Section IV.A). While a projection of a 2-D crystal can be described by a set of discrete Fourier components obtained from the diffraction spots in the Fourier transform of the projections, the description of the 3-D structure needs a continuous function (curve) of the Fourier components of each lattice line, finally represented in form of (discrete) sampling points (Section VI.D).

Fourier transforms of unit cells. In the classic Fourier approach of averaging and 3-D reconstruction of 2-D crystals (Amos *et al.*, 1982), the moduli and phases are obtained from the reflections in the Fourier transform of the *whole images*. The hybrid real space/Fourier approach of Saxton *et al.* (1984) described here, however, does not start with the Fourier transform of the whole image, but with the *unit cells* extracted from the averages of each projection. The corresponding Fourier transforms do indeed contain the complete structural information, i.e. the sampling data of the lattice lines. If the unit cells are resampled onto a *square* lattice, that is if they are, in principle, projected such that the original lattice base vectors **u** and **v** are transformed to **u**′ and **v**′, where $|\mathbf{u}'| = |\mathbf{v}'|$ and $\gamma'(\mathbf{u}', \mathbf{v}') = 90°$, thereby creating "oblique" images of the unit cells, its transforms contain the Fourier coefficients in a convenient arrangement. Since the first pixel neighbouring the central one represents the spatial frequency $1/s$, s being the image size (see above), the corresponding pixel of the Fourier transform of a resampled unit cell provides the Fourier components of the reflection of order one because here s equals the (transformed) lattice constant. Accordingly, the second pixel corresponds to the second diffraction order, and so on. Thus, the lattice line data can be obtained from the transformed unit cells by reading out the values of the appropriate pixels (Engel and Massalski, 1984; Saxton *et al.*, 1984).

B. Symmetries and space groups of two-dimensional crystals

Molecules may be packed in various fashions in 2-D crystals. Out of the 80 two-sided plane groups which describe all possible 2-D crystals with different sides ("top" and "bottom"; Vainshtein, 1981), only 17 are "allowed" for proteins, lipids, nucleotides, etc. (Fig. 4). Crystal forms which contain mirror planes, glide (mirror) planes or centres of inversion cannot be built from the chiral biological molecules (thus displaying handedness). In natural crystals (e.g. bacteriorhodopsin, bacterial surface layers, various photosynthetic and outer membranes), representing the molecules in their functional state, only those plane groups may be realized which obey the vectorial features of the proteins (e.g. transport, inside-outside properties), i.e. crystals not containing molecules with inversed orientation with respect to the layer plane ($p1, p21, p3$, $p4$ and $p6$). In artificial 2-D crystals, however, any of the 17 plane groups may be realized.

Information on the symmetry of specimens is required for the symmetrization of averages, Fourier (window) filtering and 3-D reconstruction. The plane groups can be determined by inspection of the averaged $0°$ projection of the unit cell and/or the Fourier transform or power spectrum (light-optical diffractogram) of the corresponding micrograph. The $0°$ *projection* of crystals containing molecules with inversed orientation with respect to the layer plane (i.e. rotated by 180° around an axis within the layer) show mirror symmetries over mirror axes and/or glide (mirror) axes, in addition to rotational symmetry in many cases, indicative of a certain two-sided plane group (Fig. 5). The plane groups are identified by:

- the existence and distribution of axes of rotational symmetry (point symmetry) mirror axes and glide axes (line symmetry) in the $0°$ projection of the unit cell ($\rightarrow$ image);
- the relation between the two axes, i.e. the lengths of the lattice base vectors, identical lengths or not ($\rightarrow$ image, power spectrum);
- the angle between the two axes: $60°, 90°, 120°$ or other than these angles ($\rightarrow$ image, power spectrum);
- the symmetry of the diffraction pattern ($\rightarrow$ power spectrum, Fourier transform);

- systematic absence of reflections ($\rightarrow$ power spectrum, Fourier transform);
- phases ($0°$ or $180°$) of certain reflections ($\rightarrow$ Fourier transform)—here the unit cell needs to be correctly centred ("phase origin").

See Table III for a compilation of the lattice properties, Fig. 4 for the projection symmetries, and Fig. 5 for the location of symmetry and mirror axes.

Correlation averaging (see Section V) does not need *a priori* information on the symmetry properties and is therefore a convenient tool for collecting information on the unit cell symmetry and lattice type without the necessity of determining and refining the amplitudes and phases of the reflections in the Fourier transform. In most cases a good guess of the crystal symmetry is possible. If the inspection of reflections in the Fourier transform is desired (e.g. for an independent guess of symmetry properties or in the course of averaging via Fourier techniques not described here in detail), proceed as follows (for a detailed description of the procedures briefly outlined below see Amos *et al.*, 1982):

- Obtain a correctly centred image of the $0°$ projection of the crystal (via cross-correlation).
- Compute the Fourier transform.
- Determine the reciprocal lattice base vectors $\mathbf{u}^*$ and $\mathbf{v}^*$ as precisely as possible, using the power spectrum rather than the Fourier transform (see Section V.D).
- Read out the amplitudes and phases of the reflections at their ideal lattice points. Reflection spots normally occupy more than one pixel; if so, the amplitude peak may lie somewhere between the pixels and can be interpolated (e.g. by taking the square root of the sum of the squared amplitude values). The phase values do not vary among the peak pixels except for experimental error and can simply be averaged.
- Refine the lattice base vectors using the interpolated peak positions of the high order reflections in addition and return to the previous point to improve the determination of amplitudes and phases if necessary.

If a high symmetry case (e.g. *p*222 or *p*6) cannot be assigned with certainty, it is recommended that a lower symmetry is applied (i.e. *p*21—or even *p*1—or *p*3, respectively) for symmetrization, Fourier filtration or 3-D reconstruction purposes. Doing so allows for a higher symmetry, but does not force it erroneously. Small deviations of the relation of the axes, the symmetry angle and phase values from the (expected) ideal values may be created by accidentally tilting the specimen by some degrees in the microscope, by lattice distortions, etc. Inspection of a few micrographs with apparent $0°$ projections may help to distinguish between accidental and systematic effects.

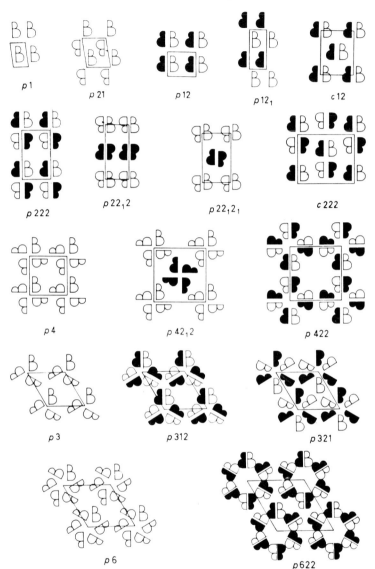

Fig. 4. Vertical projections of 2-D crystals from chiral molecules (symbolized by a B), representing the 17 two-sided plane groups. Black symbols represent molecules rotated by 180° around an axis in the crystal plane ("bottom view"). The unit cell boundaries are indicated, the positions of the axes of rotational and mirror symmetries are given in Fig. 5. The nomenclature of the plane groups follows the rules of Holser (1958). The cell type is indicated by a p (primitive) or c (centred), the rotational symmetry around the z-axis (perpendicular to the crystal plane) is described by the first number following the cell type (from Hovmöller, 1986, with permission).

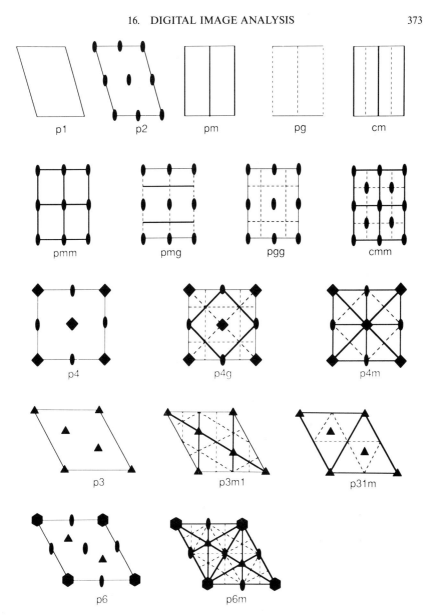

Fig. 5. Scheme of the centres of rotational symmetry and the axes of mirror symmetry in *vertical projections* of 2-D crystals belonging to the 17 two-sided plane groups. The arrangements of molecules in the unit cells are given in Fig. 4 (same order). The nomenclature describes the *projection symmetries* of the crystal types (Holser, 1958). Symbols: faint lines identify unit cell boundaries, ● centre of 2-fold, ▲ 3-fold, ◆ 4-fold and ⬡ 6-fold symmetry, respectively. Solid lines represent mirror axes and dotted lines glide (mirror) axes. Scheme adapted from Vainshtein (1981).

TABLE III

Crystallographic properties of the 17 two-sided plane groups (reproduced from Hovmöller, 1986, with permission)

No.	Symbol	Relations between a and b axes	Equivalent positions	Systematically absent reflections	Symmetry of diffraction pattern	Reflections with phase 0 or 180
1	$p1$	—	(x, y, z)	—	$\lvert F(hkl)\rvert = \lvert F(\bar{h}\bar{k}\bar{l})\rvert$	—
2	$p2_1$	—	$(x, y, z),\ (\bar{x}, \bar{y}, z)$	—	as 1 and	$(hk0)$
3	$p12$	$\gamma = 90$	$(x, y, z),\ (\bar{x}, y, \bar{z})$	—	$\lvert F(hkl)\rvert = \lvert F(\bar{h}\bar{k}\bar{l})\rvert$ as 1 and	$(h0l)$
4	$p12_1$	$\gamma = 90$	$(x, y, z),\ (\bar{x}, \tfrac{1}{2}+y, \bar{z})$	$(0k0),\ k$ odd	$\lvert F(hkl)\rvert = \lvert F(\bar{h}\bar{k}\bar{l})\rvert$ as 3	$(h0l)$
5	$c12$	$\gamma = 90$	$(x, y, z),\ (\tfrac{1}{2}+x, \tfrac{1}{2}+y, z),\ (\bar{x}, y, \bar{z}),\ (\tfrac{1}{2}-x, \tfrac{1}{2}+y, \bar{z})$	$(hkl),\ h+k$ odd	as 3	$(h0l)$
6	$p222$	$\gamma = 90$	$(x, y, z),\ (\bar{x}, \bar{y}, z),\ (\bar{x}, y, \bar{z}),\ (x, \bar{y}, \bar{z})$	—	mm-symmetry. i.e. $\lvert F(hkl)\rvert = \lvert F(\bar{h}\bar{k}\bar{l})\rvert =$ $\lvert F(h\bar{k}\bar{l})\rvert = \lvert F(\bar{h}kl)\rvert =$ $\lvert F(\bar{h}k\bar{l})\rvert = \lvert F(h\bar{k}l)\rvert =$ $\lvert F(\bar{h}\bar{k}l)\rvert = \lvert F(hk\bar{l})\rvert$	$(hk0)$ $(h0l)$ $(0kl)$
7	$p22_12$	$\gamma = 90$	$(x, y, z),\ (\bar{x}, \tfrac{1}{2}+y, \bar{z}),\ (\bar{x}, \bar{y}, z),\ (x, \tfrac{1}{2}-y, \bar{z})$	$(0k0),\ k$ odd	as 6	$(hk0)$ $(h0l)$
8	$p22_12_1$	$\gamma = 90$	$(x, y, z),\ (\bar{x}, \bar{y}, z),\ (\tfrac{1}{2}-x, \tfrac{1}{2}+y, \bar{z}),\ (\tfrac{1}{2}+x, \tfrac{1}{2}-y, \bar{z})$	$(h00)\ h$ odd $(0k0)\ k$ odd	as 6	$(hk0)$
9	$c222$	$\gamma = 90$	$(x, y, z),\ (\bar{x}, \bar{y}, z),\ (\bar{x}, y, \bar{z}),\ (x, \bar{y}, \bar{z}),$ $(\tfrac{1}{2}+x, \tfrac{1}{2}+y, z),\ (\tfrac{1}{2}-x, \tfrac{1}{2}-y, z),$ $(\tfrac{1}{2}-x, \tfrac{1}{2}+y, \bar{z}),\ (\tfrac{1}{2}+x, \tfrac{1}{2}-y, \bar{z})$	$(hkl),$ $h+k$ odd	as 6	as 6

No.	Symbol	Cell	Equivalent positions	Conditions	Amplitude relationships	Indices
10	p4	$a=b$, $\gamma=90$	(x, y, z), (ȳ, x, z) (x̄, ȳ, z), (y, x̄, z)	—	$\|F(hkl)\|=\|F(\bar{h}\bar{k}l)\|=$ $\|F(h\bar{k}l)\|=\|F(\bar{h}\bar{k}l)\|=$ $\|F(\bar{k}hl)\|=\|F(k\bar{h}l)\|=$ $\|F(k\bar{h}l)\|=\|F(\bar{k}hl)\|$	(hk0)
11	p422	$a=b$, $\gamma=90$	(x, y, z), (ȳ, x, z) (x̄, ȳ, z), (y, x̄, z) (y, x, z̄), (x̄, y, z̄) (ȳ, x̄, z̄), (x, ȳ, z̄)	—	as 10 and mm-symmetry: $\|F(hkl)\|=\|F(\pm h\pm k\pm l)\|=$ $\|F(khl)\|=\|F(\pm k\pm h\pm l)\|$	(hk0) (h0l) (0kl) (hh̄l) (hhl)
12	p42₁2	$a=b$, $\gamma=90$	(x, y, z), (ȳ, x, z) (x̄, ȳ, z), (y, x̄, z) (½+y, ½−x, z̄), (½−x, ½+y, z̄), (½+x, ½−y, z̄)	(h00) h odd (0k0) k odd	as 11	(hk0) (hh̄l) (hhl)
13	p3	$a=b$, $\gamma=120$	(x, y, z), (ȳ, x−y, z) (y−x, x̄, z)	—	6-fold symmetry $\|F(hkl)\|=\|F(k,\bar{h}+\bar{k},l)\|=$ $\|F(\bar{h}+\bar{k},h,l)\|=\|F(\bar{h}\bar{k}l)\|=$ $\|F(\bar{k},h+k,l)\|=$ $\|F(h+k,\bar{h},l)\|$	—
14	p312	$a=b$, $\gamma=120$	(x, y, z), (ȳ, x−y, z) (y−x, x̄, z), (y, x, z̄) (x−y, ȳ, z̄), (x̄, y−x, z̄)	—	as 13 and mirror symmetry across (110)-diagonal: $\|F(hkl)\|=\|F(khl)\|$etc.	(hhl) (h, 2h, l) (2k̄, k, l)
15	p321	$a=b$, $\gamma=120$	(x, y, z), (ȳ, x−y, z) (y−x, x̄, z), (y, x, z̄) (x̄, y−x, z̄), (x−y, ȳ, z̄)	—	as 13 and mirror symmetries across h and k axes: $\|F(hkl)\|=\|F(kh\bar{l})\|$ etc.	(hh̄l) (h, 2h. l) (2k̄, k, l) (hk0)
16	p6	$a=b$, $\gamma=120$	(x, y, z), (ȳ, x−y, z) (y−x, x̄, z), (x̄, ȳ, z) (y, y−x, z), (x−y, x, z)	—	as 13 and $\|F(hkl)\|=\|F(hk\bar{l})\|$ etc.	(hk0)
17	p622	$a=b$, $\gamma=120$	(x, y, z), (ȳ, x−y, z) (y−x, x̄, z), (x̄, ȳ, z) (y, y−x, z), (x̄, ȳ, z) (y, x, z̄), (x−y, ȳ, z̄) (x̄, y−x, z̄), (ȳ, x−y, z̄) (x−y, ȳ, z̄), (y−x, x, z̄)	—	as 16 and mirror symmetries across h and k axes and across (110)-diagonal $\|F(hkl)\|=\|F(khl)\|=$ $\|F(hkl)\|=\|F(k,\bar{h}+\bar{k},l)\|$ etc.	(hk0) (hhl) (hh̄l) (h, 2h, l) (h, 2h, l)

Occasionally layers or membranes (e.g. collapsed vesicles) may form double sheets with the lattices being exactly in register but attached in a face-to-face orientation (see Engel *et al.*, 1985; Baumeister and Hegerl, 1986). Here mirror symmetry is obtained in 0° projections with a characteristic absence of handedness in the diffraction pattern. Apparent mirror symmetry may thus indicate double layers of 2-D crystals not containing molecules with inversed orientation. The application of a lower symmetry (*p*1, *p*21, *p*3, *p*4 or *p*6) in 3-D reconstructions, however, will reveal the correct situation (Engel *et al.*, 1985; Dickson *et al.*, 1986). Multicomponent systems such as the photosynthetic membranes of some bacteria, displaying apparent (six-fold) symmetry, may contain subunits which are not arranged in a symmetric or even a systematic fashion (Engelhardt *et al.*, 1985). Imposing symmetry here, or even applying non-selective averaging only, will obscure the asymmetric structure and the significant variability of the particular components.

IV. Obtaining images for averaging and 3-D reconstruction

A. Tilt series

Aiming at a 3-D reconstruction with (nearly) isotropic resolution and minimum artificial aberrations, a data set as complete as possible has to be recorded. While the structural information in plane (x,y-direction) is "completely" obtained in 0° projections, i.e. there is no limitation due to the projection geometry, the data in the z-direction can only be partly recovered since tilting to 90° is impossible for practical reasons. Despite other claims, tilting far beyond 60° is necessary in order to minimize the range of missing information ("missing cone" problem, see Section VII) and to avoid serious reconstruction artefacts (Baumeister *et al.*, 1986a; Baumeister and Engelhardt, 1987).

Tilting scheme. A 3-D reconstruction of a certain resolution requires an appropriate series of tilted views in order to obtain enough sampling points of the lattice lines corresponding to the resolution sought. The minimum tolerable separation between the sample values at the resolution limit is $1/t$ for a specimen of thickness t. Saxton *et al.* (1984) derived a formula providing the approximate optimum tilt angles required for a specimen of thickness t at a resolution of r. The (largest usable) tilt increment $\Delta\psi$ (in radians) following the tilt angle ψ is given by

$$\Delta\psi = (r/t) \cdot \cos\psi \qquad (5)$$

This scheme leads to a set of angular spacings proportional to $\cos\psi$ (while being independent of the lattice constant) and results in non-identical tilt increments (Fig. 3) which ensure a suitable sampling of data in Fourier space while regular tilt angles, which are commonly applied, would not. In practice, it is desirable to obtain a somewhat closer sampling to permit some overdetermination of noisy data. This is accomplished by a smaller ratio r/t in equation (5). If the specimen thickness does not exceed 9 nm (valid for many protein layers and membranes) and the expected resolution is ≥ 1.5 nm (which appears to be the practical lower limit for negatively stained preparations), the following tilting scheme is recommended: $0°$, $9.5°$, $19.0°$, $28.0°$, $36.4°$, $44.1°$, $51.0°$, $57.0°$, $62.2°$, $66.6°$, $70.4°$, $73.6°$, $78.6°$, $80.5°$, ($81.5°$, ...). In practice, it is preferable to record a $0°$ projection first, followed by the high tilts descending to a second $0°$ projection. This sequence ensures an unobstructed view at all tilt angles. The two $0°$ projections are needed to compare their diffractograms or the resolutions achieved by (correlation) averaging, in order to assess the degree of radiation damage that may have occurred in the course of a continuous tilt series. Since it is hardly possible to determine the absolute *tilt angles* in the microscope, it is important for the subsequent determination of the tilt parameters to adjust the *tilt increments* precisely.

Tilting direction. Tilting in one direction only (from $0°$ to positive *or* negative angles) is sufficient if the internal symmetry of the unit cell is at least three-fold. Here the monomers are variously oriented with respect to the tilt axis and different projected views are simultaneously obtained, providing sufficient sampling points in Fourier space. Specimens with $p2$ or $p1$ symmetry, however, must be tilted over the full range of angles from negative to positive values, thus doubling the number of projections required.

Reliability of high tilt projections. Tilting to high angles has the consequence that the projected stain becomes very thick. This may have implications for the validity of a simple linear image formation theory. Saxton *et al.* (1984) give an assessment of the maximum values treatable. If a specimen of 10 nm thickness is tilted to $80°$ and projected at 80 or 100 kV, the resolution usable is about 1 nm which is normally sufficient for negatively stained preparations†).

Primary magnification. Magnification factors in the range of about $20\,000 \times$ to $50\,000 \times$ are appropriate for most biological macromolecules. Since high magnification means also increased irradiation (proportional to the square of the magnification factor), lower magnifications are to be preferred.

Specimen holder. High tilting ($> 60°$) is normally not possible without some technical modifications to the usual equipment. The commercial specimen

---

† The condition for the resolution r is $4\lambda \cdot t \leq r^2$, where t is the projected thickness and λ the electron wavelength (≈ 0.0038 nm for 100 kV and ≈ 0.0042 nm for 80 kV). The specimen thickness then is $t \cdot \cos \psi$, ψ being the tilt angle.

holders do not allow tilting beyond 60° due to the thickness of the tip. Shaw and Hills (1981) developed a specimen holder for top entry systems which enables tilting to 75°, and Chalcroft and Davey (1984) devised a simple "unlimited" tilt holder for side entry systems which is now commercially available (GATAN). Here only the mesh geometry of the grid itself limits the view and the maximum tilt angle applicable. Tilting up to 80° is conveniently achieved by means of 100×400 mesh grids, inserted into the specimen holder such that the tilt axis is perpendicular to the longitudinal axis of the rectangular meshes, and if the 2-D crystal lies approximately in the centre and not close to an edge. However, tilts up to 80° and beyond have also been obtained with thin hexagonal 600 mesh grids. A quite simple trick to obtain high tilt views, using the commercial holder, is to pre-bend the grids.

In most microscopes the goniometer stage is locked, prohibiting tilting beyond $\pm 60°$, which is very sensible for normal specimen holders that are too wide for the gap between the pole pieces and would seriously damage them upon high tilting. Make sure that the high tilt holder can be tilted without any problems when the safety lock has been removed!

Minimal dose conditions. Tilted views may be collected from a continuous tilt series or from various patches of the preparation, each tilted by another angle (discontinuous series). The latter procedure is necessary if the specimen, particularly unstained or, for example, glucose-embedded ones, is radiation-sensitive and only a few projections can be taken from one region. Although negatively stained preparations are relatively radiation-resistant, they, too, may suffer from extensive irradiation, resulting in reorganization of the stain and loss of resolution (Stark *et al.*, 1984). Thus, imaging conditions are recommended which minimize the electron dose suffered by the specimen; i.e. apply a low magnification, and an even lower one for screening the grid, and shift the beam to another area for focusing (by means of a low dose unit), etc. Amos *et al.* (1982) give some useful information on electron dose–imaging characteristics.

B. Image quality

Two aspects are important for the selection of micrographs for subsequent image processing: the quality of the crystal and the quality of the contrast transfer function (CTF), i.e. the focus conditions, etc., which prevailed in the microscope. Both image features are judged by inspection of the optical diffractograms using a laser diffractometer (Horne and Markham, 1972).

The Fourier transform, originating from the specimen structure, is multiplied in the course of imaging in the microscope by the CTF, weighting the spatial frequencies in addition (e.g. Misell, 1978). If the micrograph was taken under poor focus conditions, significant spatial frequencies may be suppressed, reduced to zero or, even more serious, displayed with inversed contrast via the CTF, thus limiting the usable resolution.

Quality of the 2-D crystal. The highest order of reflections obtained and the shape of the diffraction spots provide information on the crystallinity of the specimen. The crystallographic resolution (which corresponds to the resolution that can be obtained by averaging techniques relying on the crystallinity of the specimen, i.e. the classic Fourier approach) is given by

$$r_{cryst.} = 1/r^*_{cryst.} = 1/(a^* \cdot h_{max}) = (a \cdot \sin \gamma^*)/h_{max} \qquad (if \ k = 0) \qquad (6)$$

where h_{max} denotes the highest index of $h, 0$-reflections. The corresponding equation applying b and k_{max} (if $h = 0$) furnishes the resolution in the direction of the second independent lattice base vector. If $h \neq 0$ and $k \neq 0$ apply,

$$r_{cryst.} = (ab \cdot \sin \gamma^*)/\sqrt{(a^2k^2 + b^2h^2 + 2abhk \cdot \cos \gamma^*)} \qquad (7)$$

The degree of lattice disorder can be qualitatively assessed by inspection of the power spectrum. Ideal lattices produce very sharp reflections, while the diffraction spots of crystals with substantial lattice disorder (lateral disorder) are blurred and often the high frequency spots are too weak to be detectable. The consequence is a loss in crystallographic resolution. Rotational disorder (long-range disorder) is identified by circularly elongated reflections (Fig. 6). Superimposed crystals which are substantially rotated with respect to each other (by an angle of ψ), or lattices possessing corresponding discontinuities

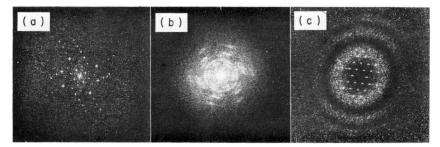

Fig. 6. Power spectra (quasi-optical diffractograms) of 2-D crystals. (a) An apparently undistorted tetragonal 2-D crystal imaged at appropriate focus conditions. (b) A badly distorted crystal showing appreciable rotational disorder; the low frequency reflections are radially elongated while the higher diffraction spots appear separated. (c) Diffractogram of a micrograph taken at strong underfocus. The first gap in the contrast transfer function eliminates a part of the reflections while high order diffraction spots can be seen in the first ring beyond the gap; the corresponding spatial frequencies in the image are displayed with reversed contrast. Some drift is indicated by the incomplete rings (in the x^*-direction).

within the area inspected, create two sets of reflections in reciprocal space on lattices rotated by ψ. The information derived from the two lattices may be separated in Fourier space by windowing one set of diffraction points and omitting the other one (Fourier filtration). Consequently, re-transformed images only show one of the two lattices (see Section V.B).

Quality of the contrast transfer function. The "pure" CTF of a micrograph may be obtained from areas free of biological material (carbon film). The diffractogram normally shows dark gaps (Thon rings; Thon, 1966; Misell, 1978) where the sign of the phase contrast changes from negative to positive values and vice versa due to the focus conditions. The corresponding spatial frequencies are weighted by factors fading out to zero, thus being eliminated in the micrograph. The first gap in the CTF therefore should be beyond the expected resolution, i.e. sufficiently far beyond the highest diffraction order, otherwise significant spatial frequencies are eliminated or obtained with inversed phase contrast (Fig. 6). For most conventionally negatively stained preparations a defocus providing a continuous CTF to a resolution of 1 to 1.5 nm is appropriate. Micrographs of tilted specimens display a focus gradient which is steepest perpendicular to the tilt axis and can easily be detected by laterally moving the micrograph through the laser beam. Apparent $0°$ projections accidently tilted in the microscope may be identified (see Section V.E dealing with focus gradients). Micrographs suffering substantially from astigmatism or drift (Fig. 6) with respect to the resolution desired should be discarded. Select only the best areas from the best micrographs for image reconstruction purposes. Any information missing in the micrographs because of unfavourable imaging conditions, can hardly be expected to be recovered in the computer! Finally, note the coordinates (e.g. relative to the top left corner of the negative as a reference point) of the area to be densitometered.

C. Digitization

Pixel size. To make a preselected area of the micrograph available to the computer, it is digitized by means of a two-dimensional scanning microdensitometer. Two variables have to be set: the spot size (common apertures are (5, 10 µm), 15, 20, 25 µm or above for most flat bed or rotating drum densitometers) and the size of the area to be digitized, preferably in dimensions of 2^n pixels (for image processing reasons), e.g. 256, 512, 1024, etc. The pixel size (on the specimen level) determines the resolution limit (r_{lim}) of the densitometered image:

$$r_{lim} = 2 \cdot \text{spot size/magnification} \qquad (8)$$

To ensure a usable resolution of 1.5 nm a pixel size of $r_{lim}/2$, i.e. ≤ 0.75 nm, is required. In practice, a lower pixel size ($\approx r_{lim}/3$) will be preferred in order not to be limited by undersampling in the course of image processing.

Image size. The size of the digitized area depends on how many unit cells are to be included. As a rule of thumb, at least 300–600 unit cells should be available for averaging. The signal-to-noise ratio, and hence the resolution of the average, increases statistically with the number n of motifs averaged (proportional to the square root of n). Images of 512 or 1024 pixels square are the most common formats, larger ones only required for high resolution work or crystals possessing a very large lattice constant (≥ 30 nm).

Mechanical alignment. Make sure that all plates of a *continuous* tilt series are identically aligned in the microdensitometer, i.e. they must not be rotated with respect to each other. Although averaging and the extraction of unit cells (see Section VI.C) are not impaired by accidental rotations, the tilt parameters will be calculated with a larger degree of error, resulting in scattering of data in Fourier space and a loss of accuracy in lattice line fitting (see Section VI.B).

V. Correlation averaging—image restoration

A. Averaging procedure

Unlike the Fourier approach (Amos *et al.*, 1982), correlation averaging does not rely on the crystallinity of the specimen, but determines the position of each unit cell via cross-correlation of the image with a representative reference, comprising one or more unit cells (Frank, 1982; Saxton and Baumeister, 1982a). In principle the reference is centred on each pixel of the source image, and the correlation coefficient between the reference and the corresponding subframe of the image is calculated. The value of the coefficient is written into the corresponding pixel of a field of source image size (the *cross-correlation function*, CCF). Now the reference is shifted to the next pixel, and so on. Thus, the CCF contains high correlation coefficients where the coincidence of the reference and the image was high (correlation *peaks*), thereby identifying the positions of the unit cells found. In practice the CCF is obtained by multiplying the Fourier transform of the image (its conjugate complex) by the Fourier transform of the reference (embedded in a "structureless" image of identical size); having done so, the product is re-transformed into real space.

Noise of any source (statistical noise, contributions of non-periodic, uncorrelated structures like the carbon film, etc.) impairs the CCF insofar as non-periodic (high-frequency) correlations are added to the correlation of the periodic structure.* The position of the apparent maximum value of a correlation peak suffering from noise may thus be different from the virtual peak maximum, identifying the "true" position of the unit cell. There are two ways partly to overcome the influence of noise and to improve the positioning of the unit cells in the CCF. Firstly, by optimizing the signal-to-noise (S/N) ratio in the CCF itself (see equation (11)) via noise reduction in the reference and/or in the source image, and by applying large references; secondly, by fitting a two-dimensional function to each peak to approximate the apparently "true" maximum position. Both possibilities are exploited in the course of correlation averaging (Fig. 7).

Once the unit cell positions have been determined and compiled in a "peak list", the preliminarily estimated lattice base vectors u and v are refined by fitting an ideal lattice to the unit cell positions found. Unit cells that are largely displaced from their corresponding lattice points can be expected to suffer substantially from displacement and may be eliminated from the peak list in order not to average distorted molecules. Further criteria for peak selection exist (see Section V.E). Besides the determination of unit cell positions, selective averaging is the second important step (and the benefit of correlation methods) to improve the resolution of the final average. For averaging, small subframes of the image are extracted at peak positions and superposed by addition. Two independent averages are calculated (derived from "odd" and "even" numbered peaks) to assess the resolution via one of the resolution criteria in use (see Section V.F). The independent sums are added to obtain the final average. The latter is either applied as a new, better defined reference for a refinement pass, or it is symmetrized, if a $0°$ projection of a symmetric specimen is processed, thereby improving the signal-to-noise ratio by another factor of $\sqrt{n}$ (n being the rotational symmetry). The whole averaging procedure is displayed schematically in Fig. 7 and discussed below in more detail.

B. Preparing a reference

Original reference. If a micrograph of a negatively stained specimen is to be processed, possessing sufficient contrast to render the lattice and the unit cells visible, it is normally sufficient to extract a small area from the source image with a unit cell (i.e. a morphological complex) centred and to apply it as a preliminary reference ("Route I" in Fig. 7). Since the CCF will still suffer from noise, the preliminary average obtained is used as a new reference (with an

---

* Let picture P and reference R consist of the signal $S(S_P, S_R)$ and the uncorrelated noise $N(N_P, N_R)$. The CCF P * R then is a sum of four independent terms, three of them describing the correlations added by noise: $P * R = (S_P * S_R + S_P * N_R + N_P * S_R + N_P * N_R)$.

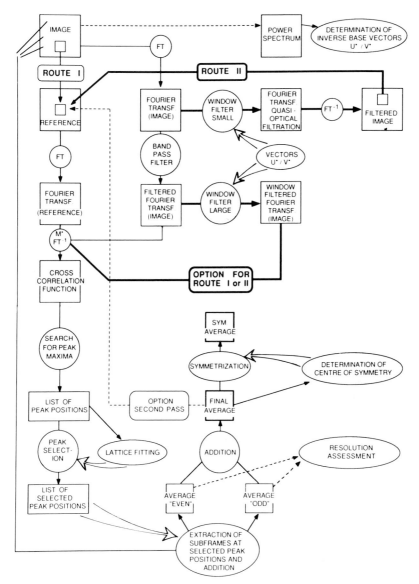

Fig. 7. Scheme of the correlation averaging procedure. FT, Fourier transformation; FT^{-1}, inverse transformation; M*, multiplication by the complex conjugate.

improved signal-to-noise ratio) for a refinement pass, resulting in a significantly improved resolution of the final average (Saxton and Baumeister, 1982). Additional refinement passes may be applied but normally the improvement, if any, is not dramatic.

Fourier filtered reference. Images with little contrast, particularly those from low dose exposures of unstained material, either require a large reference or a reference with an appropriately improved signal-to-noise ratio for cross-correlation.† If the specimen displays an apparently good crystallinity, i.e. reflections to a resolution of, say, < 3 nm and relatively sharp diffraction spots according to Fig. 6, the noise of an image is most effectively reduced by Fourier filtration (quasi-optical filtration which, in principle, is equivalent to some sort of averaging (e.g. Misell, 1978)). Here a mask is applied, with windows where the reflections in the Fourier transform are located. All the information between the reflections, originating from undesired noise and structures, is removed. To obtain an appropriate mask, the reciprocal lattice vectors $\mathbf{u}^*$ and $\mathbf{v}^*$ are determined in the power spectrum as precisely as possible. The positions of the reflections are most accurately determined by means of peak search algorithms (see Section V.D). The window size required depends on the area occupied by the reflections, i.e. it should be sufficiently large in order not to cut off the higher order reflections that may be blurred due to lattice distortions of the crystal.‡ Ideal crystals, ideally sampled by densitometry, produce reflections one pixel in diameter only; in practice a window size of two or more pixels is required, largely depending on the crystal quality. After filtration (multiplication of the Fourier transform with the mask consisting of values of "0" and "1") the real image is obtained by inverse transformation and an appropriate reference area is extracted (according to "Route II" in Fig. 7).

References of superimposed crystals. If superimposed crystals are rotated with respect to each other and distinct reflections owing to the two lattices exist in the power spectrum, the two lattices can be separated by Fourier filtration and accordingly independent references are obtained. It has been variously shown that correlation averaging is capable of coping with superimposed lattices (Pum and Kübler, 1984; Kessel *et al.*, 1985; Radermacher *et al.*, 1986) and that independent 3-D reconstructions of 2-D crystals lying upon each other can be performed (Cejka *et al.*, 1987; Chalcroft *et al.*, 1987). Theoretically, even minor differences between the lattice base vectors $\mathbf{u}_1$ and $\mathbf{u}_2$ ($\mathbf{v}_1$ and $\mathbf{v}_2$, respectively; length or orientation) are sufficient to

---

† The peak height in the CCF is proportional to the S/N ratio of either image to be correlated and to the reference diameter (Saxton and Frank, 1977).

‡ Fourier filtration of non-ideal lattices entails omitting high frequency information anyway. The consequence is a broadened correlation peak, but not necessarily a shift in its maximum position.

separate the lattices by correlation techniques (Radermacher *et al.*, 1986). In practice a difference in orientation of $> 15°$ is desirable for effective separation and convenient performance.

Reference size. According to the considerations outlined above, the reference should be large enough to obtain a significant signal-to-noise ratio in the CCF. With respect to the lattice disorder, however, a reference as small as one unit cell would be advantageous in order to follow the unit cell displacements closely; references larger than the area of apparently "perfect" local crystallinity do not find the unit cell positions exactly. Dependent on the actual signal-to-noise ratio and the crystallinity, an optimum reference size exists, producing minimal error in motif positioning. Quantitative information on the displacements, however, can only be obtained *a posteriori* (Zane, 1985) and is not available for a first attempt. Our experiences with negatively stained or heavy metal shadowed preparations suggest the following reference diameters to start with (given in lattice spacing units):

- references extracted from original images   3 (2.5–4)
- references obtained from averages or quasi-optical filtrations 2 (1.5–3)

For (extreme) low dose images larger (Fourier filtered) references are required. Values from 3 to 9 were reported (Henderson *et al.*, 1986; Rachel *et al.*, 1986), the latter only being efficient with highly ordered crystals.

C. Cross-correlation

Before the image and the ("embedded", see Fig. 7) reference are cross-correlated, two more possibilities to reduce noise exist (Fig. 7): by band-pass filtration (e.g. Misell, 1978) and by another application of a Fourier filter.

Band-pass filter. Low- and high-pass filtration should generally be applied to remove low frequency information up to the first reflection order (owing to modulations introduced by staining or focus gradients, density ramps, etc.) and high frequency noise beyond the expected resolution. The Fourier transform of the image (or of the embedded reference) is circularly masked, the edges smoothly fading to a constant value, i.e. zero (see Misell, 1978, for example, for the consequences of sharp cuts in the Fourier transform).

Fourier filter. If apparently high local crystallinity exists, i.e. neighbouring unit cells are strongly correlated in position, the image can be treated by a Fourier filter with windows large enough not to disturb the genuine lattice disorder. Thus, "noise" of any spatial frequency can be reduced. The area of local order is (artificially) determined by the size of the reference applied (see above), and so window size and reference size should coincide (Zane, 1985). (If the reference diameter is *n* times the unit cell size, the corresponding window size is a^*/n, a^* being the lattice constant in reciprocal space.) Windowing according to 1/2 of the lattice constant, i.e. corresponding to a reference size of two unit cells in diameter, "eliminates" 75% of the space between the reflections. If the window size is too restrictive, the resolution will decrease. This filter should be carefully applied since any error concerning the reciprocal lattice vectors, required for producing it, will seriously disturb the CCF. It is, however, a valuable tool for low contrast images in particular (Henderson *et al.*, 1986; Rachel *et al.*, 1986). Figure 8 illustrates the improvement of the CCF by noise reduction.

B. Peak determination

The peak search algorithm needs a global threshold value, i.e. a minimum correlation coefficient to be identified as a peak. If the threshold level is too

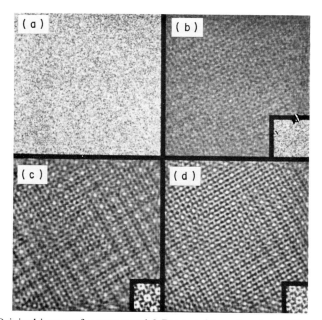

Fig. 8. Original image of a tetragonal 2-D crystal (a, subregion of the whole area; corresponding power spectrum in Fig. 6a) and cross-correlation functions (CCF) obtained from the image in (a) and the references shown as insets (b–d). (b) CCF with an original reference, three unit cells in size, extracted from the source image. (c) Improved CCF obtained by applying the average as a reference, now two unit cells in size. (d) Improved CCF obtained by a subframe of a quasi-optical filtration as a reference and by Fourier filtration of the source image with a window size of one half of the lattice constant. Low- and high-pass filtering was applied throughout (b–d).

low, many side-peaks will be found in addition. If it is too high, significant positions will be ignored. Values commonly applied are two times the standard deviation above the mean level (zero in this case) of the correlation coefficients in the CCF. If too many positions are found, there are further tools to eliminate false peaks later (see below).

The simplest and theoretically the best approach to find the "centre" of a correlation peak is to determine its maximum position. This criterion, however, is sensitive to noise. If noisy CCFs are to be treated, more robust peak search algorithms are required.

Centre of mass. Correlation peaks usually occupy an appreciable portion of the unit cell area. Since peaks are symmetric, an apparently good and noise-insensitive estimate for its maximum position is the centre of mass (centre of gravity). It is determined within a given radius around the peak position initially found, and refined by iteration. The initial radius should not be chosen too small to allow for significant shifts if the "true" peak maximum is some distance away. Appropriate values to start with appear to be 0.1 to 0.3 times the lattice spacing.

Paraboloid fitting. A very precise method for noise-deficient CCFs is to fit a two-dimensional function, i.e. a paraboloid, to an appropriate environment of each peak found (3 × 3 pixels, for example) and to determine its maximum position. This method allows the interpolation of positions to fractions of the interpixel distance. This is particularly valuable with high resolution work, fitting spots in diffractograms, or CCFs of very restrictively sampled images.

E.   Peak selection—selective averaging

Local threshold.   If the global threshold was too low or noise reduction was insufficient, it may be that more than one peak per unit cell area is found. By comparing the heights or masses, respectively, of all peaks located within a certain area (radius to be given, e.g. 1/2 times the lattice spacing) the most significant is selected, deleting the less significant from the peak list. This method is particularly suitable for treating peaks closely located together.

Peak significance. A quite simple criterion for selective averaging is to use the most significant peaks only. This is, in principle, nothing more than global thresholding *a posteriori*. Occasionally the benefit of averaging more homogeneous unit cells may be overcompensated by the loss of statistical significance if the number of unit cells selected for averaging is reduced drastically.

Masking of peak lists.   Peaks located near the edges of either the image or the crystalline area, if it does not fill the whole image, are omitted from the peak list in order not to average over sharp edges or regions free of specimen. How many peaks are to be ignored depends on the particular size of the average (usually two times the unit cell in diameter or, if it will be used as a reference subsequently, at least as large as the reference should be).

Lateral displacement.   Once a complete peak list has been obtained, the actual peak positions can be used to calculate an ideal lattice, closely fitting to the real crystal (by least-squares techniques, using the preliminary estimates of the lattice base vectors **u** and **v**, and the lattice offset vector **w** as starting values; Saxton and Baumeister, 1982a). The *displacements* of the individual unit cells from their expected, ideal lattice points (Fig. 9) may be used as another criterion to eliminate positions, assuming that largely displaced peaks are indicative of some distortions of the corresponding unit cells or of "false" correlations, particularly if they are located at crystal edges.

Local strain.   Displacement maps (Fig. 9) often show patches of about 10 to 20 unit cells strongly correlated and displaced together. The local order within these patches is apparently good, but the unit cells between them are particularly strained by displacements with respect to their nearest neighbours. Saxton and Baumeister (1982b) introduced the local strain criterion to eliminate substantially strained, i.e. potentially distorted,

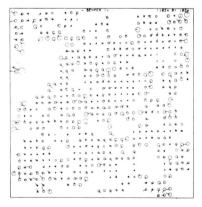

Fig. 9.   Displacement and strain map obtained from the CCF of a tetragonal lattice (according to that of Fig. 8c). The displacement vectors (magnified five times) indicate the positions of the fitted, i.e. ideal lattice points. The circles represent a measure of strain at a given site. Note particularly that the unit cells between the patches of relatively good local order are strongly strained, probably indicating structural distortions.

molecules from averaging. As a simple measure of strain at a given site, the root mean square displacement change between the site and its immediate neighbours (at least one), normalized by the lattice vector length, is used (Saxton and Baumeister, 1982b; see also Engelhardt *et al.*, 1985, for a similar approach).

Rotational disorder—long-range distortions. Besides lateral displacements, rotational deviations from one unit cell to another may occur, albeit largely restricted by molecular interactions within the crystal. With apparently well-ordered crystals orientational variability appears not to be limiting, even at a resolution level of 0.8 nm (Rachel *et al.*, 1986). If, however, large crystals are to be averaged, and that is inevitably the case for high resolution work, slight lattice bending may accumulate to significant angular deviations from one end of the crystal to the other, identified by curved lattices in displacement maps. Two strategies have been applied to compensate for long-range distortions: *lattice unbending* by means of spline algorithms (Henderson *et al.*, 1986) to obtain perfect crystallinity, and *local averaging* of a few (e.g. four) subregions, followed by rotational alignment and final addition of the correctly oriented averages (Rachel *et al.*, 1986).

Badly distorted crystals displaying substantial lateral and rotational disorder within small areas of, say, about 20 to 50 unit cells are hardly suitable for lattice unbending techniques. Here local averaging of smaller crystal patches, correction for the angular deviations and final addition of the aligned averages is to be preferred (if better preparations are not available). Hegerl and Baumeister (1988) found that even the normal correlation averaging procedure is safe enough to exclude seriously rotated unit cells from averaging.

Correspondence analysis. While peak selection, as outlined above, is predominantly suited to selecting positions which belong to a lattice, and to reducing the heterogeneity among the unit cells to be averaged, correspondence analysis is capable of *analysing* the differences between them and sorting the unit cells with respect to their similarity, or differences, respectively, in multidimensional space (Frank and van Heel, 1982; van Heel, 1984; Bretaudiere and Frank, 1986). Subsequent classification and selective averaging of images combined in the various classes reveal the different features. The peak positions obtained from the CCF are used to extract either small images comprising one morphological complex, neighbouring complexes masked off (Engelhardt *et al.*, 1985), or to extract the unit cells resampled onto a square lattice (Engelhardt *et al.*, 1986b; Bingle *et al.*, 1987). The single images are then applied to correspondence analysis. If systematic differences do not exist, classification can be used to exclude those unit cells from averaging, exhibiting particularly strong variability with respect to the majority of the unit cells. Related techniques, aiming at identifying "atypical" molecules, have been described recently (Unser *et al.*, 1986).

Averaging of tilted projections. In principle, averaging of tilted views is performed in the same way as averaging of untilted projections, except that symmetrization cannot be applied here. However, projections of the highly tilted specimen exhibit a significant focus gradient which has to be taken into account. Averaging should, therefore, be restricted to stripes parallel to the tilt axis azimuth ranging from minimum to maximum tolerable underfocus.† The CTF, which depends on the imaging conditions and parameters of the electron microscope (Misell, 1978), should not be limiting with respect to the expected resolution. With tilts above 50° to 60° (or even less, depending on the image size) the images or the peak lists have, therefore, to be masked off for regions outside the tolerable focus conditions. The focus gradient may be judged by means of the power spectra of small subframes extracted from the image. Reconstructions of unstained material, imaged at appreciable defocus to produce sufficient contrast, and/or highly tilted views of large crystal patches that are not to be masked off particularly aiming at high resolution, require a *restoration* of the CTF, correcting for the contrast reversal of certain spatial frequencies (here the reader is referred to Lepault and Pitt, 1984, and Henderson *et al.*, 1986, for recent applications).

F. Resolution assessment

The two independent averages ("odd" and "even", Figs 7 and 10) are extracted and smoothly masked to comprise one morphological complex (unit cell) each. Corresponding Fourier coefficients of the two transforms representing identical spatial frequencies (on rings) are summed according to the function of the resolution criterion applied (see below) and displayed as a function of the spatial frequency. The intersection of the experimental curve with a threshold level function is taken as a measure of resolution (Fig. 11). Recently a resolution criterion for 3-D reconstructions was also suggested by van Heel and Harauz (1986).

Phase residual. This criterion compares the phase differences $\delta\theta$ between the Fourier coefficients F_1 and F_2 of the two independent averages (Frank *et al.*, 1981a). The coefficients of each spatial frequency under scrutiny are combined

† If a focus gradient Δf perpendicular to the tilt axis azimuth is tolerable, the maximum diameter d of the projected specimen suitable for averaging and reconstruction without correction of the CTF is $d = \Delta f/\tan\psi$, ψ being the tilt angle. Assuming that a gradient of ± 200 nm, i.e. $\Delta f = 400$ nm, applies, d equals 230 nm at 60°, 145 nm at 70° and 70 nm at 80°, respectively.

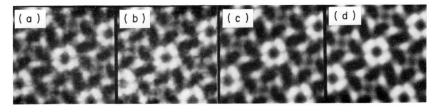

Fig. 10. Correlation averages of a negatively stained bacterial surface protein array. (a, b) Averages of "even" and "odd" numbered unit cells as used for resolution assessment. (c) Cumulative (final) average of (a) and (b). (d) Four-fold symmetrized final average (note the improvement of perceptibility of structural detail, i.e. resolution, in c and d).

by summation according to

$$\Delta\theta = ([\sum (|F_1| + |F_2|)\,\delta\theta^2]/[\sum (|F_1| + |F_2|)])^{1/2} \qquad (9)$$

A phase residual of $45°$ is normally used as a measure for resolution.

Fourier ring correlation. The radial correlation function (Saxton and Baumeister, 1982a; van Heel, 1987b) compares the Fourier components (F_2^* is the complex conjugate), summed over rings of identical spatial frequencies, by cross-correlation according to

$$RCF = [\sum F_1 F_2^*]/[\sum |F_1|^2 \cdot \sum |F_2|^2]^{1/2} \qquad (10)$$

A value of $2/\sqrt{n}$, n being the number of independent pixels per Fourier ring, is normally used as the significance level of resolution.

Radermacher *et al.* (1986) and van Heel (1987b) discuss the properties of these two and some closely related resolution criteria. Generally the radial

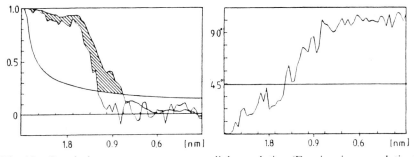

Fig. 11. Resolution assessment curves, radial correlation (Fourier ring correlation) function (left) and phase residual criterion (right). The hatched area corresponds to a statistical improvement of the resolution as achieved from averaging of about 2000 molecules instead of *c.* 500. (Courtesy of R. Rachel, Martinsried.)

correlation function suggests somewhat more optimistic resolution values (Fig. 11); this, however, depends on the threshold level applied. It is hardly possible to give an absolute resolution measure; but whatever criterion is used, it should be stated to make the comparison of resolution values more reliable.

Resolution of symmetric averages. The correctly centred, cumulative (final) average of $0°$ projections of symmetric specimens may be applied to another resolution test, now comparing the original average with a copy rotated about the symmetry angle (Saxton and Baumeister, 1982a). This comparison checks the symmetry assumed, and if it applies, the isotropy of the resolution (here the number of averaged unit cells has to be taken into account). Accidentally but seriously tilted "$0°$ projections" may be identified by a loss of resolution with respect to the initial assessment.

Signal-to-noise ratio. The signal-to-noise ratio (S/N) is calculated according to

$$S/N = \rho/(1 - \rho) \qquad (11)$$

where ρ denotes the cross-correlation coefficient between the independent averages (Frank and Al Ali, 1975). Accordingly, the radial correlation curve is a measure of the signal-to-noise ratio as a function of the spatial frequency.

Resolution gaps. Occasionally resolution assessment curves indicate a "temporary" loss of resolution significance. Frequently one of the following explanations applies:

• Staining gradients (differently accumulated in the two images) creating a characteristic resolution gap at spatial frequencies $\geq 1/a*$. Such gradients are removed by high-pass filtering (Section V.C).

• Radiation damage, occasionally observed with negatively stained membranes, causing a loss of resolution in the range of 2–3 nm (Stark *et al.*, 1984).

• Appreciable underfocus (Fig. 6), creating a resolution loss according to the gaps in the CTF. Beyond the gap(s) significant Fourier components, now possessing reversed phases, may correlate again.

G. Symmetrization

According to the symmetry properties of the average, symmetrization is performed by rotation about the symmetry centre (for $p2$, $p3$, $p4$ and $p6$ symmetries) and/or by reflection at the mirror axes (all the other symmetry groups, except for $p1$). For symmetrization purposes the axis of symmetry (the mirror axis) must be correctly centred; the axes can be conveniently determined by correlation techniques (Frank, 1980).

Rotational symmetry. To determine the centre of centrosymmetric structures, proceed as follows:

- obtain a copy of the image to be centred, rotated about 180° if $p2$-related symmetries apply, or rotated about 120° in case of $p3$ symmetries;
- mask smoothly off regions outside the apparently central morphological complex;
- cross-correlate the rotated with the original copy;
- and find the position of the correlation peak maximum (x, y) by means of peak search algorithms as outlined in Section V.D.

The symmetry centre (x_c, y_c) with respect to the image centre is given by

$$x_c = x/2 \qquad y_c = y/2$$

With $p3$ symmetric images the symmetry centre is calculated by

$$x_c = (x + y\sqrt{3})/2 \qquad y_c = (y - x\sqrt{3})/2$$

or by

$$x_c = (x - y\sqrt{3})/2 \qquad y_c = (y + x\sqrt{3})/2$$

if the copy was rotated to $-120°$.

The rotated copies are produced from the correctly centred average and finally added. In order not to accumulate interpolation errors upon successive rotation, each rotated copy is derived from the original (centred) average. Symmetrization can also be performed in Fourier space by averaging the Fourier components (see Crowther and Amos, 1971).

Mirror symmetry. If rotational symmetry exists in addition to mirror symmetry, centre and symmetrize the image first. Now find the axes of mirror symmetry:

- Rotate the image such that one of the lattice base vectors becomes parallel to the x-axis ("source image"). The orientations of the mirror axes are now easily identified (Fig. 8).

- Reflect the source image about an axis parallel to the mirror axis and mask the copies smoothly off outside a radius of about 1.5 times the lattice spacing.

- Cross-correlate the reflected with the source image and find the position of the correlation peak (x, y).

- Centre the mirror axis by shifting the source image about $x/2$ and $y/2$.

Now obtain the reflected images and add the copies. Alternatively, symmetrization may be done in Fourier space (e.g. Fuller, 1981).

VI.   3-D reconstruction—the hybrid real space/Fourier space approach

A.   Reconstruction procedure

The hybrid real space/Fourier space approach (Engel and Massalski, 1984; Saxton *et al.*, 1984) starts with averages of the tilted views instead of (Fourier transforms of the) original images (Amos *et al.*, 1982). Thus, the benefit of correlation averaging is exploited, too, for 3-D reconstructions. As outlined in Section III.A, the lattice line data are obtained from the Fourier transforms of the appropriately resampled unit cells. The lattice base vectors as well as the tilt parameters are required to extract and resample the unit cells and to insert the Fourier components into the lattice line data set. Data of symmetry-related lines are combined. A quasi-low-pass filtering is applied by excluding all lattice lines beyond an expected resolution. Aiming at an almost isotropic resolution, high frequency information of the lattice lines are "cut off" accordingly. The unit cells must be aligned with respect to each other, i.e. they have to be centred to their common origin. Correlation of successive projections, starting with the untilted view and proceeding to the next in turn, a method that is commonly used, fails with tilt angles $>60°$. Saxton *et al.* (1984) describe an improved procedure developed for highly tilted views ("iterative alignment cycle" in Fig. 12). At this step the lattice line data, particularly the moduli of the Fourier coefficients, still scatter appreciably. This may be due to density differences among the various projections introduced by variations in exposure or development, etc.; a normalization procedure partly corrects for these variations. The "raw" (aligned and normalized) data are fitted and interpolated by least-squares techniques to obtain continuous, regularly sampled lattice line "curves" for the moduli and phases. Now the most important steps of a 3-D reconstruction have been completed, the following procedures are only necessary to convert the reciprocal data into real space and into interpretable representations. The lattice line data are three-dimensionally back-transformed to a "unit cell cube" and *horizontal slides* and *vertical sections* are generated for documentation (Fig. 12). For a final representation a 3-D model is built either by means of the computer ("surface shading"; Radermacher and Frank, 1984; Saxton, 1985) or by hand (using balsa-wood and a saw). Both modelling approaches require the layer slices thresholded, i.e. a distinct border between protein and negative stain, or whatever the specimen is surrounded with, must be defined. Unfortunately, this information is lost due to the "missing cone" effect, which

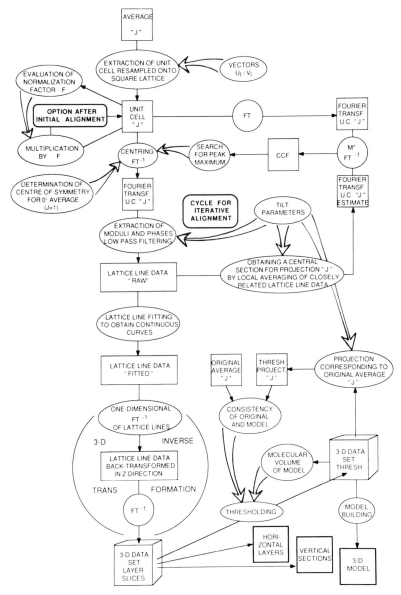

Fig. 12. Scheme of the 3-D reconstruction procedure, the hybrid real space/Fourier space approach. "J", number of projections from a continuous tilt series; CCF, cross-correlation function; other symbols as in Fig. 7.

removes the Fourier components of, at the least, the 0, 0 lattice line, which would otherwise provide the threshold levels. Thresholding is therefore an attempt to *assess* the borders, using constraints such as the molecular volume, the thickness of the reconstruction, etc., as guides. In the following the various steps of a 3-D reconstruction are described.

B. Tilt parameters

The actual lattice base vectors $\mathbf{u_i}$, $\mathbf{v_i}$ are used to extract the unit cells from the corresponding averaged projections and to calculate the tilt parameters required: (i) the original, untilted vectors $\mathbf{u_0}$, $\mathbf{v_0}$; (ii) the tilt axis azimuth (ϕ); (iii) the tilt angle (ψ); and (iv) the inclination of the specimen around an axis perpendicular to the tilt axis (α). With a continuous tilt series, the parameters can be determined particularly accurately by means of a least-squares fit, if the tilt angle increments are exactly known from the goniometer readings. Saxton *et al.* (1984) derived a matrix (R_{ij}) which describes the transformation of a lattice base vector upon tilting and projection in the microscope. The vector $\mathbf{u_i}$ with respect to the specimen is transformed to $\mathbf{u_i'} = R_{ij} \cdot \mathbf{u_j}$ with respect to the microscope, i.e. the projection. Matrix R_{ij} is obtained from

$$R_{ij} = T_{ij} I_{ij} \qquad (11)$$

and

$$I_{ij} = \begin{pmatrix} \cos^2 \phi \cos \alpha + \sin^2 \phi & \sin \phi \cos \phi (\cos \alpha - 1) & -\cos \phi \sin \alpha \\ \sin \phi \cos \phi (\cos \alpha - 1) & \sin^2 \phi \cos \alpha + \cos^2 \phi & -\sin \phi \sin \alpha \\ \cos \phi \sin \alpha & \sin \phi \sin \alpha & \cos \alpha \end{pmatrix}$$

$$T_{ij} = \begin{pmatrix} \sin^2 \phi \cos \psi + \cos^2 \phi & \sin \phi \cos \phi (1 - \cos \psi) & -\sin \phi \sin \psi \\ \sin \phi \cos \phi (1 - \cos \psi) & \cos^2 \phi \cos \psi + \sin^2 \phi & \cos \phi \sin \psi \\ \sin \phi \sin \psi & -\cos \phi \sin \psi & \cos \psi \end{pmatrix}$$

Before calculating the tilt parameters it is useful to check the actual vectors of a continuous tilt series for consistency. Figure 13 shows a plot of the vector components that should lie on straight lines perpendicular to the tilt axis. If a vector (pair) substantially deviates from the lines (see Section IV.C for possible reasons), it is excluded from the least-squares fit of the tilt parameters, and the corresponding unit cell is handled like an independent projection.

A single pair of vectors is sufficient to calculate the apparent values of ϕ and ψ (containing the information on α which cannot be separated) and the corresponding untilted vectors $\mathbf{u_0}$ and $\mathbf{v_0}$, using the equations of Shaw and Hills (1981), provided that the (symmetry) angle between the vectors and their relative or absolute lengths are known. The estimation of the parameters from low tilt projections ($\leq 20°-30°$), however, suffers from small errors concerning $\mathbf{u}$ and $\mathbf{v}$ (Amos *et al.*, 1982; Engel and Massalski, 1984). Aiming at high

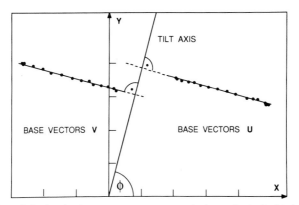

Fig. 13. Plot of the lattice base vector components (y against x) obtained from the images of a continuous tilt series; ϕ, tilt axis azimuth in plane.

resolution, say, clearly better than 1.5 nm, it is recommended that a short continuous series comprising 3 to 4 tilts is recorded in order to determine the tilt parameters precisely for this single projection, going into the 3-D reconstruction afterwards. To make independent projections compatible with a given set of tilts, the equivalent basic untilted vectors are aligned by rotation, and the tilt axis azimuths are corrected for the rotation angle.

C. Extraction of unit cells

The unit cells are extracted from the pre-centred (by hand) averages using the original lattice base vectors $\mathbf{u_i}$, $\mathbf{v_i}$. Each unit cell is resampled onto a square lattice as outlined in Section III.A (Fig. 14). To obtain an appropriate representation of the unit cells in Fourier space, the real space vectors $\mathbf{u_i}$, $\mathbf{v_i}$ should derive from untilted vectors 120° apart for $p6$, 90° for $p4$ and 60° for $p3$ symmetry, respectively. The size of the resampled unit cell is of some importance with respect to the accuracy needed for the subsequent alignment. If the correlation peak position is not interpolated to fractional intervals, the correlation is only of the accuracy to a spatial frequency according to the lattice line of order $n \leq s/8$, where s denotes the unit cell size (in pixels) in the plane, i.e. a resolution of $a \cdot \sin \gamma^*/n$.†

D. Alignment, normalization and lattice line fitting

Transformation into lattice line form.   The Fourier components are read out from the Fourier transforms of the resampled unit cell projections and

† The accuracy necessary is approximately $\pm a \cdot \sin \gamma^*/16n$ (i.e. a phase accuracy of $\pm 22.5°$), a being the unit cell dimension and n the highest lattice line order to be recovered (Saxton *et al.*, 1984).

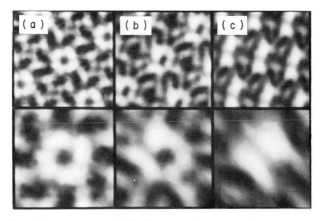

Fig. 14. Correlation averages of tilted views of a tetragonal protein crystal and the corresponding (enlarged) unit cells resampled on a square lattice. Tilt angles 0°, 51° and 70°, respectively.

compiled in a lattice line data set. At this step the plane group characteristics are introduced by selecting only those lattice lines corresponding to a certain plane group (Table III), and by taking all symmetry-related components. (Alternatively, lattice lines may be selected later, when the structure is to be inversely transformed.) The positions z^* of the various Fourier components are determined by geometrical considerations, i.e. by means of the lattice line order (h, k), the (untilted) reciprocal lattice base vectors $\mathbf{u}_0^*, \mathbf{v}_0^*$, and the tilt parameters α, ϕ, ψ_i, using the transformation via equation (12) (Saxton *et al.*, 1984). A quasi-low-pass filter is applied by considering only lattice lines to an expected resolution of $1/(a^* \cdot h_{\max})$.

Alignment to a common origin. As well as the accurate superposition of the unit cells in the course of correlation averaging of each projection, the exact alignment of the unit cells of a series to a common origin is one of the most important steps in the course of a 3-D reconstruction. Saxton *et al.* (1984) studied various procedures (centre of mass, cross-correlation of all images with the untilted projection, cross-correlation of successive projections with each other) and recommended the following improved method for series comprising tilts beyond 60° in particular (Fig. 12):

(1)   centre the apparently untilted and resampled unit cell ($i = 1$) first (see Section V.G) and transform it into lattice line form;

(2)   remove the sampling data from the next projection ($i + 1$) from the lattice line data set;

(3)   obtain a central section only for this projection ($i + 1$) by local

averaging of closely related Fourier components in the z^*-direction in the various lattice lines ("estimate", Fig. 12);

(4) cross-correlate it with the image to be aligned, find the correlation peak maximum (see below) and centre the image $(i+1)$;

(5) now transform it into lattice line form and exchange the former Fourier components from the projection $(i+1)$ by the new values;

(6) proceed to Step 2 until all projections are aligned;

(7) refine the alignment by a new cycle, again starting with the lowest tilt $(\neq 0°)$ until the result remains consistent (2–4 cycles are normally sufficient).

The cross-correlation function will usually show a broad peak. To improve the peak profile the moduli of one or both Fourier transforms are set to 1, which gives a uniform weighting of all spatial frequencies and therefore a sharp correlation peak (Saxton et al., 1984). This approach is noise-sensitive and should only be applied to substantially noise-limited images (like the averaged unit cells and the estimated projections here).

Normalization.   The relative normalization of the unit cells, i.e. the moduli of their Fourier transforms, reduces the scattering of the lattice line sampling data in the z^*-direction. The transform modulus of each projection is matched to the modulus of an estimated lattice line section such as was used for alignment. The factor F_i is given by

$$F_i = \sum (|f_i| \cdot |f_i'|)/\sum |f_i'|^2 \qquad (13)$$

where f_i denotes the Fourier components of the estimated section and f_i' those of the projection to be normalized. The sum is taken over all components existing simultaneously in both transforms, excluding the central $(0,0)$ component (Saxton et al., 1984). The procedure in detail:

(1) obtain the Fourier transforms of the projection and the estimated section and determine the common origin as outlined above;

(2) take the moduli of the Fourier transforms $(|f_i|, |f_i'|)$ and set the central pixels to zero;

(3) make a copy of the estimated section $(|f_i|)$ and create a *mask* selective for components $\neq 0$ by thresholding (minimum value $= 0$, maximum value $= 1$);

(4) calculate the products $|f_i| \cdot |f_i'|$ and $|f_i'|^2$ and multiply each by the mask;

(5) evaluate the mean values of the products and calculate the normalization factor using equation (13);

(6) now multiply the aligned image by F_i and proceed to the next projection $(i+1)$;

(7) refine the normalization by a new cycle if necessary (2–4 cycles are sufficient).

The recording and refinement of the lattice line data is now completed; Fig. 19 illustrates how densely the reciprocal space is filled with measured data and which lines contain the dominant Fourier components.

Smoothing and interpolating the lattice line data. To recover the specimen density distribution by 3-D inverse transformation, the lattice lines are smoothed in addition to give a *continuous line* interpolated so as to provide regularly spaced sampling points. Smooth lines may either be drawn by hand or automatically fitted by least-squares methods. Often the hand-fitted curves are very good estimates of the solution found by automatic algorithms, and the structures are apparently indistinguishable (Baumeister *et al.*, 1986a). The more objective, constrained least-squares approach of Shaw (1984) is useful for lattice lines where there are sampling gaps or appreciable scattering of data (particularly with less significant lines). The algorithm minimizes the second derivative of the estimate and, therefore, normally finds curves that are smooth enough to avoid artificial high frequency modulations in the 3-D structure. Occasionally it may be necessary to smooth wildly oscillating curves by hand as well. Lattice line data are frequently displayed in *modulus/phase form*, but for fitting purposes the *real/imaginary form* (Fig. 15) has proved to be advantageous (Saxton *et al.*, 1984) since the two (independent) parts may be treated separately. The lattice lines are now interpolated so as to provide a fine enough sampling in real space for data documentation afterwards.

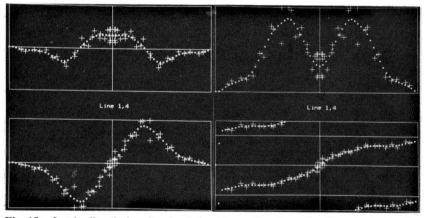

Fig. 15. Lattice line (indexed as $h = 1$, $k = 4$) sampling data of a tilt series (*p4* lattice) comprising tilts from $0°$ to about $75°$. Left, real/imaginary form; right, modulus/phase form (the phase is given in the range of $-3\pi/2$ to $+3\pi/2$). The sampling data of the phases scatter significantly and characteristically less than the moduli components. Crosses, sampling data of an aligned and normalized lattice line data set; small dots, regularly spaced, fitted lattice line data. The horizontal axis corresponds to the z^*-direction.

E. Calculation of the structure

Three-dimensional inverse transformation from the lattice line data can either be performed in one step, producing a unit cell cube (Fourier transform/real space representation), or the lattice lines can be individually back-transformed followed by the inverse transformation of each layer slice in turn ("horizontal sections"). The procedures are equivalent and depend on the basic data organization and programs used in the image processing system. At this point the appropriate lattice lines corresponding to a certain space group must be selected (see Section VI.4). *Horizontal layers* (Fig. 16) and *vertical sections* (Fig. 19) are produced at reasonable spacings to follow the density variations in adjacent layers closely. Interlayer spacings in the range of 0.2 to 0.6 nm are suitable for an informative representation of 3-D models (models produced via surface shading techniques require 20–40 layer slices to give a reasonable representation of fine detail). The layer slices, containing one or a few unit cells in a plane, are now inversely resampled (if necessary) to obtain the original "undistorted" representation of the structure. The inverse vectors $\mathbf{u}', \mathbf{v}'$ are given by

$$\mathbf{u}' = (v_y/d, -u_y/d) \qquad \mathbf{v}' = (-v_x/d, u_x/d), \qquad \text{where} \qquad d = (u_x \cdot v_y - u_y \cdot v_x) \tag{14}$$

Absolute handedness of structure. There is a wealth of possibilities to invert the handedness of a structure in the course of a 3-D reconstruction: from the orientation of the grid in the microscope to the production of copies for documentation. All these steps may be controlled; an important point is to ensure that the tilt angles are recorded with the correct sign. Amos *et al.* (1982) describe a suitable procedure to check by means of the focus gradient in the micrograph in which direction the specimen was originally tilted. Alternatively, the absolute handedness may be derived from independent experiments such as surface relief reconstruction starting with heavy-metal shadowed preparations (Guckenberger, 1985; see Section VIII).

F. Thresholding and model building

Thresholding, i.e. finding distinct density values that represent the apparent border between the structure and the surrounding medium in each horizontal layer, is the most critical and least objective step. The 3-D model built from thresholded layers stacked upon each other, however, is the result that is normally paid most attention to. Constraints used for model building, thus, should be stated clearly.

Kittler *et al.* (1985) reviewed automatic thresholding algorithms and introduced a novel one designed for any kind of image. The criteria normally used are local and global image statistics, grey level histograms, algorithms for edge detection, etc., which in principle are also valid for the individual layer slices of a 3-D reconstruction; the Kittler algorithm indeed produced results similar to those found by constraints listed below (Woodcock *et al.*, 1986). Since the layer slices are not independent of each other but show the density distribution of a structure continuing in the z-direction, and since often additional information on its molecular structure is available, some useful additional constraints may be introduced.

Thickness of the reconstruction. The layer slices do have significant contrast where the structure is contained, fading out to a nearly uniform density distribution above and below the unit cell. Frequently used measures for the density variations are: (i) the difference between the maximum and minimum pixel value; and (ii) the standard deviation of the density distribution, each plotted against the distance of the corresponding layer from the central plane (Fig. 16).

Steepness of density gradient. A criterion that is reminiscent of "edge detection" considers the steepness of density gradients, adequately displayed

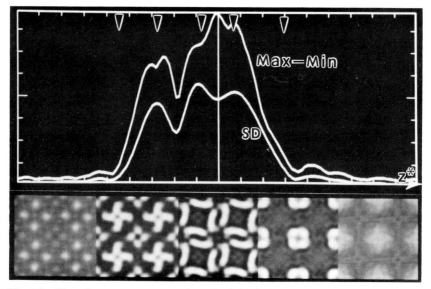

Fig. 16. Plot of contrast measures (standard deviation, difference of maximum and minimum density values) of the horizontal layer slices of a 3-D reconstruction to identify the approximate thickness of the reconstruction. The arrows indicate the positions of the horizontal layers displayed below (two unit cells in size); bright features denote the structure (bacterial surface layer).

in contour plots of the layers (see Fig. 19). The threshold level is chosen so as to fit into the region of steepest gradient. This approach is particularly successful for layers with high contrast and provides results similar to those offered by automatic procedures, but it is quite ambiguous for the top and bottom layers comprising an almost uniform density distribution and very flat gradients (Fig. 16). Here even small differences in the threshold levels may create dramatic changes in the resulting structure.

Connectivity.   Significant structural domains known to be continuous should appear so in the model. "Floating" masses either indicate a very high threshold level, peeling off the structure, or "reconstruction-induced blobs" (Baumeister *et al.*, 1986a) due to artefacts introduced by low tilt reconstructions in particular (see Section VII). Those blobs need not be connected to the structure proper.

Molecular volume.   The apparent volume of the 3-D model is calculated (by counting the pixels representing the structure in each layer and calculating the occupied volume) and compared with the expected molecular volume, derived from molecular weight determinations. The apparent molecular weight (M_r) is estimated from

$$M_r\,[\text{kD}] \approx 0.6 \cdot d \cdot \text{Volume}\,[\text{nm}^3] \qquad (15)$$

where d denotes the value of the relative density of the biological material ($d \approx 1.3$ for partially hydrated to $\approx 1.37\,[\text{g cm}^{-3}]$ for apparently dry proteins; Richards, 1977). The proportion of the mass represented may be used as a guide to correct the tendency of thresholding. Most 3-D models present only 50–75% of the expected volume (Baumeister and Engelhardt, 1987) and are sometimes not far from a "backbone" representation. This, however, often gives the misleading impression of a relatively open structure (Fig. 17). Thus, it is mandatory to state clearly the proportion of mass occupied by the model.

Threshold curve.   Although no general rules exist concerning the shape of a threshold curve (i.e. the threshold values plotted as a function of the layer "number"), some features are obvious. Threshold curves: (i) do not possess discontinuities; (ii) do not oscillate wildly, particularly if the reconstruction was oversampled in the z-direction with respect to the resolution achieved; and (iii) threshold curves often appear to be cup-shaped with high values where the structure fades out (to obtain a smoothly bounded model).

Consistency of model and projections.   A valuable tool is the consistency test, i.e. appropriate projections through the thresholded 3-D data set are calculated and compared with the original views (Fig. 18; Engelhardt *et al.*,

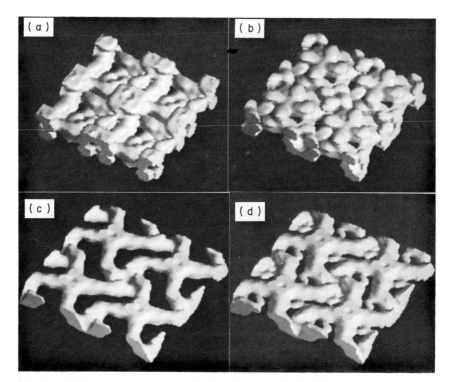

Fig. 17. 3-D representations obtained by "surface shading" in the computer. (a, b) Views of the inner and the outer face of the surface protein layer of *Sporosarcina ureae* (for details see Engelhardt *et al.*, 1986a). (c, d) 3-D models of the *Azotobacter vinelandii* surface protein layer displayed with 68% and about 90% of the expected molecular volume. While the gross morphology remains similar small domains are removed in the 3-D model on the left, creating the impression of a relatively open structure. (For details see Bingle *et al.*, 1987.)

1986a). If structural domains were accidentally overemphasized or deleted by thresholding, the comparison enables the experimenter to identify inconsistent image regions and to correct for the aberrations. Comparing the projections, one must keep in mind that image features introduced by the CTF are largely missing in the projected 3-D reconstruction. This approach has been incorporated into an automatic thresholding procedure, using the global cross-correlation coefficient as a measure for similarity which is to be maximized (Guckenberger, unpublished).

Experimental constraints. With negatively stained preparations, only those structural details can be expected to be recovered in the 3-D reconstruction that are, at least partially, embedded in the staining (contrast-producing)

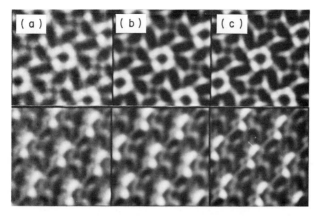

Fig. 18. Projections of a 2-D crystal untilted (top row) and tilted to 70° (bottom row) obtained by correlation averaging (a) and from the 3-D reconstruction non-thresholded (b) and thresholded (c). The projection in (c) appears more slender (only 68% of the volume is represented), but each structural domain is displayed with approximately the same relative intensity as in (a) and (b).

material. Hence proteins or molecular domains buried in the lipid phase of a membrane are *invisible* and should not appear in the reconstruction (see Section VII). Only if they contribute to the image contrast, e.g. in unstained, frozen-hydrated or density-matched preparations, can they be reconstructed.

VII. Reconstruction artefacts

A. The missing cone problem

Besides possible artefacts introduced by the preparation of the specimen (e.g. flattening upon adsorption on the specimen support) and electron microscopy (e.g. radiation damage), which are briefly discussed elsewhere (Baumeister and Engelhardt, 1987), the quality of the reconstruction itself largely depends on the completeness of the recorded data. The larger the gap of missing data in Fourier space, i.e. the smaller the range of tilting, the more aggravating are the artefacts in the 3-D reconstruction (Fig. 19). Three sources of problems are due to the missing cone: (i) anisotropy of resolution; (ii) artificial density modulations; and (iii) missing mean values in the layer slices. The consequences for the 3-D reconstruction and the interpretability of the model are numerous.

Anisotropy of resolution. The "resolution element" (image point; Hoppe and

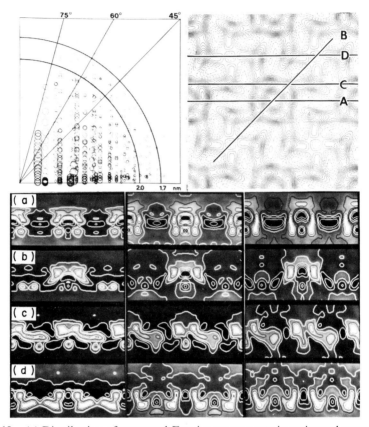

Fig. 19. (a) Distribution of measured Fourier components in reciprocal space. The area of each circle indicates the modulus of the Fourier coefficient. The abscissa corresponds to the distance r^* of the lattice lines from the image centre ($r^* = \sqrt{(x^{*2} + y^{*2})}$), and the ordinate is the z^*-axis. The resolution spheres at 2 nm and 1.7 nm are indicated as well as the missing data (lines) if tilting was restricted to 60° or 45°, respectively. (b) Vertical projection (contour lines plot) through the 3-D reconstruction of the specimen as shown in Fig. 14; the positions of the vertical sections (below) are indicated. (c) Vertical section obtained from 3-D reconstructions comprising tilts to 75° (left), 58.5° (middle) and 47° (right), respectively. Note the structural alterations introduced by low tilting. (From Baumeister and Engelhardt, 1987, with permission of the publisher).

Hegerl, 1980) and concomitantly the whole structure is elongated in the z-direction, whereby the elongation is an inverse function of the highest tilt angle. As an obvious consequence the reconstruction becomes thicker (Fig. 19) and the apparent reconstructed molecular volume is closer to the expected value. Structural details which are not or only weakly connected in the z-direction tend to become continuous (Fig. 19). Negatively stained domains of integral membrane proteins become elongated into the blind region of the membrane matrix and, in particularly bad cases, are rigorously made continuous across the membrane. Though this may happen to "reflect" the real situation, the "masses" obtained are by no means reconstructed but artificially created and must not be taken as a real structure. Serious consequences are to be expected if the first lattice line $(1,0)$, which suffers particularly strongly from limited tilting, contained the most dominant Fourier components (Baumeister et al., 1986a), and if the crystal possesses a large lattice constant (for geometric reasons†).

Artificial density modulations. Density modulations in Fourier space introduced by restricted tilting ("clutter"; Hoppe and Hegerl, 1980) may appear so strongly in the reconstruction that they cannot be separated from the real structure by thresholding. These spurious features appear to be particularly dominant on symmetry axes in the z-direction, creating the "reconstruction-induced blobs" that occasionally float over pores, for example (frequently situated in symmetry centres), or close them altogether. Evidently artificial density modulation is a more serious problem than resolution anisotropy, and it produces significant aberrations even with tilting up to 60°! (Fig. 19; Baumeister et al., 1986a).

Missing mean values in layer slices. The Fourier components responsible for the mean density values in the layer slices are located on the $0, 0$ lattice line and, hence, are completely missing, even with high tilts. Thus, all layer slices have the same mean density (i.e. zero) and it is left to the experimenter to find suitable threshold densities in calculating a 3-D model (Section VI.F).

B. Approaches to overcome the missing cone problem

A pragmatic though not yet completely successful approach is to collect an as complete data set as possible, i.e. comprising tilted views far beyond 60° (75° to $\geq 80°$). This will significantly reduce the problems of artificial density modulations and elongation of the point response, but cannot solve the thresholding problem. Some additional information, however, can be obtained from other experiments. Thus, Henderson and Unwin (1975) used X-ray diffraction patterns to assess (some of) the missing data in the $0, 0$ lattice line. Amos et al. (1982) suggest that profile projections from folded membranes are included (an approach that suffers, however, from distortions introduced by folding), or vertical sections through the negatively stained specimen, a technique that was experimentally explored by Jesior (1982).

---

† Assuming that one lattice base vector is (nearly) perpendicular to the tilt axis, the lattice line of order $1,0$ comprises data to a resolution of $r_{1,0} \geq a^*/\tan \psi_{max}$, where a^* denotes the lattice constant and ψ_{max} the highest tilt angle.

Recent developments in image processing deal with *data extrapolation* using constraints such as the specimen thickness (Agard and Stroud, 1982) or, even better, surface reliefs derived from heavy-metal shadowing experiments (Baumeister *et al.*, 1986a; Barth, 1987). Approaches based on optimization via the maximum entropy principle (Skilling and Bryan, 1984) and recursive projection algorithms (Gerchberg, 1974) are currently explored (Barth, 1987; Carazo and Carrascosa, 1987) and promise (partly) to fill the data gap left by the experiment.

VIII.   Surface relief reconstruction

A complementary though not completely alternative method to yield (quasi) 3-D information on a (regular) specimen is surface relief reconstruction (Smith and Kistler, 1977; Guckenberger, 1985) of *unidirectionally* heavy-metal shadowed preparations such as freeze-etched biological objects. Although the resolution achievable is limited to about 2 nm, for experimental reasons rather than by image processing relief reconstruction possesses some valuable features:

- The surface reconstructed is not adsorbed to the carbon support and, hence, can be assumed to be undistorted (Baumeister *et al.*, 1986b).

- The two surfaces are reconstructed individually and independently, which is advantageous if substantial deviations from regularity or symmetry must be taken into account (disabling a straightforward 3-D reconstruction even by means of correlation techniques) (Engelhardt *et al.*, 1985, 1986b).

- Only one untilted view of each surface is required for specimens with symmetries higher than two-fold, or (at least) two for *p*1 or *p*2 symmetric objects. Thus, relief reconstruction is very fast (one day from digitization to documentation).

The rationale of the method is that the density distribution in the micrograph of a shadowed specimen can be described in terms of the first derivative of the two-dimensional surface profile function and that the relief can, therefore, be recovered by integration. Smith and Kistler (1977) developed a method based on integration in real space, while Guckenberger (1985) introduced the Fourier space approach, which possesses some advantages for periodic specimens in particular. Here the reconstruction is performed by applying a Fourier (Wiener) filter. The theory, basic algorithms and special features, along with various applications to periodic specimens, are described in detail

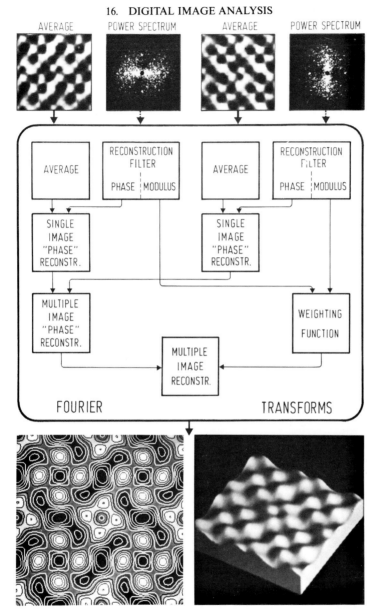

Fig. 20. Scheme of surface relief reconstruction from unidirectionally heavy-metal shadowed specimens. Top: averages (original and a copy rotated by 90°) and the corresponding power spectra. Bottom: surface relief (contours indicate heights, black areas are valleys) and 3-D representation by "surface shading" in the computer (reproduced with permission of the publisher). The relief represents the outer face of the surface layer of *Sporosarcina ureae*, whose 3-D reconstruction is shown in Fig. 17(b) (from Baumeister *et al.*, 1986b, with permission of the publisher).

elsewhere (Engelhardt *et al.*, 1985; Guckenberger, 1985; Baumeister *et al.*, 1986b). The basic strategy, illustrated in Fig. 20, is as follows:

- The image of an untilted view of the shadowed specimen is applied to correlation averaging as outlined in Section V.

- The *shadowing direction* with respect to the 2-D crystal (lattice base vectors) is determined by means of the power spectrum of the image and/or the average, identified as a "zero line" perpendicular to the shadowing direction (Fig. 20).

- The *asymmetry centre* of the average, which corresponds to the symmetry centre of the non-shadowed specimen, is determined and the image centred accordingly.

- The Fourier filter function is created (Wiener filter) such that the real part of the Fourier transform of the average is suitably weighted (Fourier components on the "zero line" are eliminated) and the imaginary part is corrected by a "phase filter" upon multiplication.

- The average is rotated about the symmetry angle (if $|\gamma| \leq 120°$) and the filtration is performed again. The rotated and individually phase-filtered Fourier transforms are combined, thereby filling the zero line with data from the rotated copies, and multiplied by the weighting function. The "multiple image reconstruction" (Fig. 20) is now re-transformed.

The relief may be calibrated in the z-direction, provided that areas free of heavy metal (true shadow casts) are available (Engelhardt *et al.*, 1985; Guckenberger, 1985).

Acknowledgements

I thank all the authors who have contributed published and unpublished figures, in particular W. Baumeister, S. Hovmöller, R. Rachel and I. Wildhaber. Special thanks are due to R. Hegerl for valuable discussions and critically reading the manuscript, and to W. Baumeister for continuous support.

References

Aebi, U., Fowler, W. E. and Smith, P. R. (1982). *Ultramicroscopy* **8**, 191–206.
Aebi, U., Fowler, W. E., Buhle, E. L. and Smith, P. R. (1984). *J. Ultrastruct. Res.* **88**, 143–176.
Agard, D. A. and Stroud, R. M. (1982). *Biophys. J.* **37**, 589–602.
Amos, L. A., Henderson, R. and Unwin, P. N. T. (1982). *Prog. Biophys. Mol. Biol.* **39**, 183–231.

Baker, T. S. (1981). In *Electron Microscopy in Biology* (J. D. Griffith, ed.), Vol. 1, pp. 189–290. John Wiley and Sons, New York.

Barth, M. (1987). "Näherung experimentell nicht zugänglicher Strukturdaten in der Elektronentomographie raumbegrenzter Objekte", Thesis, Technische Universität München, München.

Baumeister, W. and Engelhardt, H. (1987). In *Electron Microscopy of Proteins, Vol. 6: Membranous Structures* (J. R. Harris and R. W. Horne, eds), pp. 109–154. Academic Press, London.

Baumeister, W. and Hegerl, R. (1986). *FEMS Microbiol. Lett.* **36**, 119–125.

Baumeister, W., Barth, M., Hegerl, R., Guckenberger, R., Hahn, M. and Saxton, W. O. (1986a). *J. Mol. Biol.* **187**, 241–253.

Baumeister, W., Guckenberger, R., Engelhardt, H. and Woodcock, C. L. F. (1986b). *Ann. N.Y. Acad. Sci.* **483**, 57–76.

Bingle, W. H., Engelhardt, H., Page, W. J. and Baumeister, W. (1987). *J. Bacteriol.* **169**, 5008–5015.

Bretaudiere, J.-P. and Frank, J. (1986). *J. Microscopy (Oxford)* **144**, 1–14.

Carazo, J. M. and Carrascosa, J. L. (1987). *J. Microscopy (Oxford)* **145**, 23–43.

Cejka, Z., Hegerl, R. and Baumeister, W. (1987). *J. Ultrastruct. Mol. Struct. Res.* **96**, 1–11.

Chalcroft, J. P. and Davey, C. L. (1984). *J. Microscopy (Oxford)* **134**, 41–48.

Chalcroft, J. P., Engelhardt, H. and Baumeister, W. (1987). *FEBS Lett.* **211**, 53–58.

Cormack, A. M. (1964). *J. Appl. Phys.* **35**, 2908–2913.

Crowther, R. A. (1984). *Ultramicroscopy* **13**, 295–304.

Crowther, R. A. and Amos, L. (1971). *J. Mol. Biol.* **60**, 123–130.

Crowther, R. A. and Luther, P. K. (1984). *Nature* **307**, 569–570.

Dickson, M. R., Downing, K. H., Wu, W. H. and Glaeser, R. M. (1986). *J. Bacteriol.* **167**, 1025–1034.

Engel, A. and Massalski, A. (1984). *Ultramicroscopy* **13**, 71–84.

Engel, A., Massalski, A., Schindler, H., Dorset, D. L. and Rosenbusch, J. P. (1985). *Nature* **317**, 643–645.

Engelhardt, H., Guckenberger, R., Hegerl, R. and Baumeister, W. (1985). *Ultramicroscopy* **16**, 395–410.

Engelhardt, H., Saxton, W. O. and Baumeister, W. (1986a). *J. Bacteriol.* **168**, 309–317.

Engelhardt, H., Engel, A. and Baumeister, W. (1986b). *Proc. Natl Acad. Sci. USA* **83**, 8972–8976.

Frank, J. (1973). In *Advanced Techniques in Biological Electron Microscopy* (J. K. Koehler, ed.), pp. 215–274. Springer-Verlag, Berlin, Heidelberg and New York.

Frank, J. (1980). In *Computer Processing of Electron Microscope Images*, Topics in Current Physics, Vol. 13 (P. W. Hawkes, ed.), pp. 187–222. Springer-Verlag, Berlin, Heidelberg and New York.

Frank, J. (1982). *Optik* **63**, 67–89.

Frank, J. and Al Ali, L. (1975). *Nature* **256**, 376–379.

Frank, J. and van Heel, M. (1982). *J. Mol. Biol.* **161**, 134–137.

Frank, J., Verschoor, A. and Boublik, M. (1981a). *Science* **214**, 1353–1355.

Frank, J., Shimkin, B. and Dowse, H. (1981b). *Ultramicroscopy* **6**, 343–358.

Frank, J., Radermacher, M., Wagenknecht, T. and Verschoor, A. (1986). *Ann. N.Y. Acad. Sci.* **483**, 77–87.

Frank, J., McEwen, B. F., Radermacher, M., Turner, J. N. and Rieder, C. L. (1987). *J. Electron Microscopy Tech.* **6**, 193–205.

Fuller, S. D. (1981). *Meth. Cell Biol.* **22**, 251–296.

Gerchberg, R. W. (1974). *Optica Acta* **21**, 709–720.

Gilbert, P. (1972). *J. Theor. Biol.* **36**, 105–117.
Glaeser, R. M. (1985). *Ann. Rev. Phys. Chem.* **36**, 243–275.
Gordon, R., Bender, R. and Herman, G. T. (1970). *J. Theor. Biol.* **29**, 471–481.
Guckenberger, R. (1985). *Ultramicroscopy* **16**, 357–370.
Harauz, G. and Ottensmeyer, F. P. (1984). *Ultramicroscopy* **12**, 309–320.
Harauz, G. and van Heel, M. (1986). *Optik* **73**, 146–156.
Hawkes, P. W. (ed.) (1980). *Computer Processing of Electron Microscope Images*, Topics in Current Physics, Vol. 13. Springer-Verlag, Berlin, Heidelberg and New York.
Hegerl, R. and Altbauer, A. (1982). *Ultramicroscopy* **9**, 109–116.
Hegerl, R. and Baumeister, W. (1988). *J. Electron Microscopy Tech.* (in press).
Henderson, R., Baldwin, J. M., Downing, K. H., Lepault, J. and Zemlin, F. (1986). *Ultramicroscopy* **19**, 147–178.
Henderson, R. and Unwin, P. N. T. (1975). *Nature* **257**, 28–32.
Holser, W. T. (1958). *Z. Kristallogr.* **110**, 266–281.
Hoppe, W. and Hegerl, R. (1980). In *Computer Processing of Electron Microscope Images*, Topics in Current Physics, Vol. 13 (P. W. Hawkes, ed.), pp. 127–185. Springer-Verlag, Berlin, Heidelberg and New York.
Horne, R. W. and Markham, R. (1972). In *Practical Methods in Electron Microscopy* (A. M. Glauert, ed.), Vol. 1, Part II, pp. 327–435. North-Holland, Amsterdam and London.
Hovmöller, S. (1986). In *Techniques for the Analysis of Membrane Proteins* (C. I. Ragan and R. J. Cherry, eds), pp. 315–344. Chapman and Hall, London.
Jesior, J.-C. (1982). *EMBO J.* **1**, 1423–1428.
Kessel, M., Radermacher, M. and Frank, J. (1985). *J. Microscopy (Oxford)* **139**, 63–74.
Kittler, J., Illingworth, J. and Föglein, J. (1985). *Comp. Vision Graph. Image Process.* **30**, 125–147.
Lepault, J. and Pitt, T. (1984). *EMBO J.* **3**, 101–105.
Mellema, J. E. (1980). In *Computer Processing of Electron Microscope Images*, Topics in Current Physics, Vol. 13 (P. W. Hawkes, ed.), pp. 89–126. Springer-Verlag, Berlin, Heidelberg and New York.
Misell, D. L. (1978). *Practical Methods in Electron Microscopy, Vol. 7: Image Analysis, Enhancement and Interpretation* (A. M. Glauert, ed.). North-Holland, Amsterdam, New York and Oxford.
Pum, D. and Kübler, O. (1984). In *Proc. 8th Eur. Congr. Electron Microscopy* (A. Csanady, P. Röhlich and D. Szabo, eds), Vol. 2, pp. 1331–1340. Budapest.
Rachel, R., Jakubowski, U., Tietz, H., Hegerl, R. and Baumeister, W. (1986). *Ultramicroscopy* **20**, 305–316.
Radermacher, M. and Frank, J. (1984). *J. Microscopy (Oxford)* **136**, 77–85.
Radermacher, M., Frank, J. and Manella, C. A. (1986). In *Proc. 44th Ann. Meet. Electron Microscopy Soc. America* (G. W. Bailey, ed.), pp. 140–143. San Francisco Press, San Francisco.
Richards, F. M. (1977). *Ann. Rev. Biophys. Bioeng.* **6**, 151–176.
Saxton, W. O. (1978). *Computer Techniques for Image Processing in Electron Microscopy*. Academic Press, New York.
Saxton, W. O. (1985). *Ultramicroscopy* **16**, 387–394.
Saxton, W. O. and Baumeister, W. (1982a). *J. Microscopy (Oxford)* **127**, 127–138.
Saxton, W. O. and Baumeister, W. (1982b). In *Developments in Electron Microscopy and Analysis* (M. Goringe, ed.), Inst. Phys. Conf. Ser. No. 61, pp. 333–336.
Saxton, W. O. and Frank, J. (1977). *Ultramicroscopy* **2**, 219–227.
Saxton, W. O., Pitt, T. J. and Horner, M. (1979). *Ultramicroscopy* **4**, 343–354.

Saxton, W. O., Baumeister, W. and Hahn, M. (1984). *Ultramicroscopy* **13**, 57–70.

Shaw, P. J. (1984). *Ultramicroscopy* **14**, 363–366.

Shaw, P. J. and Hills, G. J. (1981). *Micron* **12**, 279–282.

Skilling, J. and Bryan, R. K. (1984). *Mon. R. Astr. Soc.* **211**, 111–124.

Smith, P. R. (1978). *Ultramicroscopy* **3**, 153–160.

Smith, P. R. and Kistler, J. (1977). *J. Ultrastruct. Res.* **61**, 124–133.

Smith, P. R., Peters, T. M. and Bates, R. H. T. (1973). *J. Phys. A: Math., Nucl. Gen.* **6**, 361–382.

Stark, W., Kühlbrandt, W., Wildhaber, I., Wehrli, E. and Mühlethaler, K. (1984). *EMBO J.* **3**, 777–783.

Stewart, M. (1986). In *Ultrastructure Techniques for Microorganisms* (H. C. Aldrich and W. J. Todd, eds), pp. 333–364. Plenum Press, New York and London.

Thon, F. (1966). *Z. Naturforsch.* **21a**, 476–478.

Unser, M., Steven, A. C. and Trus, B. L. (1986). *Ultramicroscopy* **19**, 337–348.

Vainshtein, B. K. (1978). In *Advances in Optical and Electron Microscopy* (V. E. Cosslett and R. Barer, eds), Vol. 7, pp. 281–377. Academic Press, London.

Vainshtein, B. K., Fridkin, V. M. and Inderbom, V. L. (1982). *Modern Crystallography Structural Crystallography*. Springer-Verlag, Berlin, Heidelberg and New York.

Vainshtein, B. K., Fridkin, V. M. and Inderbom, V. L. (1982). *Modern Crystallography II. Structure of Crystals*. Springer-Verlag, Berlin, Heidelberg and New York.

van Heel, M. (1984). *Ultramicroscopy* **13**, 165–184.

van Heel, M. (1987a). *Ultramicroscopy* **21**, 111–124.

van Heel, M. (1987b). *Ultramicroscopy* **21**, 95–100.

van Heel, M. and Harauz, G. (1986). *Optik* **73**, 119–122.

van Heel, M. and Keegstra, W. (1981). *Ultramicroscopy* **7**, 113–130.

Woodcock, C. L. F., Engelhardt, H. and Baumeister, W. (1986). *Eur. J. Cell Biol.* **42**, 211–217.

Zane, B. (1985). "Korrelationsmittelung elektronenmikroskopischer Bilder von zweidimensionalen biogenen Kristallen mit schwach gestörtem Gitter bei schlechtem Signal-Rausch-Verhältnis." Diploma Thesis, Universität München, München.

Index